T0320703

Thermophysical Properties of Lithium Hydride, Deuteride, and Tritide and of Their Solutions with Lithium

Thermophysical Properties of Lithium Hydride, Deuteride, and Tritide and of Their Solutions with Lithium

É. É. Shpil'rain
K. A. Yakimovich
T. N. Mel'nikova
A. Ya. Polishchuk

Translated by
Stephen J. Amoretty

Originally published as *Teplofizicheskie svoĭstra gidrida, deĭterida, tritida litiya i ikh rastvorov s litiem: Spravochnik*

Printed in the United States of America.

Library of Congress Cataloging-in-Publication Data

Thermophysical properties of lithium hydride, deuteride, and tritide and of their solutions with lithium.

Bibliography: p.
1. Lithium hydride--Thermal properties--Handbooks, manuals, etc. 2. Solution (Chemistry)--Thermal properties--Handbooks, manuals, etc. 3. Solids--Thermal properties--Handbooks, manuals, etc. I. Shpil´raĭn, Ė. Ė. (Ėval´d Ėmil´evich) II. Title: Deuteride, and tritide and of their solutions with lithium. III. Series.
QC176.8.T4T46 1987 620.1'98 87-12601
ISBN 0-88318-532-6

Contents

Chapter 1. Physical properties of lithium hydride and of its isotopic modifications

Chapter 2. Thermodynamic properties of the isotopic modifications of lithium hydride in the condensed state

Chapter 3. Thermodynamic properties of Li-LiH (LiD, LiT) systems

Chapter 4. Transport properties

Chapter 5. Tables of the thermophysical properties of lithium hydride, deuteride, and tritide and their solutions with lithium

Preface

The properties of the isotopic compounds of lithium with hydrogen have recently attracted considerable interest and rapidly increasing attention. Many ways in which the isotopic modifications of lithium hydride and lithium-containing solutions of these modifications can be used in new branches of science and technology have been discussed extensively in the literature and at conferences and symposia.

This interest stems from the fact that these substances have a striking combination of physical, thermal, and mechanical properties, making them promising candidates for use in relevant engineering projects.

Many papers dealing with various characteristics of lithium hydride have been published in the scientific literature. Considerably less information is available on the isotopic modifications of lithium hydride and on lithium-containing solutions of these modifications. At the same time, the discrepancies in the data of various authors can be attributed to either the inherent flaws of the experimental methods or to insufficient purity of the samples, or to both factors. Lithium hydride is not easily amenable to experimental study, because it easily decomposes, thereby releasing part of the hydrogen, which changes the composition and, hence, the properties of the sample. Discrepancies in the data have produced some caution among scientists, engineers, and designers who routinely use these data.

A book by Shpil'rain and Yakimovich,[7] entitled *Lithium Hydride*, which was published in 1972, was the first attempt to compile data on the thermophysical properties of lithium hydride in a more systematic form. That book provides a rather comprehensive analysis of the scientific literature on lithium hydride and lithium-containing solutions published before 1970. It does not, however, review the properties of the isotopic modifications, lithium deuteride and tritide and their solutions with lithium.

In the present book, we analyze and systematically classify reference data obtained at high temperatures on the thermophysical properties of lithium hydride, deuteride, and tritide and their solutions with lithium. Some of the properties of the substances considered here are the density, heat capacity, enthalpy, entropy in the solid and liquid states, pressure and composition of the vapor phase when the liquid is in equilibrium with the vapor, surface tension, elastic properties in the crystalline state, and transport properties in the condensed and vapor phases. In sum, we consider those properties which are of specific use in some branches of the new technology (some of which are briefly discussed in this book) and are useful in the theoretical analysis of the substance. The thermal regime governing the performance of these materials in the apparatus sets the limiting temperature at 1300 K for the calculated tables on thermophysical properties.

The reference data are based on experimental and theoretical studies carried out either by the authors or with their participation. Additionally, this book includes a comprehensive review of the literature published in this field. Special attention is given to the presentation of complete original information on the topics discussed in this book. Sometimes we had access to data reported in rare or little-known publications from research centers, proceedings of international conferences, and dissertations. In these cases we strove to make the reviews more comprehensive and to make the data more readily available.

The experimental data are subjected to a critical statistical analysis in terms of the thermal properties (density, thermal expansion coefficient, elastic properties) in the solid phase and the caloric properties in the solid and liquid states. The results of this analysis are then used, along with the present understanding of solid state physics and the techniques of thermodynamic scaling in the liquid state, to determine these properties for all isotopic compositions of lithium hydride.

Special attention is given to the thermodynamics of the Li-LiH (LiD, LiT) systems, in which the liquid is in equilibrium with the vapor, since systems of this type are particularly nonideal due to dissociation. This dissociation, as measured from pressures of the dissociation products in the equilibrium vapor, compositions of the vapor and liquid, and other parameters, must be taken into account in the determination of the thermodynamic properties.

In calculating the transport coefficients in the gas phase, extensive use is made of the asymptotic method of calculating the interaction potentials and collision integrals. This method is now being widely used.

The methods suggested in this book for determination of the thermophysical properties of lithium hydride and its isotopic modifications are, as a rule, of a more general nature and can also be used for other isotopically similar compounds.

In the last chapter, a list of comprehensive tables on the thermophysical properties of lithium hydride, deuteride, and tritide and their solutions containing lithium is presented. These tables are doubly significant in that they give this publication the characteristics of a handbook. These tables have been certified as recommended reference data by the All-Union Scientific-Research Government Service Center for Standard Reference Data. We did not consider it necessary to standardize the notation and dimensions of the original data given in the review sections of this book. The recommended data in Chap. 5 are given in the International System (S.I.) of Units.

The Introduction, Chap. 3, and Secs. 1 and 2 of Chap. 4 were written by K. A. Yakimovich; Chaps. 1 and 2 and Sec. 3 were written jointly by K. A. Yakimovich and T. N. Mel'nikova, and Sec. 4 was written by A. Ya. Polishchuk. The tables presented in Chap. 5 were compiled by the authors in accordance with the thematic outline given above. The work on this book was carried out under the guidance, and with the direct participation, of É. É. Shpil'rain.

This book is intended for scientific workers, engineers, and designers who use lithium hydride, deuteride, and tritide and the solutions containing lithium in their specific work, and also for graduate and undergraduate students studying the techniques of experimental and theoretical determination of the thermophysical and physicochemical properties of substances, including the isotopy.

We welcome any comments or suggestions.

Introduction

The uses of lithium hydride, deuteride, and tritide and their solutions with lithium in the new technology

Lithium hydride as a moderator and a neutron shield

Lithium hydride is a light material which is suitable for use as a neutron shield in nuclear reactors.[196,252] The large mass fraction of hydrogen in lithium hydride, its high chemical, thermal, and radiation stability, and its adequate strength and relatively low density make it a promising material for use as a neutron moderator[186,190] or a material for shielding an airborne reactor against meteorites.[211] Welch[252] studied the feasibility of using lithium hydride as a neutron shield in space vehicles. He noted that the LiH molecule is a stable combination of two elements that moderate and absorb neutrons most efficiently. Calculations carried out for nuclear reactors for use in space, with thermal power outputs[172] of up to 1000 MW, have shown that the relative mass (ton/m^3) of a reactor in which different materials are used as a moderator is: 19.4 for BeO, 11.4 for Be, 1.8 for ^7LiH, and 1.8 for H_2O. Clearly, the use of lithium hydride and water reduces the reactor mass substantially. In practice, however, water cannot be used for this purpose, since the high saturated-steam pressure must be dealt with in reactors under thermal stress.

Concrete, water, and other hydrogenous materials are used for building land-based stationary reactors whose total mass of the neutron shield is of no particular importance. However, materials with a high hydrogen content and low density must be used as neutron shields in transport vehicles in which the mass is an important factor of their efficiency.

After an analysis of the properties of many substances (within the scope of the programs for developing nuclear aviation and space reactors), lithium hydride was chosen as the most suitable material. The following were key factors in selecting lithium hydride as a potentially useful reactor shield:

(1) The large relative hydrogen content (12.68 wt.%) and a large number of hydrogen atoms per cubic centimeter (5.85×10^{22}) make it a highly efficient neutron shield for man and equipment.

(2) The high melting point (~690 °C) makes it possible to use this material in the thermally stressed zone without fear of melting.

(3) The low dissociation pressure at the melting point of commercially pure lithium hydride (4×10^3 Pa) makes it possible to melt and freeze this material without decomposition and evolution of hydrogen at relatively low pressures.

(4) The low density (775 kg/m^3 at room temperature) makes it an attractive candidate for use in space vehicles.

On the negative side, the relatively low thermal conductivity and low mechanical strength of solid lithium hydride cause problems involving the structural design of space vehicles.

Source of rocket fuel

Lo[178] has noted that lithium hydride is a promising source of hydrogen for rocket engines. The necessary amount of hydrogen can be obtained from the reaction

$$LiH + LiOH \cdot H_2O \rightleftarrows 2LiOH + H_2,$$

in which lithium hydride is a lightweight solid fuel source.

When lithium hydride is used directly as a source of rocket fuel, the specific volume flash can last as long as 360 s, as indicated by Sarner,[53] upon the combustion of LiH in ClF_3.

Thermal-energy storage systems

Lithium hydride has many advantages when used as an energy storage system. These advantages are particularly important for transport vehicles.

Because of its large specific heat and large latent heat of fusion, lithium hydride can be used effectively as a thermal-energy storage system, which is particularly important in the case of solar-powered space vehicles when they are in the region of a planet that is shaded from the sun. Cazeneuve[116] carried out a special study of promising energy sources for transport vehicles, in particular, for ships. He found that lithium hydride is one of the most promising materials. He also found that the accumulation and storage of energy is an important problem for ships when the classical thermodynamic methods of energy production cannot be used for physical or logistical reasons and nontraditional energy sources, produced, for example, by acoustic or electromagnetic methods, must instead be used. In this case, lithium hydride can provide an energy density ten times greater than that of the classical (lead-acid) electrical storage batteries.

In his study, Cazeneuve[116] presents some interesting data obtained as a result of analysis of the specific mass energy (W h/kg) of the most typical heat-accumulating substances. The results were calculated from the heat capacity in the liquid phase and from the latent heat of transition:

Substance	W h/kg	(°C)
Water (pressurized)	630	(300)*
Aluminum	100	(600)
Lithium hydride	850	(685)
Lithium fluoride	290	(847)
Eutectic alloy of beryllium with silicon	360	(1089)
Sodium fluoride	220	(1092)
Beryllium	360	(1283)
Silicon	500	(1420)

*The phase transition temperatures (°C) are enclosed in the parentheses.

It is evident that lithium hydride has a high specific mass energy and a phase-transition temperature more favorable than water from the standpoint of energy conversion in heat engines.

By way of example, Cazeneuve notes that 15 tons of molten lithium hydride could provide the mechanical equivalent of 5000 kW h, or enough fuel to power a 350-kW motor for approximately 15 h.

The use of lithium hydride in new methods of electric energy production

Among the new methods of direct electric energy production, the development of electrochemical converters as fuel cells merits attention. In these systems lithium hydride is one of the more promising components.

Several methods of using lithium hydride for regenerative electrochemical cells were considered in Argonne National Laboratory and in several other research centers in the U.S.A.[143,167] The basic electrochemical cell uses the electrochemical reaction between lithium and hydrogen to produce lithium hydride at a temperature T_2. The electromotive force produced between the lithium and hydrogen electrodes is the voltage source for the external load. The reaction product enters the regenerative cell, which is maintained at a temperature T_1 ($T_1 > T_2$). This heat causes the hydride to break down into the original components, which are returned to the electrochemical cell, producing a closed thermodynamic cycle.

Based on the experimental data, the following values were obtained from a comparison of the efficiency of these converters and the efficiency of thermoelectric and thermoelectronic converters: The efficiency of the cycle with a regenerative electrochemical cell is 8%–12% at temperatures between 1200 and 1500 K, while the efficiency of the thermoelectric cycle reaches 5% and that of thermoelectronic systems 8% at a maximum temperature of 1600 K.

Johnson and Heinrich[167] presented the results of an experimental study for several operating points of lithium hydride fuel cells. In one case, an emf of 0.3 V at an electrode current density of 2.15 kA/m^2 was measured at a temperature of 570 °C. In another experimental setup, the electrode current density was greater than 10 kA/m^2 at a hydrogen pressure of 100 kPa and a temperature of 549 °C.

Heat shield

Lithium hydride has a large effective heat capacity if its nominal heat capacity and the heats of phase transitions and the heat of dissociation are taken into account. Accordingly, it can be used as a heat shield for systems with a large heat flux and high temperatures. For example, liquid lithium hydride was used as a heat-transfer agent in the cooling systems of space vehicles.[145,198] Additionally, the feasibility of using lithium hydride as an active heat shield for hypersonic aircraft, space vehicles, reentry vehicles, and solid-fuel rocket engines was considered. It was noted that lithium hydride has no competition with respect to heat absorption capacity in the temperature range from 25 to 800 °C, over which the heat capacity and the heat of fusion account for the consumption of 1810 kcal/kg. Dissociation may account for an additional consumption of up to 2500 kcal/kg. According to the data from Refs. 40 and 190, lithium hydride begins to break down at ~800 °C and a pressure of 100 kPa. However, only a small fraction of the original material breaks down under these conditions. Keeping the temperature constant at a reduced pressure of 27 kPa causes up to 80% of lithium hydride to break down, while a further reduction of the pressure to 1 kPa extends the breakdown to 92%. Feasibility studies of using lithium hydride as a heat shield at temperatures up to 1100 °C and heat fluxes of up to $\sim 4\times 10^5$ kcal/(m^2 h) have largely confirmed their initial estimates.

Fusion reactors and tritium production

Lithium hydride, its isotopic compounds (LiD, LiT), and solutions of hydrogen isotopes in lithium are attracting considerable interest in new technology because lithium is one of the most suitable blanket materials for fusion reactors (see Refs. 106, 181, 183, 193, 216, and 247). Lithium is not only an ideal heat-sink material but also a fast-neutron moderator and a tritium production source. The neutrons released in the fusion reaction,

$$D + T \rightarrow {}^4He + n + 17.6\ \text{MeV},$$

interact with lithium in the blanket:

$$^6Li + n \rightarrow {}^4He + {}^3T + 4.8\ \text{MeV};$$

$$^7Li + n \rightarrow {}^4He + {}^3T - 2.5\ \text{MeV}.$$

As a result, dilute solution of tritium in lithium is produced, giving rise to the possibility of extracting the tritium for use in the fusion reaction.

From a safety standpoint, lithium must be purified of tritium to the maximum extent in order to prevent leakage of tritium into the surrounding medium. A very thorough purification, however, appreciably increases the cost of the system.

The tritium produced in the lithium blanket, which is maintained at a temperature of about 400 °C, produces lithium tritide (LiT). The problem of extracting tritium from lithium is therefore related principally to the problem of separating LiT from lithium. Since this is a costly process, a relatively large concentration of LiT must be reached to make this process economically feasible.

Two methods of tritium recovery are considered promising:

(1) Cooling a hypereutectic Li-T solution to separate solid LiT (tritium is further separated through the dissociation of LiT at a temperature of 800 °C and higher).

(2) Cooling an ~700 °C equilibrium vapor over the solution to 300 °C; this condenses lithium vapor while tritium remains in the gas phase. This approach, however, involves the recombination of Li and T_2 to LiT during the cooling, so that the relationship between the rate of recombination, $Li + \frac{1}{2}T_2 \rightleftarrows LiT$, and the cooling rate is important.

Tritium can also be separated by diffusing it through a niobium diaphragm. The efficiency of this method, i.e., the rate of diffusion through the diaphragm depends essentially on the tritium concentration in the solution.

An alternative method of tritium separation involves a multiple distillation of the solution to increase the tritium concentration in the vapor in combination with the subsequent use of a niobium (hydrogen-permeable) wall.

Maroni *et al.*[181] noted that large lithium liquid-metal systems used as blankets in D-T fusion electric power plants will contain a significant amount of hydrogen and tritium. Successful development of efficient methods for the recovery (extraction) of tritium and the development of systems for the removal of hydrogen isotopes from the blanket require a thorough knowledge of the behavior of these isotopes in solution with liquid lithium.

In laser fusion systems, the frozen D-T fuel pellet, which is shot into the chamber at the velocity of about 1000 m/s, can be replaced by a room-temperature LiD-LiT pellet.[13,105] This method has several advantages.

The composition of the vapor phase in the core of a controlled fusion reactor must be known. The results calculated by Smith *et al.*[219] show that the partial pressure of LiT molecules in the reaction zone may be quite large at certain temperatures and may even be higher than the partial pressure of the principal component—the T_2 molecules. In the list below, the first column of numbers corresponds to a temperature of 800 K and the second column corresponds to a temperature of 1300 K; the range of values for P_{T_2} and P_{LiT} corresponds to different compositions of the condensed equilibrium phase:

P_{Li}, Pa	0.95	690 0
P_{T_2}, MPa	0.001–01	0.001–0.1
P_{LiT}, MPa	0.000 64–0.006 4	1.6–16

The examples of promising directions of the new technology cited in this review by no means represent all of the practical uses of lithium hydride and the lithium-lithium hydride system (in different isotopic combinations). The demand for hydrogen in various types of apparatus has led to a study of the feasibility of simple and rapid production of hydrogen from lithium hydride.[264] It should be noted, however, that the introduction of lithium hydride, its isotopic modifications, and lithium-containing solutions of these compounds into technology is limited by the lack of reliable information about the thermophysical and physicochemical properties of these materials, particularly in the high-temperature region.

This monograph presents a study of several physicochemical properties and basic thermophysical properties of the isotopic modifications of lithium hydride in the condensed and vapor phases and a study of the lithium-containing solutions of these modifications when the liquid phase is in equilibrium with the vapor.

Chapter 1
Physical properties of lithium hydride and of its isotopic modifications

1. Structure of the crystal lattice

Lithium hydride crystallizes to a NaCl-type lattice. Figure 1.1 shows the lattice of the LiH crystal. This crystal has a face-centered cubic structure and its unit cell is a parallelepiped built up of basis vectors $\mathbf{a}_1$, $\mathbf{a}_2$, and $\mathbf{a}_3$. Each unit cell contains a Li^+ ion and a H^- ion. The volume of the unit cell is $v = l^3/4$, where l is the lattice constant, and it has six nearest neighbors.

The crystal structure of LiH and its parameters, including the absorption spectra, have been studied experimentally by many investigators (see Refs. 24, 95, 96, 113, 119, 203, 214, 221, 228, and 259–262, and also, Chap. 2, Sec. 1) by both x-ray diffraction and neutron-diffraction methods over broad temperature and pressure ranges. In all cases, lithium hydride retained the NaCl structure up to 12 GPa of pressure.[199]

According to the current theory, LiH is an ionic crystal with the formula $Li^+ H^-$. The ionicity of LiH has been determined by various means. For example, Pretzel *et al.*[204] have shown that the physical characteristics of lithium hydride (compressibility, crystal structure, lattice constant, thermal expansion, melting point, ionic conductivity) justify the assumption that the bond in crystalline LiH is ionic. From Pauling's[200] empirical relationship between a bond's ionicity and the difference between atomic electronegativities, these authors concluded that the bond in lithium hydride is 87.5% ionic. From an analysis of the phonon spectra of LiD measured from neutron scattering, Verbl *et al.*[243] found the ionicity to be 0.878. Calder *et al.*[115] used both x-ray diffraction and neutron diffraction studies of lithium hydride to determine the electron density of the hydrogen ion. According to their data, the ionicity of LiH is in the range 0.8–1.0, in good agreement with the conclusions of the studies cited above. Using the correlation between the differences $(z - e^*)$ and $(\beta_+ - \beta_-)$ (z is the ion charge, e^* is the effective Szigeti charge, and β_+ and β_- are the ionic polarizabilities) obtained by Hanlon and Lawson[137] for alkali-metal halides, these investigators found that $z - e^* \approx 0.32$ for LiH. It follows that $z \approx 0.85$, since $e^* = 0.53$ (see Refs. 111 and 123). The results obtained by Calder *et al.*[115] are also in good agreement with the experimental study of Pinsker and Kurdyumova,[46] who concluded from electron diffrac-

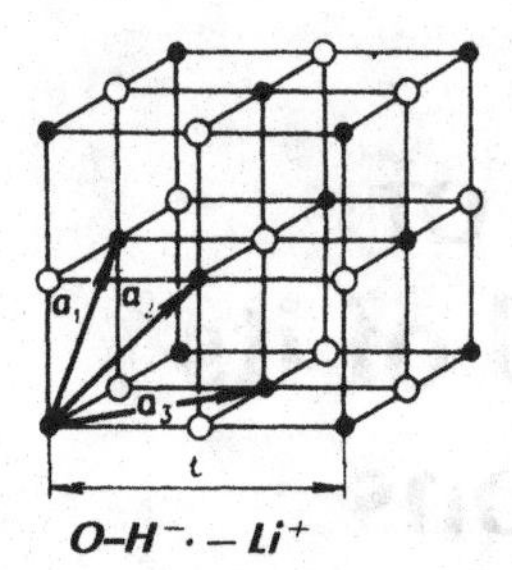

Figure 1.1. Structure of the crystal lattice of lithium hydride (a_1, a_2, a_3 **are the basis vectors, and l is the lattice constant).**

tion studies of LiH powder that the bond in crystalline lithium hydride is largely ionic.

The results of experimental and theoretical studies and calculations thus show that lithium hydride forms an ionic crystal. Various model-based ionic potentials, which give reasonably good agreement with experiment, are now widely used for calculating the physical properties of lithium hydride. The results of a study by Ahmed,[95] in which he found the ionic charge of a LiH crystal to be 0.25 by making use of x-ray diffraction data, were subsequently found to be erroneous (see Ref. 101).

2. Potential energy of lithium hydride

Classification of LiH as an ionic crystal makes it possible to construct potentials which describe the interaction of ions in LiH. The potential energy of interaction of two ions, i and j, with charges $\pm ze$, separated by a distance r_{ij}, can be written in the form

$$U_{ij}(r_{ij}) = U_{ij}^{C}(r_{ij}) + U_{ij}^{nC}(r_{ij}), \quad U_{ij}^{C} = (-1)^{i+j}\frac{(ze)^2}{r_{ij}} \tag{1.1}$$

if the central forces are the only active forces between the charges (which is usually the case for ionic crystals). The function $U_{ij}^{C}(r_{ij})$ contains the Coulomb part of the interaction and the function $U_{ij}^{nC}(r_{ij})$ contains the part attributable to the non-Coulomb central forces (the repulsion of ion cores, van der Waals forces).

The total Coulomb energy of a crystal

$$U^C = \sum_{ij} u_{ij}^{C}(r_{ij})$$

depends on the distance r between the nearest ions in the crystal (Li^+ and H^-) in the following way[22]:

$$U^C = Nu^C = N\frac{\Omega(ze)^2}{r}, \tag{1.2}$$

where u^C is the Coulomb energy of a single cell, z is the ion charge (typically, $z = 1$ or 0.878 for lithium hydride), Ω is Madelung's constant (which is 1.748 for the NaCl structure), and N is the number of ion pairs. The van der Waals forces are usually ignored in the calculations of various properties requiring the knowledge of the potential, since these forces contribute only small corrections to the thermodynamic functions of lithium hydride.

Since the non-Coulomb repulsive energy decreases with distance quite rapidly (the potential is effective only at distances on the order of the ionic radii), it is usually assumed that this interaction involves only the nearest neighbors. For LiH, for example, the distance between the nearest ions is ~0.204 nm, whereas the radius of an Li^+ ion is ~0.068 nm and that of an H^- ion is ~0.136 nm (Ref. 204). The incorporation of the non-Coulomb repulsion of the second nearest neighbors of the H^-–H^- ions, for example, gives a correction no greater than 0.05% to the binding energy, the ionic spacing, and the compressibility.[124]

In earlier studies, the non-Coulomb repulsive potential was written as A/r^n (the Born potential), where r is the distance between the nearest ions, and A and n are the potential parameters. More recent quantum-mechanical calculations have shown, however, that a more accurate way to write this potential would be in the Born–Mayer form $Be^{-r/\sigma}$, where B and σ are the Born–Mayer parameters. Such quantum-mechanical calculations are generally variational calculations in which trial wave functions are used.

The most complete calculation of the repulsive potential with a wide class of trial wave functions was carried out by Fischer *et al.*[124] with use of the Heitler–London method and the Slater ion wave functions (which were used previously for He) with different variational parameters δ (the wave functions of the hydrogen ion were selected in the form $e^{-\delta r}$). The repulsive potential was written in the form of the Born–Mayer potential and its parameters were then determined. From this result and the Coulomb energy, the equilibrium distance between the ions $\tilde{r}$, (corresponding to the position of the potential-energy minimum), the equilibrium static compressibility $\tilde{\kappa}$ (in terms of the second derivative of the potential at the point $\tilde{r}$), and the binding energy of the crystal $\tilde{W} = -\tilde{U}(\tilde{r})$ were determined.

The quantum-mechanical calculations of the potential of lithium hydride and of the quantities $\tilde{r}$, $\tilde{\kappa}$, and $\tilde{W}$ were also carried out in other studies (see Refs. 28, 98, 129, 140, 151, 152, 180, and 197). Each of these calculations has shown from first principles that the equilibrium distance between ions, the compressibility, and the binding energy of the lattice cannot simultaneously be in good agreement with experiment. If the wave functions which give the best agreement in terms of energy and ion spacing are used, the compressibility will be much lower (by 20%–30%).

The use of noncentral forces by Wilson and Johnson[255] and by Gerlich and Smith[125] improved the results only slightly. Using the parameters of the potential given in Ref. 124 and the noncentral forces, Wilson and Johnson calculated $\tilde{r}$, $\tilde{\kappa}$, and $\tilde{W}$ for ^6LiD. The values they obtained for $\tilde{r}$ and $\tilde{\kappa}$ are in reasonably good agreement with the experimental values, but the energy $\tilde{W}$ is too high. Consequently, the following procedure is generally used to select the potential. Phenomenologically, the potential per cell (at $z = 1$) can be written in the form

$$U(r) = (-\Omega e^2/r) + Be^{-r/\sigma} \tag{1.3}$$

(if only the repulsive forces of the nearest neighbors are taken into account). The parameters B and σ contained in this potential can be determined from

Table 1.1. Comparison of the coefficients of linear thermal expansion, α, of lithium hydride and lithium deuteride at room temperature obtained from experimental and calculated data, 10^{-6} deg^{-1}.

Isotopic modification	α_{calc}	α_{exp}
^{7}LiH	31.9 (Ref. 108)	35.22 (Ref. 113);
	33.76 (Ref. 220)	34.8 (Ref. 5); 32–33 (Ref. 196); 32.0 (Ref. 138)
^{7}LiD	37.22 (Ref. 108)	41.22 (Ref. 113); 40.4 (Ref. 5); 36 (Ref. 138); 41.4 (Ref. 214)
	39.04 (Ref. 220)	

Note: Although the experiment was carried out with the use of compounds containing natural ^{n}Li (92.5% ^{7}Li, 7.5% ^{6}Li), the values of α for the compounds ^{n}LiH and ^{n}LiD are essentially the same as those for ^{7}LiH and ^{7}LiD, respectively

experimentally measured values of the compressibility κ and equilibrium distance $\bar{r}$ between the ions (Li^+ and H^-) at a given temperature by using the equations

$$(\Omega e^2/\bar{r}^2) - (B/\sigma)e^{-\bar{r}/\sigma} = 0;$$
$$-(2\Omega e^2/\bar{r}^3) + (B/\sigma^2)e^{-\bar{r}/\sigma} = (18\bar{r}/\kappa), \tag{1.4}$$

which are obtained from the relations

$$\left(\frac{dU}{dr}\right)_{r=\bar{r}} = 0; \quad \frac{1}{18\,\bar{r}}\left(\frac{d^2U}{dr^2}\right)_{r=\bar{r}} = \frac{1}{\kappa}.$$

In this semi-empirical approach it is assumed that the position of the potential minimum coincides with the true equilibrium distance $\bar{r}$ between the ions for the given isotopic composition and temperature. The parameters of the potential determined in this manner generally depend on both temperature and isotopic composition and differ slightly from parameters of the true potential found from first principles. A phenomenological potential of this sort includes anharmonicity expressed in terms of $\bar{r}$ and κ. In dealing with such a potential, the use of a harmonic approximation is sufficient in all cases and a further allowance for the anharmonicity is not necessary.

The semi-empirical Born–Mayer potential has recently been used extensively in calculating the properties of lithium hydride and of its isotopic modifications. The results of such calculations are in good agreement with the experimental data. Using this potential, Bowman[108] and Srivastava and Saraswat[220] have calculated the thermal expansion coefficient of several isotopic modifications of lithium hydride at room temperature. The results obtained by them are in fairly good agreement with the experimental results (Table 1.1). The Born–Mayer potential was also used by Dass and Saxena[119] and by Stephens and Lilly[222] to calculate the binding energy of ^{7}LiH, and by Bowman[107] to calculate the binding energy of ^{7}LiH and LiD. The results obtained by these authors are in good agreement with the experimental data.

Table 1.2. Heat of formation of lithium hydride.

$\Delta_f H°(298.15)$ (cal/mole)	Method	References
− 21 600	Calorimetric	Guntz (Refs. 134, 135)
− 21 600 ± 250	Calorimetric	Moers (Ref. 194)
− 22 000 ± 1000	Measurement of saturated vapor pressure	Hurd and Moore (Ref. 150)
− 22 900	Spectroscopic, calorimetric measurements	Kapustinskii *et al.* (Ref. 20)
− 21 340 ± 210		Messer and Fasolino (Ref. 188)
− 21 790 ± 290	emf	Johnson *et al.* (Ref. 163)
− 20 940 (800 K)		
− 21 663 ± 52	Calorimetric	Gunn and Green (Ref. 131)
− 21 784 ± 21 (for LiD)		Gunn (Ref. 132)

3. Heat of formation and binding energies

The heat of formation of lithium hydride in the solid phase was measured extensively. The basic measurements are summarized in Table 1.2. Gunn and Green[131] have also measured the heat of formation of lithium deuteride. The heat of formation of lithium tritide has not been measured.

Guntz[134,135] determined the heat of formation of solid lithium hydride from solid lithium and hydrogen gas at room temperature, as the difference between the heats of hydrolysis of lithium and lithium hydride, which are − 53 200 and − 31 600 cal/mole, respectively.

Moers[194] determined experimentally the heats of hydrolysis of both lithium and lithium hydride at room temperature and then calculated the heat of formation of lithium hydride. The heat of hydrolysis was measured at the temperatures indicated in Table 1.3. At room temperature (the exact temperature is not given), the heat of hydrolysis is scaled to infinite dilution. The heat of formation of lithium hydride was then calculated as − 21 600 ± 250 cal/mole by Moers. The analyzed results were corrected for the heat of vaporization of water. The samples were weighed in a vacuum.

Hurd and Moore[150] studied the equilibrium pressure of hydrogen above lithium hydride. Making use of the dependence, $\log P_{H_2} = f(1/T)$, they calculated the heat of the reaction $2LiH_S \rightleftarrows 2Li_S + H_{2\,gas}$ and found $\Delta_f H = -22\,000 \pm 1000$ cal/mole of LiH. Analysis of the method and the tolerance error indicated by the investigators suggest that this value is acceptable for the heat of formation of LiH.

Kapustinskii *et al.*[20] measured the wavelength of lithium hydride corresponding to the absorption-band maximum in the ultraviolet (wavelength $\lambda_{LiH} = 251.7$ nm). The authors used the following equation to relate this quantity to the heat of formation of lithium hydride at room temperature:

$$\frac{h}{\lambda_{LiH}} = \Delta_f H(LiH) + \Delta_S H(Li) + 0.5\,D_0(H_2), \qquad (1.5)$$

where $\hbar$ is Planck's constant, $\Delta_f H(LiH)$ is the heat of formation of lithium

Table 1.3. Experimental data obtained by Moers (Ref. 194).

Hydrolysis reaction of lithium Temperature °C	Heat of reaction cal/mole	Hydrolysis reaction of lithium hydride Temperature °C	Heat of reaction cal/mole
9.8	− 52 095	10.5	− 30 715
12.2	− 52 553	11.1	− 31 106
18.2	− 52 553	13.0	− 31 113
Room temperature	− 52 723 ± 200	Room temperature	− 31 110

hydride, $LiH_S \rightleftarrows Li_S + 0.5\ H_{2\ gas}$, $\Delta_S H(Li)$ is the heat of sublimation of lithium, $D_0(H_2)$ is the dissociation energy of hydrogen, and $\Delta_f H(LiH)$ is calculated to be − 22 900 cal/mole.

Messer and Fasolino[188] determined the heat of formation of lithium hydride at room temperature (298.15 K) from experimentally determined values of the heats of hydrolysis of lithium and lithium hydride at temperatures very close to this temperature. Lithium hydride was synthesized by passing hydrogen through molten metal for 24 h at $T = 720$ °C and $P = 100$ kPa. The authors give the following values for the heat of hydrolysis, calculated for infinite dilution, and the heat of formation of lithium hydride (cal/mole):

Heat of hydrolysis of lithium .... − 53 100 ± 100
Heat of hydrolysis of lithium hydride .... − 31 760 ± 100
Heat of formation of lithium hydride .... − 21 340 ± 210

Gun and Green[31] determined the heats of reaction for the formation of both lithium hydride and lithium deuteride by a method similar to that used in the preceding study. The heat of hydrolysis was determined experimentally at a temperature of 25 ± 0.4 °C. Special attention was given to control the purity of the samples and special care was taken in preparing the samples. Specifically, the metallic lithium designated for direct testing and for the production of LiH was obtained by vacuum melting and subsequent distillation. Four lithium samples were tested. The hydrogen used to prepare the LiH was purified using a liquid nitrogen trap to remove water vapor. The heat of formation of lithium hydride at 298.15 K, determined by Gunn and Green,[131] is − 21 666 ± 26 cal/mole.

Corrections were later introduced by Gunn,[32] who took into account the solubility of hydrogen in water (and the thermal effect associated with it) and found a systematic error resulting from the calibration of the calorimeter. This error was 0.05%–0.20%. The corrected heat of formation of lithium hydride is − 21 663 ± 52 cal/mole. Although the net value of the heat of formation of LiH was refined by 3 cal/mole, the error tolerance increased by a factor of 2 in comparison with the previous value. For LiD, Gunn[132] gave the value $\Delta_f H^0(298.15) = -21\ 784 \pm 21$ cal/mole.

Johnson *et al.*[163] determined the heat of formation of lithium hydride from liquid lithium and hydrogen gas at a temperature of 800 K by measuring the emf of the chemical circuit. The emf between the hydrogen and lithium electrodes, $\mathscr{E}$, which were immersed into melts of various salts with lithium

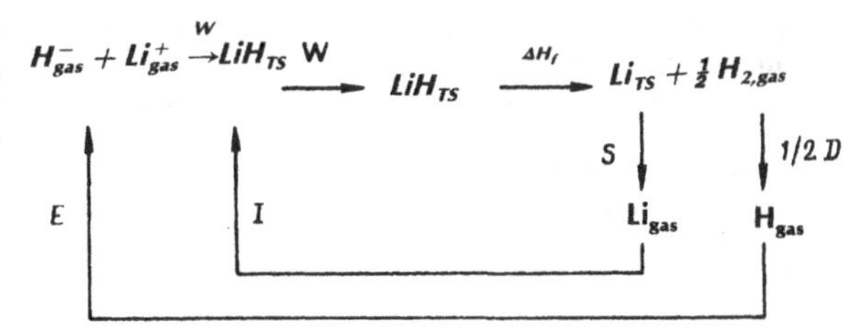

Figure 1.2. The Born-Haber cycle for determining the binding energy of solid lithium hydride (see text for notation).

hydride, was measured over the temperature range 675–885 K. They obtained the following experimental dependence of $\mathscr{E}$ (volts) on temperature T (K):

$$\mathscr{E} = 0.9081 - 7.699 \times 10^{-4}\, T. \tag{1.6}$$

Using the standard relation for the heat of formation

$$\Delta_f H = nF\left[T\left(\frac{d\mathscr{E}}{dT}\right)_P - \mathscr{E}\right] \tag{1.7}$$

(where F is the Faraday number and n is the ion valence), they obtained $\Delta_f H = -20\,940$ cal/mole at a temperature of 800 K.

Using data on the heat capacities of lithium and hydrogen[170] and lithium hydride[246] and on the heat of fusion of lithium,[223] Johnson *et al.*[163] corrected their results to $T = 298$ K and found $\Delta_f H = -21\,790 \pm 290$ cal/mole. As can be seen from the magnitude of the error, this value is considered satisfactory by the authors.

From the preceding analysis of the experimental studies of the heat of formation of LiH, the most complete study appears to be that of Gunn and Green,[131] along with the corrections of Gunn.[132] The remaining data are either approximate in nature or have large errors (considering the magnitude of the tolerable error) and are essentially the same, within the tolerable error limits, as the values given in these studies.

The binding energy W of an ionic crystal is defined as the difference between the internal energy of the crystal and the energy of the individual ions in the free state. The binding energy of a crystal, usually determined at a temperature of 0 or 298 K, can be found thermodynamically from the Born–Haber cycle (for a lithium hydride crystal at $T = 0$ K; see Fig. 1.2):

$$W = -(-\Delta_f H^0 + \Delta_S H^0 + I_0 + \tfrac{1}{2} D_0 + A_0), \tag{1.8}$$

where $\Delta_f H^0$ is the enthalpy of formation of solid lithium hydride from solid lithium and hydrogen gas, $\Delta_S H^0$ is the enthalpy of sublimation of lithium, I_0 is the ionization potential of lithium, D_0 is the dissociation energy of molecular hydrogen, and A_0 is the electron affinity of the hydrogen atom.

The following values of the binding energy (kcal/mole) were determined from Eq. (1.8) by using the most recent data in the literature (Table 1.4): $W_{^7\mathrm{LiH}}(0) = -216.738$ and $W_{^7\mathrm{LiD}}(0) = -217.914$.

The slight difference in the properties of the lithium isotope ^{7}Li and natural lithium ^{n}Li was ignored in the calculations.

Equation (1.8) cannot be used, however, to determine the binding energy of lithium tritide ^{7}LiT and other isotopic modifications of lithium hydride,

Table 1.4. Input data for calculation of the binding energy of the isotopic modifications of lithium hydride in the crystalline state.

Thermodynamic function	Value (kcal/mole)	References
$\Delta_s H°(\text{Li},0)$	37.700	Ref. 16
$I_0(\text{Li})$	124.336	Ref. 16
$D_0(H_2)$	103.267	Ref. 15
$D_0(D_2)$	105.070	Ref. 15
$D_0(T_2)$	105.870	Ref. 15
$A_0(\text{H,D,T})$	− 17.392	Ref. 15
$\Delta_f H°(\text{LiH}, 298.15)$	− 21.663	Ref. 132
$\Delta_f H°(\text{LiH},0)$	− 20.460	Recalculation
$\Delta_f H°(\text{LiD}, 298.15)$	− 21.784	Ref. 132
$\Delta_f H°(\text{LiD},0)$	− 20.735	Recalculation

principally because of the absence of data on the heats of formation of these compounds. This problem can be resolved by making use of the available results on ^{7}LiH and ^{7}LiD and the standard theoretical relation used to calculate the binding energy of a crystal, W_{calc} (see Ref. 31, for example),

$$W_{calc} = U(l) + E_{vib}(T), \tag{1.9}$$

where $U(l)$ is the potential energy of the crystal, $E_{vib}(T)$ is the vibrational energy at the temperature T, and l is the lattice constant of the crystal at the given temperature.

We can accordingly write

$$E_{vib}(T) = E_{vib}(0) + E_{therm}(T), \tag{1.10}$$

where $E_{vib}(0)$ is the vibrational energy at $T=0$ (zero-point energy), and $E_{therm}(T)$ is the thermal vibrational energy at temperature T. At low pressure, the thermal-vibrational energy of a crystal may be replaced by the enthalpy; i.e., we have

$$E_{therm}(T) \approx H(T) - H(0). \tag{1.11}$$

This component is not required for the calculation of the binding energy of a crystal at 0 K, so that

$$W_{calc}(0) = U(l) + E_{vib}(0). \tag{1.12}$$

The potential energy of a crystal can be expanded as the series

$$U(l) = \tilde{U}(\tilde{l}) + f\frac{(l-\tilde{l})^2}{2!} + g\frac{(l-\tilde{l})^3}{3!} + \dots, \tag{1.13}$$

where l is the lattice constant of the crystal in the harmonic approximation as determined by the position of the interaction-potential minimum, $\tilde{U}(\tilde{l})$ is the minimum interaction potential, and f and g are, respectively, the derivatives of the interaction potential evaluated at the minimum on the curve:

$$f = \left.\frac{d^2U}{dl^2}\right|_{\tilde{l}}; \quad g = \left.\frac{d^3U}{dl^3}\right|_{\tilde{l}}. \tag{1.14}$$

Higher terms of the expansion are, as estimates show, unimportant for lithium hydride. The quantities $\tilde{U}(\tilde{l})$, $\tilde{l}$, f, and g, which are determined by the interaction potential for the atoms in the crystal, do not depend on the isotopic composition of the compound. According to the harmonic theory of the crystal, we have (per mole)

$$\left.\frac{d^2U}{dl^2}\right|_{\tilde{l}} = \frac{1}{4}\left.\frac{d^2U}{dr^2}\right|_{\tilde{r}} = \frac{N_A}{4}\frac{18\tilde{r}}{\tilde{\kappa}}, \tag{1.15}$$

where $\tilde{r} = \tilde{l}/2$, κ is the compressibility of the crystal in the harmonic approximation, and N_A is Avogadro's number.

The coefficient g can be estimated in a quasiharmonic approximation from a known expression (see Ref. 22, for example),

$$a = -\frac{c_V}{2\tilde{r}}\left.\frac{d^3U/dr^3}{(d^2U/dr^2)^2}\right|_{\tilde{r}}, \tag{1.16}$$

where a is the coefficient of linear thermal expansion of a crystal and c_V is the heat capacity of a crystal at constant volume. We can thus write the relation (per mole)

$$\left.\frac{d^3U}{dl^3}\right|_{\tilde{l}} = \frac{1}{8}\left.\frac{d^3U}{dr^3}\right|_{\tilde{r}} = -\frac{N_A a 2\tilde{r}}{8c_V}\left(\frac{18\tilde{r}}{\tilde{\kappa}}\right)^2. \tag{1.17}$$

In Chap. 2, Sec. 2.2, we find the values $\tilde{l} = 0.39947$ nm and $\tilde{\kappa} = 2.84\times10^{-11}$ m^2/N, and in Tables 2.18, 2.39, and 5.5, we give the lattice constant $\tilde{l}$, the heat capacity c_V, and the thermal expansion coefficient a, as functions of temperature, for various isotopic compositions of lithium hydride. Using these data, we find $f = 4.561\times10^{17}$ kcal/mole cm^2 and $g = -8.75\times10^{25}$ kcal/mole cm^3.

The zero-point-vibration energy $E_{vib}(0)$ in Eq. (1.10) obeys the following relation in the harmonic approximation of the Debye theory[31]:

$$E_{vib}(0) = \frac{9}{4}\hbar\sqrt{\frac{5}{2}\frac{\tilde{l}}{\tilde{\kappa}}\frac{1}{\mu}} = A\sqrt{\frac{1}{\mu}}, \tag{1.18}$$

where $\hbar$ is Planck's constant, $\mu = Mm/(m+M)$ is the reduced mass, M is the atomic weight of the lithium isotope, and m is the atomic weight of the hydrogen isotope.

Clearly, the quantity A does not depend on the isotopic composition. A direct calculation of this quantity and the subsequent determination of $E_{vib}(0)$ from Eq. (1.18) produce, however, a large error, principally because of the inaccurate value of $\tilde{\kappa}$.

The error in determining A and hence $E_{vib}(0)$ for the isotopic compositions of lithium hydride can be substantially reduced by using the data on the binding energy and the lattice constants of ^{7}LiH and ^{7}LiD crystals. Using Eqs. (1.12) and (1.13), at a temperature of 0 K, we find

$$W_{calc} = U(\tilde{l}) + f\frac{(l-\tilde{l})^2}{2} + g\frac{(l-\tilde{l})^3}{3!} + E_{vib}(0). \tag{1.19}$$

Equating $W = W_{calc}$, we can write the following expression for ^{7}LiH and ^{7}LiD:

$$\Delta W = W(^7\text{LiH}) - W(^7\text{LiD})$$
$$= \frac{f}{2}[\Delta(l-\tilde{l})^2] + \frac{g}{3!}[\Delta(l-\tilde{l})^3] + \Delta E_{vib}(0),$$

where Δ is the difference between the pertinent quantities for ^{7}LiH and ^{7}LiD.

Retaining in this approach the condition that $E_{vib}(0) \sim 1/\sqrt{\mu}$ for the isotopic modification and that the proportionality factor A in Eq. (1.18) does not depend on the type of isotopic modification, we first find $\Delta E_{vib}(0)$ from the previously determined values of $W(^7\text{LiH})$, $W(^7\text{LiD})$, f, and g and from the lattice constants of the ^{7}LiH and ^{7}LiD crystals [from the data of Ref. 96 (see Chap. 2, Sec. 2)] and then find A, since

$$\Delta E_{vib}(0) = A\left(\sqrt{\frac{1}{\mu(^7\text{LiH})}} - \sqrt{\frac{1}{\mu(^7\text{LiD})}}\right). \tag{1.21}$$

ΔE_{vib} was found to be 1.129 kcal/mole and the factor $A = 4.249$ kcal/mole.

We then find $E_{vib}(0)$, $U(\tilde{l})$, the binding energy of isotopic modifications of lithium hydride and the heat of formation $\Delta_f H(^7\text{LiT})$. For example,

$$E_{vib}(^7\text{LiH}, 0) = A\sqrt{\frac{1}{\mu(^7\text{LiH})}}. \tag{1.22}$$

Using the data on ^{7}LiH [W, $E_{vib}(0)$, l], we find

$$U(\tilde{l}) = W - f\frac{(l-\tilde{l})^2}{2} - g\frac{(l-\tilde{l})^3}{3!} - E_{vib}(0). \tag{1.23}$$

In a similar way, we can find other quantities.

The following calculated values (kcal/mole) of the energy parameters of the isotopic compositions of lithium hydride were obtained from the relations derived above:

$E_{vib}(^7\text{LiH}, 0)$. . .4.526 $U(\tilde{l})$. . . − 221.379
$E_{vib}(^7\text{LiD}, 0)$. . .3.398 $W(^7\text{LiT}, 0)$. . . − 218.401
$E_{vib}(^7\text{LiT}, 0)$. . .2.927 $\Delta_f H^0(^7\text{LiT}, 0)$. . . − 20.822

Calculations have shown that the isotopic substitution of ^{6}Li for ^{7}Li in lithium hydride does not cause the binding energies and the heats of formation to differ appreciably.

4. The melting points of lithium hydride, lithium deuteride, and lithium tritide

Only data on the melting point of lithium hydride are given in the literature. The differences in the reported melting points of lithium hydride from the literature (see Table 1.5) stem primarily from the presence of impurities in the test samples. Unfortunately, sample composition was not monitored in every study. The principal impurities in lithium hydride are metallic lithium (because of a partial decomposition of the hydride) and lithium oxide (because

Table 1.5. Reported melting points of lithium hydride found in the literature.

Melting (solidification) point, °C	Purity of sample, %	References
680	98.7	Guntz (Refs. 134, 135)
680	—	Akulichev and Klyachko (Ref. 1)
686 ± 2	—	Hellman and Salmm (Ref. 144)
688 ± 1	99.8	Messer *et al.* (Ref. 40)
688	—	Messer (Ref. 189)
688.6 ± 0.3	—	Messer *et al.* (Ref. 191)
690.2 ± 0.7	99.6–99.9	Messer and Levy (Ref. 192)
690.8 ± 0.2	—	Messer and Levy (Ref. 192)
692 ± 2	100	Messer and Levy (Ref. 192)
686.15	—	Johnson *et al.* (Ref. 164)
686.35	—	Johnson *et al.* (Ref. 165)
686.4	99.8	Johnson *et al.* (Ref. 163)

of the exposure of the hydride to air during various procedures involving the production and loading of LiH). The presence in the lithium hydride of each component indicated above and of other impurities lowers the melting point of the sample, as pointed out by Messer *et al.*,[191,192] for example.

The first attempts to determine the melting point of lithium hydride were made at the end of the last century. In 1896 Guntz[134,135] found a relatively low value (680 °C), clearly indicating the presence of an appreciable fraction of impurities in the sample. The next attempts to determine the melting point of lithium hydride yielded higher values (if the calculations of Akulichev and Klyachko[1] are disregarded). From the law of conservation of momentum in rotational motion, Akulichev and Klyachko[1] derived an equation for the dependence of the absolute melting point of a binary compound on the melting point of the component parts, on the atomic numbers, and on several other factors. Specifically, they obtained a temperature of 680 °C for lithium hydride. Heuman and Salman[144] found the melting point of lithium hydride to be (686 ± 2) °C. Unfortunately, they did not disclose the method they used to measure this value or the purity of the lithium hydride they studied.

A group of investigators headed by Messer have reported in a number of studies[40,187,189,191,192] the melting point of LiH, which varies from 688 to 692 °C. These values were determined by analyzing the diagrams for the melting of binary LiH mixtures containing Li, Li_2O, CaH_2, and other components and then extrapolating the melting point of the mixture to 100% LiH. Their experiments were performed carefully both in terms of the measurement methods and in terms of the chemical monitoring of sample composition. The solidification temperatures of lithium hydride, measured experimentally by Messer and Levy,[192] are 690.8 ± 0.2 °C (for eight samples) and 690.2 ± 0.7 °C (for two samples) with a standard deviation within 0.3°. The molar concentration of LiH in the samples ranged from 99.6 to 99.9%. Extrapolating these results to 100% lithium hydride, Messer and Levy found the solidification temperature to be 691.3 °C. They attributed the higher value, as compared with the preceding results,[40,189,191] to the use of advanced experimental techniques and to higher purity samples.

Upon analyzing the results from phase diagrams of the binary LiH system with other components, Messer and Levy concluded that the most accurate determination of the solidification temperature of pure lithium hydride is 692 ± 2 °C.

Johnson *et al.*[164,165] carried out similar studies of the behavior of binary mixtures with other components—NaCl, LiBr, and LiI. They focused more attention, however, on the thermodynamics of dissolution of the components of the mixture and less attention on the melting point of pure lithium hydride and on the monitoring of minute quantities of impurities in it. Furthermore, using the emf method to study the thermodynamic properties of lithium hydride, Johnson *et al.*[163] made a special effort to determine the melting point of the original lithium hydride sample. The values of the solidification temperature and of the melting point of lithium hydride obtained in all of the most recent studies are lower than those obtained in the studies of the Messer school, conceivably because of the presence of a certain uncontrolled fraction of impurities in the samples.

Analysis of the studies of the melting point of lithium hydride shows that the most thorough studies are those of Messer and Levy, so that the melting point of 965 ± 2 K obtained by them for pure lithium hydride may therefore be assumed correct. With increasing lithium impurity, the melting point of lithium hydride decreases to the temperature of the monotectic mixture at a lithium content on the order of 0.01 mole fraction (see Chap. 2, Sec. 3.1).

Lithium hydride generally used in industry contains this amount and sometimes even larger amounts of lithium, so that industrial lithium hydride melts at the monotectic temperature. In this case, there exists, in fact, a solution of lithium in lithium hydride. In Chap. 2, Sec. 3.1, we examine the phase diagram of a Li–LiH system and find the temperature of the monotectic mixture in the Li–LiH(D, T) systems from the data on vapor pressure. The functional dependence $P_{2D} = f(T)$, where P_{2D} is the vapor pressure near the monotectic temperature (the so-called two-dimensional pressure), changes abruptly at the monotectic point. The monotectic temperature, which was found to be 961 ± 2 K for the Li–LiH system, 963 ± 2 K for the Li–LiD system, and 964 ± 3 K for the Li–LiT system, was determined from the position of the discontinuity on the curve.

The monotectic temperatures for the Li–LiD and Li–LiT systems made it possible to estimate the melting points of LiD and LiT. As was pointed out above, no special studies were conducted to determine the melting points of these compounds. Since the monotectic temperatures found for the Li–LiD and Li–LiT systems are higher by 2 and 3 K, respectively, than the monotectic temperature of the Li–LiH system, it can be assumed that this difference also applies to the melting points of the isotopic hydrides. Thus, the melting point $T_m = 967 \pm 2$ K should be used for LiD, and $T_m = 968 \pm 3$ K should be used for LiT.

Chapter 2 Thermodynamic properties of the isotopic modifications of lithium hydride in the condensed state

1. Caloric properties

1.1. Review and analysis of experimental studies of caloric properties

The principal experimental studies involving the measurement of the caloric properties of the isotopic modifications of lithium hydride are cited in Table 2.1. The first measurements of the heat capacity of solid lithium hydride were carried out in 1920 by Günther[133] and Moers[194] by a direct-heating method over the temperature range 74–95.7 K and at 292.7 K. On the basis of these data, they calculated the Debye temperature, which turned out to be 825 K. Günther indicated that the error contained in their results, including the Debye temperature, stemmed from the incomplete purity of the sample. The data of Günther and Moers were, nevertheless, used by Kelly[169] and Rossini *et al.*[209] and by other researchers to calculate the heat capacity of lithium hydride in the standard state. They obtained a value of 8.3 cal/mole deg for c_P at 298.15 K. Subsequent studies have indicated that these results are inaccurate.

The experimental data on the heat capacity of lithium hydride obtained by Günther and Moers are listed in Table 2.2.

Lang[175] published measured results of the enthalpy of lithium hydride over the temperature interval 284–1186 °F (410–910 K). The experiments were carried out by the displacement method in an apparatus with an adiabatic water calorimeter. The water mass (distillate) in the calorimeter was determined by gravimetry to within 0.1 g. A globar tube served as the heater. The temperature gradient in the working zone was no greater than 0.2 deg/cm at 910 K. The temperature of the furnace in the region in which a stainless-steel cell containing the sample was suspended was controlled by two platinorhodium–platinum thermometers which were installed in a pro-

Table 2.1. Experimental studies of the caloric properties of lithium hydride and lithium deuteride.

Property	Isotopic modification	Temperature interval, K	References
Heat capacity	^{n}LiH	74–96; 292.7	Günther (Ref. 133) Moers (Ref. 194)
	^{n}LiH	3.7–295.5	Kostryukov (Ref. 25)
	^{n}LiH	373–1173	Kite (see Ref. 250)
	^{7}LiH,^{7}LiD	30–300	Busey and Bevan (Ref. 114)
	^{7}LiH,^{7}LiD	3–320	Yates *et al.* (Ref. 263)
Enthalpy	^{n}LiH	298–673	Rodygina *et al.* (see Ref. 70)
	^{n}LiH	336–1360	Novikov *et al.* (see Ref. 70)
	^{n}LiH	413–914	Lang (Ref. 175)
	^{n}LiH	363–1023	Wilson *et al.* (Ref. 254)
	^{n}LiH	400–1325	Shpil'rain *et al.* (Ref. 69)
Heat of fusion	^{n}LiH		Kite (see Ref. 250)
	^{n}LiH		Messer *et al.* (Refs. 189, 191)
	^{n}LiH		Vogt (see Refs. 164, 165, 245)
	^{n}LiH		Hoffman (see Ref. 245)
	^{n}LiH		Vogel *et al.* (Ref. 245)

Table 2.2. Heat capacity of lithium hydride c_p [cal/(mole deg)] from the experimental data of Günther (Ref. 133) and Moers (Ref. 194).

T, K	c_p	T, K	c_p
74.0	0.674	88.4	1.095
75.2	0.734	90.5	1.217
76.6	0.749	93.0	1.34
77.6	0.722	95.7	1.42
83.2	0.925	292.7	8.177
85.9	1.017		

tective tube. A possible temperature drop between the axis of the sample and the thermocouple's position was checked in a special experiment. The temperature drop was found to be less than 1°. The temperature of the water in the calorimeter and of the external water shield was measured by specially calibrated Beckmann thermometers with a 0.005° error at the 1.5° average temperature rise of the calorimeter. The part of the total error contributed by the calorimeter was no greater than 1% according to the author.

The results of Lang's measurements are given in Table 2.3 (the enthalpy ΔH is reckoned from the value at 80 °F).

Lang used a least-squares interpolation equation for the temperature interval 284–1053 °F (410–840 K),

$$\Delta H = H(T) - H(80) = -59.17 + 847.5\times10^{-3}T + 398.9\times10^{-6}T^2, \quad (2.1)$$

where H is the enthalpy (Btu/lb) and T is the temperature (°F). He found that the temperature dependence of the enthalpy of lithium hydride increases, beginning at 1100 °F (865 K).

In addition to lithium hydride, Lang measured the enthalpies of several other substances—aluminum oxide, niobium, stainless steel, and others with the use of an apparatus described above. Comparison of these results with

Table 2.3. Experimental data on the enthalpy of lithium hydride obtained by Lang (Ref. 175).

T, °F(K)	ΔH, Btu/lb (J/mole)	T, °F(K)	ΔH, Btu/lb (J/mole)
284 (413)	212.1 (3 919)	929 (771)	1166 (21 543)
397 (476)	345.0 (6 374)	1029 (827)	1234 (22 800)
561 (567)	541.2 (10 110)	1053 (840)	1280 (23 649)
648 (615)	653.6 (12 076)	1105 (869)	1305 (24 111)
735 (663)	779.2 (14 397)	1117 (876)	1411 (26 070)
807 (703)	886.7 (16 383)	1158 (898)	1526 (28 195)
895 (752)	1026 (18 864)		
937 (776)	1080 (19 954)		

reliable data in the literature and analysis of the measurement procedure and its implementation show that these data are sufficiently accurate. Unfortunately, Lang does not indicate how the lithium hydride sample was prepared nor does he disclose its chemical composition. The scatter in the data on LiH is within $\pm 2\%$, as is the maximum relative error (which is within 1.5%–2.0%).

An equation for calculating the heat capacity c_P [kcal/(mole deg)] over the same temperature interval (which was used in Refs. 14 and 251, for example) can be obtained by differentiating Eq. (2.1):

$$c_P = 3.82 + 11.4 \times 10^{-3} T, \tag{2.2}$$

where T is the temperature in K. Lang estimated the error in calculating the heat capacity to be within 0.66%–2.90%.

In 1958, Rodygina, Gomel'skii, Lushin, and Akchurina experimentally studied the enthalpy of solid lithium hydride over the temperature interval 98–400 °C at the Mendeleev All-Union Scientific Research Institute of Metrology (see Ref. 70). They used a diathermal calorimeter with an aluminum block. Thermostatic control was implemented at 25 °C. A platinum cell was filled with the substance to be studied through a flow of nitrogen which was evaporated from a Dewar vessel. In the experiments, the cell became deformed at $T = 500$ °C because of the decomposition of lithium hydride. The investigators noted that the method of filling the cell used by them did not rule out the possibility of sample contamination. Because of decomposition and the possibility of contamination, they considered their results to be preliminary (in Table 2.4, the enthalpy is reckoned from the value at 25 °C).

An interpolation equation was derived for calculating the enthalpy, ΔH (cal/mole), of lithium hydride over the temperature interval 373–700 K,

$$\Delta H = H(T) - H(298.15) = 5.275T + 49.12 \times 10^{-4} T^2 + \frac{13.45}{T} - 2461. \tag{2.3}$$

In 1959, Novikov, Gruzdev, Kozlov, and Smirnov, at the Moscow Engineering Physics Institute, studied the enthalpy of lithium hydride over a broader temperature interval, 60–1100 °C (see Ref. 70). The measurements were carried out by the displacement method with the use of a massive copper calorimeter. The distinguishing feature of the apparatus was its ability to repeatedly drop the cell into the calorimeter. A special system was built for

Table 2.4. Enthalpy of lithium hydride ΔH (J/mole) from the experimental data of Rodygina *et al.*

T,°C	ΔH	T,°C	ΔH	T,°C	ΔH
98.46	2255	202.28	6 022	310.76	10 504
96.55	2198	309.62	10 463	390.70	14 291
94.33	2200	312.82	10 623	390.48	14 278
203.55	6084	307.58	10 382	390.78	14 259
201.39	5986				

Table 2.5. Enthalpy of lithium hydride, ΔH (cal/g), from the experimental data of Novikov *et al.*

T,°C	ΔH	T,°C	ΔH	T,°C	ΔH
63.1	52.3	690.5	1596	625.5	846.8
290.2	319.3	702.7	1640	785.2	1729
154.0	152.6	747.3	1720	815.0	1791
213.9	232.2	832.5	1857	868.1	1880
413.5	494.0	205.2	206.8	892.8	1888
496.6	620.9	274.3	294.1	911.5	1943
610.3	810.5	204.5	204.9	946.0	2007
648.3	887.0	314.7	343.6	1018	2096
663.5	939.0	400.6	459.0	1054	2140
679.2	1094	656.4	907.7	1087	2217

controlling the cell-supporting electromagnet, the shutter that screened the calorimeter from the furnace, and the calorimeter lid. The cell, made from 1X18H9T steel, had an interval volume of $\sim 3.5\ \mathrm{cm}^3$ and a wall thickness of 0.4–0.8 mm. The hermetically sealed cell held 1.1–1.4 g of lithium hydride. The composition of the sample was not analyzed.

The results of the experiment are presented in Table 2.5. An interpolation equation was derived by the authors for calculating the enthalpy, ΔH (cal/g), of solid lithium hydride as a function of temperature T (°C),

$$\Delta H = H_T - H_0 = 85.6\times10^{-2}T + 8.7\times10^{-4}T^2 - 15.3\times10^{-8}T^3\,. \quad (2.4)$$

A graphic analysis of the experimental data for solid and liquid phases led the authors to conclude that the heat of fusion of lithium hydride is 5200 ± 200 cal/mole.

Complete measurements of the heat capacity of ^{n}LiH over the temperature range 3.72–295.5 K were carried out by Kostryukov[25] by the direct-heating method. The experiments were carried out in a helium atmosphere in a stainless-steel calorimeter with an approximately 0.15-mm wall thickness. The calorimeter was calibrated using benzoic acid. The temperature was measured with a platinum resistance thermometer above 12 K and with a carbon resistance thermometer below 12 K. The sample was 99.8% pure.

The author noted that the measurements at low temperatures were not entirely stable, and that the heat capacity of the sample was slightly higher in the temperature interval 10–20 K. He attributed this behavior to a possible absorption of hydrogen above the sample. This assumption does not account,

Table 2.6. Enthalpy of lithium hydride (cal/g) from the experimental data of Wilson *et al.* (Ref. 254).

T,°C	$\dot{H}_t - H_{25}$	T,°C	$H_t - H_{25}$
Set I		Set II	
89.6	63.0	491.0	595.7
125.4	103.6	496.0	608.0
189.5	185.3	562.6	673.7
235.0	247.3	588.3	753.2
329.0	375.4	652.0	858.0
372.0	435.0	669.0	938.6
425.3	495.3	690.0	1110.3
470.5	556.1	714.5	1592.2
521.3	632.4	727.6	1620.4
585.0	740.3	753.5	1669.6
617.5	835.6		
643.0	863.7		
662.0	997.2		

however, for the fact that this phenomenon has been observed only over a relatively narrow temperature interval (see Sec. 2 of this chapter).

Wilson *et al.*[254] studied the enthalpy of lithium hydride over the temperature interval 90–750 °C by the displacement method. They followed the standard procedure using a copper calorimeter, whose temperature was measured by a platinum resistance thermometer. The calorimeter was calibrated within 0.5% with an electric heater.

The test sample was placed into a stainless-steel cell in a helium atmosphere. The cell was then sealed by arc welding. The disk-shaped cell was 12 mm high and 23 mm in diameter. The enthalpy of the sample was determined from the difference between the thermal effects in the calorimeter produced by a cell filled with hydride and by an empty cell.

Two sets of measurements were carried out. In the first set of measurements, the apparatus was filled with carbon dioxide gas, and in the second set, it was filled with helium.

The researchers did not report whether corrections were made for the thermal loss due to the insertion of the cell into the calorimeter. They reported, however, that the original sample contained 99.1% ^{n}LiH and that the cell's seal was broken at the end of the first set of measurements. They also noted that the cell was completely filled with hydride in the first set of measurements and only two-thirds filled in the second set. The experimental data obtained by Wilson *et al.*[254] are summarized in Table 2.6.

The authors attempted to estimate the melting point and the heat of fusion of lithium hydride from their data. Their estimates of these quantities turned out to be rather crude: $T_m = 690 \pm 5$ °C and $\Delta H_{\text{fus}} = 625 \pm 31$ cal/g. In determining the heat of fusion, the authors ignored the region near the melting point because of the appreciable prefusion effect.

In another study,[250] Welch presented data obtained from Kite's private communication. Using an ice calorimeter, Kite measured the enthalpy of solid-phase lithium hydride over the temperature interval 373–953 K and the enthalpy of liquid-phase lithium hydride over the temperature interval 973–

Table 2.7. Heat capacity c_p of lithium hydride [cal/(mole deg)] from the data of Kite.

T,K	c_p	*T*,K	c_p
Solid LiH		873	14.1
373	8.4	953	15.1
473	9.53	Liquid LiH	
573	10.7	973	15.7
673	11.8	1073	15.6
773	12.9	1173	15.5

1173 K and then calculated its heat capacity from these data (Table 2.7).

The studies of Shpil'rain *et al.*,[69] Busey and Bevan,[114] and Yates *et al.*[263] are but three among many recent studies dealing with the measurement of the caloric properties of lithium hydride.

The first study, which includes both the solid and liquid states, was carried out in the temperature interval 400–1325 K. The authors used the displacement method and a boiling water calorimeter built by them. This experiment was preceded by many measurements of the enthalpy of other substances with the use of the same apparatus. These measurements confirmed the high accuracy of the results.

The method used by these investigators can be described as follows. The lithium hydride sample consisting of small crystals was placed inside a stainless-steel cell in an argon atmosphere. The cell was then sealed in a vacuum by electron-beam welding. The cell was placed in the working zone of the furnace in a thermostatically controlled molybdenum block which was mounted on a long support bracket by means of an electromagnet. Upon reaching the specified temperature, the cell was dropped into the calorimeter containing the evaporating liquid, which in this case was distilled water. The additional part of the condensed vapor produced due to the heat liberated by the cell was determined by a gravimetric method. The temperature of the sample was measured with three platinorhodium-platinum thermocouples placed vertically along the thermostatically controlled molybdenum block.

In the analysis of the results of initial observations, the researchers introduced appropriate corrections for the heat capacity of the cell, for the heat of dissociation reaction of lithium hydride, for the barometric pressure, and for other quantities. Special attention was focused on the possible leakage of hydrogen through the cell walls. For this purpose, the cell containing the sample was heated in the furnace a reasonably short time (30–40 min), and the hydrogen loss was calibrated by weighing the cell on an analytical balance. Spectral and chemical analyses of the sample were also carried out in order to determine the impurity content before and after the experiment.

The latter procedures, along with the weighing, showed that at temperatures below ~1150 K the sample's composition remained constant within experimental error. At a higher temperature (1325 K), the hydrogen loss during the experiment was approximately 17 mg, corresponding to a 2.5% increase in the content of free lithium in the sample (an appropriate correction was introduced).

The authors have shown that the total mass content of the impurities

Table 2.8. Enthalpy of lithium hydride (cal/mole) from the experimental data of Shpil'rain *et al.* (Ref. 69).

T,K	*H*(*T*) − *H*(273)	*T*,K	*H*(*T*) − *H*(273)	*T*,K	*H*(*T*) − *H*(273)
412.7	861	890.5	6 466	968.6	12 993
521.0	2179	940.5	7 271	1012.7	13 907
562.0	2539	950.0	7 564	1083.8	14 784
701.4	4263	961.6	12 882	1151.6	16 038
732.7	4410	966.4	13 031	1325.5	18 500

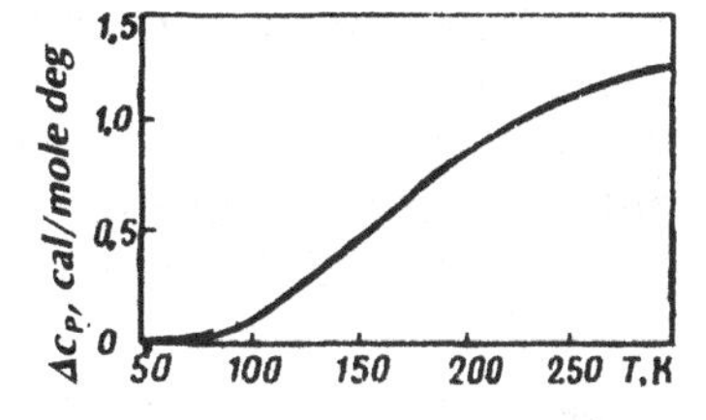

Figure 2.1. Difference between the heat capacities Δc_p versus temperature.[114]

was about 1.0%, of which 0.5% corresponds to free lithium. The internal volume of the cell was ~15 cm^3, and the mass of the sample was ~6 g. The experimental data are presented in Table 2.8. The value $H(373) - H(273) = 748$ cal/mole was taken from the data of Novikov *et al.* The authors estimate the error in the enthalpy to be in the range 0.7%–0.8%. From these results, the heat of fusion of lithium hydride was found to be 5300 ± 80 cal/mole.

Busey and Bevan[114] studied the heat capacity of ^{7}LiH and ^{7}LiD over the temperature interval 30–300 K. Unfortunately, they do not describe the measurement method used (they simply mention that the samples are of high isotopic purity) and do not give the experimental values of the heat capacity c_P obtained. They discuss in detail the difference between the heat capacities of lithium hydride and lithium deuteride (Fig. 2.1), analyze the results in terms of the data on the phonon spectra obtained by Verbl *et al.*[243] and Jaswal and Hardy,[158] and determine the Debye temperatures of these isotopic modifications.

The analysis of the heat capacities of ^{7}LiH and ^{7}LiD carried out by Busey and Bevan was based on the theory which states that the heat capacity of a crystal of constant volume, c_V, can be expressed as a sum:

$$c_V = c_{\rm ac} + c_{\rm opt}, \tag{2.5}$$

where $c_{\rm ac}$ and $c_{\rm opt}$ are the components of the acoustic and optical branches of the oscillations, respectively. The heat capacity associated with the acoustic oscillations can be approximated by the Debye function,

$$c_{\rm ac} = 3D(\Theta_{\rm ac}/T), \tag{2.6}$$

Table 2.9. The spectrum-averaged optical frequencies $\overline{\omega}$ of ^{7}LiH and ^{7}LiD crystals (cm^{-1}).

Isotopic modification	$\overline{\omega}_{LO}$	$\overline{\omega}_{TO}$	References
^{7}LiH	956	784	Verbl *et al.* (Ref. 243)
	1125	800	Jaswal and Hardy (Ref. 158)
^{7}LiD	690	554	Verbl *et al.* (Ref. 243)
	820	570	Jaswal and Hardy (Ref. 158)

where Θ_{ac} is the characteristic temperature of the acoustic oscillations. Busey and Bevan assumed that the values of Θ_{ac} and hence c_{ac} for ^{7}LiH and ^{7}LiD are the same.

The optical part of the heat capacity was analyzed using the Einstein approximation, in which the oscillation spectrum was divided into longitudinal and transverse components,

$$c_{opt} = E(\Theta_{LO}/T) + 2E(\Theta_{TO}/T), \tag{2.7}$$

where $\Theta_{LO} = \hbar\overline{\omega}_{LO}/k$; $\Theta_{TO} = \hbar\overline{\omega}_{TO}/k$; $\overline{\omega}_{LO}$ and $\overline{\omega}_{TO}$ are the spectrum-averaged frequencies of the longitudinal and transverse optical oscillations, respectively; and E is the Einstein function.

At constant pressure, the heat capacity c_P can be determined from the standard thermodynamic relation

$$c_P = c_V + 9\alpha^2 Tv/\kappa_{iso}, \tag{2.8}$$

where α^2 is the linear thermal expansion coefficient, v is the molar volume, κ_{iso} is the isothermal compressibility, and T is the absolute temperature.

Busey and Bevan found that the difference between c_P and c_V is vanishingly small up to temperatures of about 100 K, at which temperature this difference becomes noticeable, reaching about 4.5% at room temperature. They also reported that the difference* for the isotopic compositions studied by them is

$$\Delta = \left(\frac{9\alpha^2 Tv}{\kappa_{iso}}\right)_{^7\mathrm{LiH}} - \left(\frac{9\alpha^2 Tv}{\kappa_{iso}}\right)_{^7\mathrm{LiD}} \ll c_V. \tag{2.9}$$

Working from the assumption that the acoustic components of the heat capacity of ^{7}LiH and ^{7}LiD are the same and using relation (2.9), Busey and Bevan found

$$\Delta c_P = c_{P,^7\mathrm{LiD}} - c_{P,^7\mathrm{LiH}} = c_{\mathrm{opt},^7\mathrm{LiD}} - c_{\mathrm{opt},^7\mathrm{LiH}} \tag{2.10}$$

and they compared the values of Δc_P obtained by them experimentally with the values of Δc_{opt}, calculated in the Einstein model at the optical frequencies $\overline{\omega}_{TO}$ and $\overline{\omega}_{LO}$, and which were found from the data of Refs. 158 and 243. The values of Δc_P measured experimentally are in good agreement with the calculated value at the frequencies $\overline{\omega}_{TO}$ and $\overline{\omega}_{LO}$, which are the average frequencies based on the data of Refs. 158 and 243. The frequencies for the ^{7}LiH and ^{7}LiD crystals are given in Table 2.9, and their average values are

* We have estimated this difference to be 1% at room temperature, which is within the measurement error margin of the heat capacity.

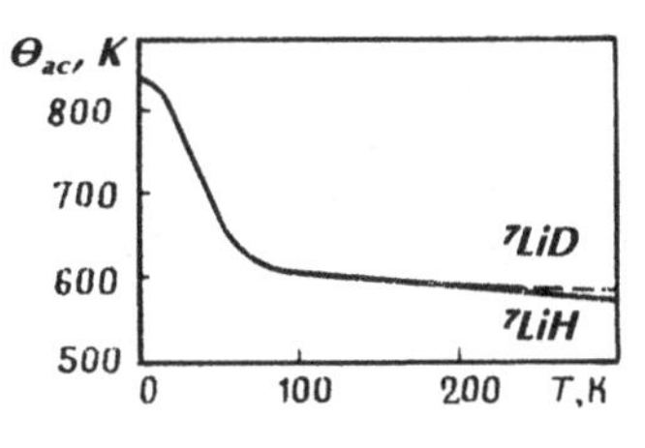

Figure 2.2. Temperature dependence of the characteristic temperature of the acoustic oscillations θ_{ac} for ^{7}LiH and ^{7}LiD (Ref. 114).

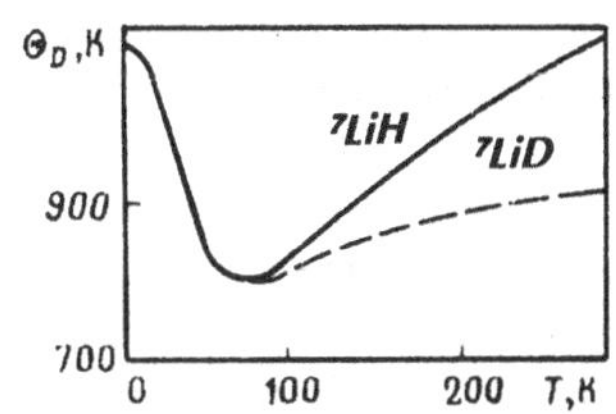

Figure 2.3. Temperature dependence of θ_D for ^{7}LiH and ^{7}LiD.

$\overline{\omega}_{TO}(^7\text{LiH}) = 792\,\text{cm}^{-1}$, $\overline{\omega}_{LO}(^7\text{LiH}) = 1040\,\text{cm}^{-1}$, $\overline{\omega}_{TO}(^7\text{LiD}) = 562\,\text{cm}^{-1}$, and $\overline{\omega}_{LO}(^7\text{LiD}) = 755\text{ cm}^{-1}$.

Using the analytical method indicated above, Busey and Bevan also found the values of c_{ac} and Θ_{ac} for these compounds. The functional dependence $\Theta_{ac} = f(T)$ in Fig. 2.2. shows that above ~100 K, Θ_{ac} changes only slightly and is ~600 K.

Busey and Bevan found that if the characteristic Debye temperature is determined in terms of the total heat capacity, as is often done, i.e.,

$$c_V = 6D(\Theta_D/T), \tag{2.11}$$

then Θ_D will depend on the temperature in a rather complicated manner. A complex functional dependence $\Theta_D = f(T)$ for the two isotopes in Fig. 2.3 shows that the Debye approximation of the whole spectrum for lithium hydride over the temperature range studied is a poor approximation, since the crystal oscillation spectrum has a large gap between the acoustic and optical branches because of the large difference in the masses of the heavy and light particles in the cell (in this case, Li and H).

The average values of the Debye temperature for the ^{7}LiH and ^{7}LiD crystals found by Busey and Bevan from the relation

$$\overline{\Theta}_D = \frac{1}{T}\int_0^T \Theta_D(T)dT \tag{2.12}$$

are listed in Table 2.10, along with other data in the literature, obtained from the analysis of the measurements of the heat capacity.

Yates *et al.*[263] studied the heat capacity of ^{7}LiH and ^{7}LiD over the temperature interval from liquid helium to 320 K. They used the direct-heating method with an adiabatic calorimeter. The main components of the calorimeter (Refs. 172 and 256) were made from silver which has a high thermal conductivity and low helium-absorption capacity. The internal space of the calorimeter, in which the test sample was placed, was filled with helium to ensure good heat transfer between the calorimeter's heater and the sample. The

Table 2.10. Average values of the Debye temperature $\overline{\Theta}_D$ (K) for lithium hydride and lithium deuteride.

Temperature interval, K	$\overline{\Theta}_{D,\ LiH}$	$\overline{\Theta}_{D,\ LiD}$	References
74–293	825	—	Günther (Ref. 133) and Moers (Ref. 194)
70–100	815	611	Ubbelohde (Ref. 229)
0–298	936	—	Kostryukov (Refs. 25, 114)
0–300	953	878	Busey and Bevan (Ref. 114)

calorimeter was surrounded by a special screen equipped with an auxiliary heater in order to prevent heat loss from the calorimeter. The temperature of the calorimeter was measured by a germanium resistance thermometer. Before and after each experiment, a control check based on the freezing point of water was carried out to determine the stability of the thermometer. The authors estimated the error in measuring the temperature of the sample to be 0.01 K.

In their experiments, the authors were aware of the anomalous behavior of the heat capacity of lithium hydride at low temperatures, which was discovered by Kostryukov.[25] Accordingly, they gave special attention to the purity of the sample and to monitoring its content during the experiment. They estimated relative error of the heat capacity temperature from room temperature to 50 K to be 2%. A further decrease in the temperature caused the error to increase to 10% because of the increase in the heat capacity of the measuring instrument, which changed the final result.

These experiments confirmed the anomalous increase in the heat capacity of lithium hydride and lithium deuteride at low temperatures, as was pointed out by Kostryukov. Heat-capacity peaks indicating a possible structural transformation were detected in ^{7}LiH at temperatures 11.1 ± 0.2 K and in ^{7}LiD at temperatures 12.8 ± 0.2 K. Attempts of other researchers to detect similar transitions in lithium hydride by compressing the sample at high pressures, however, have not been successful, as was pointed out by Yates *et al.*[263] The Debye temperature (1190 ± 80 K) obtained by Yates *et al.*[263] for ^{7}LiH at low temperature is higher than that obtained by Kostryukov (850 K). The Debye temperature for ^{7}LiD is 1030 ± 50 K. At temperatures above 50 K, the data on the heat capacity of lithium hydride obtained by Kostryukov[25] and Yates *et al.*[263] are in satisfactory agreement (within 3%). At lower temperatures, the divergence between the results of these studies increases because of the greater care taken by Yates *et al.*[263] in planning the experiment in this temperature range.

There are also several studies in which the heat of fusion of lithium hydride was determined. Messer and co-workers carried out several measurements of the heat of fusion of ^{n}LiH, generally using cryoscopy methods. In studying the LiH–Li_2O and LiH–CaH_2 systems, Messer[189] found $\Delta H_{fus} = 4900 \pm 700$ cal/mole at the melting point of ^{n}LiH (688 °C). Messer *et al.*[191] found the average value of the heat of fusion of lithium hydride to be 5095 ± 460 cal/mole at the melting point (688.6 °C) for these systems.

Vogt's data on the heat of fusion of lithium hydride have been cited by several authors in the literature. A value of $\Delta H_{fus} = 5240$ cal/mole, for example, is given by Johnson *et al.*,[164,165] in which they refer to a study of Vogt. A review by Vogel *et al.*[245] contains some additional information that seems to refer (implicitly) to the origin of these data. Specifically, in this review the authors report, with a citation to a study by Vogt, that the heat of fusion was found to be 5237 cal/mole by a group of researchers who used an adiabatic displacement calorimeter. In the same review, the authors point out, on the basis of several private communications, that Hoffman's calorimetric measurements of the heat of fusion of lithium hydride yield a value of 5518 cal/mole. The average value of ΔH_{fus}, as pointed out in this review, is 5240 ± 225 cal/mole.

Vogel and co-workers carried out an independent study of the heat of fusion of lithium hydride and of several other compounds by the thermal-analysis method. Before carrying out thermal measurements, they calibrated the system on the basis of the heat-transfer rate of substances with known thermophysical properties. The experiments with lithium hydride were performed under hydrogen pressure. The heat of fusion of lithium hydride was found to be 5240 cal/mole; i.e., it was the same as the mean value indicated above. Welch[250] gives a value of 5519 ± 25 cal/mole for the heat of fusion of ^{n}LiH. This value was obtained by Kite from measurements of the enthalpy in an ice calorimeter.

In summary, our analysis of the data in the literature on the caloric properties of lithium hydride and of its isotopic modifications shows that the properties of lithium hydride have been measured extensively, both in the solid and liquid phases. Our study also shows that a systematic statistical analysis of the data on the heat capacity and enthalpy, with allowance for the relative error of the results of each study, must be carried out. Such an analysis would make it possible to derive standard working equations for calculating the caloric functions of the heat capacity, enthalpy, entropy, and the thermodynamic potential.

In the case of lithium deuteride, only the heat capacity of this compound in the solid phase has been measured up to a temperature of 320 K. There are no experimental data on the caloric properties of LiD for higher temperatures, including the liquid phase. The caloric properties of lithium tritide have not been determined.

1.2. Analysis of experimental data on the caloric properties of lithium hydride and lithium deuteride

In the preceding section, we have discussed the principal studies in which the caloric properties of lithium hydride and lithium deuteride were measured. Only the heat capacities of solid-phase LiH and LiD have been measured over the temperature interval 0–320 K. At higher temperatures, only the enthalpy of lithium hydride has been measured. For lithium deuteride and lithium tritide, no measurements have been carried out in this temperature range.

The data on the heat capacity of solid lithium hydride, which were obtained by Kite and cited by Welch[250] in a table of fitted values, were determined from the analysis of enthalpy measurements in the temperature inter-

Table 2.11. Relative error δ of the initial data on the caloric properties of lithium hydride.

Authors	δ (%)	Authors	δ (%)
Heat capacity		*Enthalpy*	
Günther and Moers	3	Rodygina *et al.*	1.6
Kostryukov	2	Novikov *et al.*	1.0
Kite	3	Lang	1.6
Yates *et al.*	2	Shpil'rain *et al.*	1.0
		Wilson *et al.*	1.6

val 373 –1173 K in an ice calorimeter. These data are therefore a secondary source of information.

Taking into account these circumstances and the anomalous behavior of the heat capacity of lithium hydride and lithium deuteride at low temperatures, as pointed out in Refs. 25 and 263, we decided to divide the entire temperature interval from 0 K to T_m into three parts (0–50, 50–298.15, and 298.15 K–T_m) in the analysis of the solid-phase data, and to approximate each part separately (see Ref. 87). The three parts were joined on the basis of the heat capacity. This method can be used to obtain relatively simple approximate equations, without reducing the accuracy necessary for practical applications.

Since the effect of the heat capacities of lithium hydride and lithium deuteride in the initial temperature region, 0–50 K, on the caloric properties in the next principal temperature region is small, and since the dependence of the heat capacity in the initial temperature region is complex—it has a peak—we had to analyze the data graphically in the temperature region 0–50 K, following the procedure used by Yates *et al.*[263] The data in regions 2 and 3 were analyzed by the method of least squares, with a precise estimate of the relative error of each initial source. This estimate was based on the analysis of the studies discussed in Sec. 1.1. The scale for the estimates is indicated in Table 2.11.

The larger errors in the data of Günther and Moers, Rodygina *et al.*, Lang, and Wilson *et al.* stem from inadequate purity of the tested samples (as pointed out by the authors of these studies) and from the large spread in the experimental data. As for Kite's study, we have access to only the data on the heat capacity which were obtained from the measurements of the enthalpy. In the solid phase, the measurements were carried out over the temperature interval 370–950 K, and in the liquid state, they were carried out over the temperature interval 970–1170 K. Although our aim was to include only the primary experimental results in the joint analysis, we felt it inadvisable to completely ignore Kite's data, since they were obtained experimentally. These considerations account for the errors of the data in Table 2.11.

The results of Novikov *et al.* and Shpil'rain *et al.* were scaled to the temperature of 298.15 K on the basis of Ref. 70.

In the temperature interval 50–298.15 K, the heat capacity c_P (J/mole deg) can be approximated by a power series,

$$c_P = T^{1.5} \sum_{n=1}^{6} a_n T^{n/2}. \qquad (2.13)$$

Table 2.12. Coefficients in Eq. (2.13) for lithium hydride and lithium deuteride.

n	a_n LiH	a_n LiD	n	a_n LiH	a_n LiD
1	$1.963\,98\times10^{-3}$	$-4.779\,53\times10^{-4}$	4	$-3.560\,68\times10^{-5}$	$-1.854\,53\times10^{-5}$
2	$-1.496\,21\times10^{-3}$	$-4.147\,18\times10^{-4}$	5	$1.561\,42\times10^{-6}$	$8.303\,56\times10^{-7}$
3	$3.617\,84\times10^{-4}$	$1.688\,08\times10^{-4}$	6	$-2.551\,14\times10^{-8}$	$-1.344\,61\times10^{-8}$

The coefficients a_n of the power series for LiH and LiD are given in Table 2.12.

At temperatures above 298.15 K, the data on the heat capacity and enthalpy of lithium hydride were analyzed jointly.* In this case, the heat capacity c_P (J/mole deg) was expressed as a whole-number power series

$$c_P = 27.08 + \sum_{m=1}^{5} b_m (T - 280)^m. \tag{2.14}$$

The coefficients b_m of the power series are given in Table 2.13.

Equation (2.14) shows that the heat capacity tends to increase near the melting point. The scatter in the experimental data on the heat capacity and enthalpy of lithium hydride in the solid phase with respect to the curves plotted from Eqs. (2.13) and (2.14) are shown in Figs. 2.4 and 2.5. As can be seen in these figures, equations describe the experimental data satisfactorily; the divergence of most of the experimental points does not exceed the 2.5% limit.†

A specific analysis of the data on the heat capacity and enthalpy of lithium hydride at temperatures between 298.15 K and T_m showed that the analysis of the available data beginning at $T_0 = 280$ K produces the best agree-

*The method of joint analysis of the data can be summarized as follows. The temperature dependence of the heat capacity can be written as a power series $c_P(T)_{\text{calc}} = c_P(T_0) + \Sigma_{m=1}^{m} b_m x_1(m)$, where $x_1 = (T - T_0)^m$. If the heat capacity is expressed in this manner, we can write the temperature dependence of the enthalpy as $[H(T) - H(0)]_{\text{calc}} = H(T_0) - H(0) + c_P(T_0)(T - T_0) + \Sigma_{m=1}^{m} b_m x_2(m)$, where $x_2(m) = \int_{T_0}^{T} x_1(m) dT = (m+1)^{-1} (T - T_0)^{m+1}$. We can then write the functional

$$F = \Sigma_{i=1}^{k} [\sigma_i^2(c_P)]^{-1} [c_P(T_i)_{\text{calc}} - c_P(T_i)_{\text{exp}}]^2 + \Sigma_{i=1}^{n} [\sigma_i^2(H)]^{-1} \times \{ [H(T_i) - H(0)]_{\text{calc}} - [H(T_i) - H(0)]_{\text{exp}} \}$$

and minimize it by the method of least squares in order to determine the coefficients b_m of the series. Here k is the number of experimental points for the specific heat, n is the number of experimental points for the enthalpy, and σ_i is the error in determining the experimental points. The number of terms, m, in the series is determined from the condition corresponding to the best fit between the calculated and experimental points over the whole temperature interval under consideration.

† In Fig. 2.4, the divergence of the experimental data has a systematic sinusoidal nature. This behavior indicates that another approximation, which would allow the spread of data to be reduced slightly, can be found. Such an attempt, however, would inevitably lead to the formulation of a more complex equation. In the analysis of these data, this was not thought to be the best approach.

Table 2.13. Coefficients in Eq. (2.14) for lithium hydride.

m	b_m	m	b_m
1	$1.090\,78\times10^{-1}$	4	$-2.935\,47\times10^{-10}$
2	$-2.796\,15\times10^{-4}$	5	$5.595\,45\times10^{-14}$
3	$4.747\,55\times10^{-7}$		

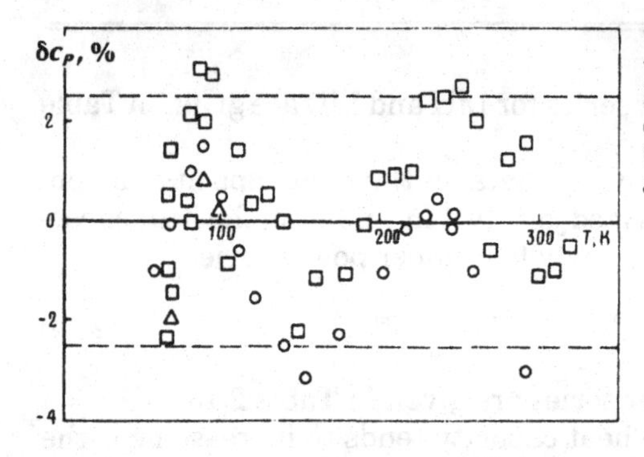

Figure 2.4. Deviation of the experimental values of the heat capacity of solid lithium hydride from the calculated values determined from Eq. (2.13). ○—Kostryukov; □—Yates et al.; △—Günther and Moers.

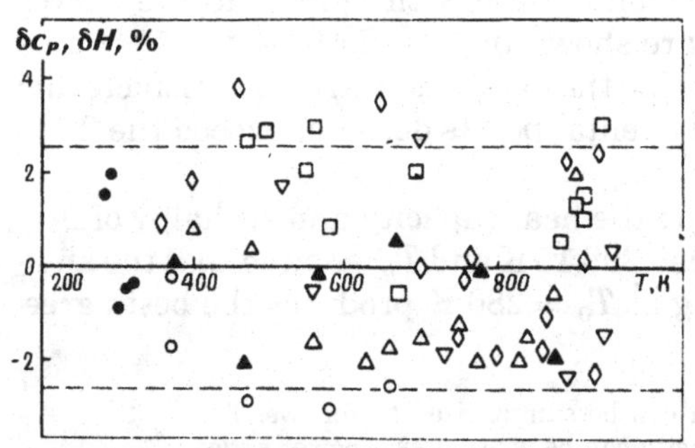

Figure 2.5. Deviation of the experimental values of the heat capacity (filled circles) and enthalpy (open circles) of solid lithium hydride from the calculated values determined from Eq. (2.14) (according to the data of different investigators). ○—Rodygina; □—Novikov *et al.*; △—Lang; ▽—Shpil'rain *et al.*; ◇—Wilson *et al.*; ●—Yates et al.; ▲—Kite.

ment between the calculated and experimental values. The discrepancy in the heat capacity at a temperature of 298.15 K found from Eqs. (2.13) and (2.14) is 0.2% in this case. Since this discrepancy is considerably smaller than the relative error of the initial experimental data, the joining of these two equations at a temperature of 298.15 K may be assumed valid. A correction for the difference between the caloric properties of hydride containing natural lithium ^{n}Li or a pure isotope ^{7}Li was not introduced in the analysis, since this difference produces a much smaller effect than the errors of the initial data or of the calculations.

For the liquid phase, the only data that merit attention are those on the enthalpy of lithium hydride which were obtained by Novikov *et al.*, Shpil'rain *et al.*, and Wilson *et al.* (see Table 2.1). These data were analyzed, without changing the relative error in Table 2.11, in the form of a polynomial (J/mole) of the type

Table 2.14. Coefficients in Eq. (2.15).

k	c_k	k	c_k
0	-1.3213×10^{5}	2	-2.9844×10^{-1}
1	4.0125×10^{2}	3	8.4659×10^{-5}

Table 2.15. Relative deviation of the experimental values of the enthalpy of liquid lithium hydride from the values calculated from Eq. (2.15).

T,K	δH,%	T,K	δH,%
The data of Novikov *et al.*		The data of Shpil'rain *et al.*	
963.6	−0.29	961.6	0.21
975.8	−0.41	966.4	0.78
1020.4	1.00	968.6	0.27
1105.6	0.88	1012.7	2.29
1058.3	−1.91	1083.8	1.80
1088.2	−1.13	1151.6	4.32
1141.2	−0.73	1325.5	5.42
1165.9	−2.17	The data of Wilson *et al.*	
1184.6	−0.86	987.7	−2.83
1219.1	−0.30	1000.8	−2.52
1291.1	−1.36	1026.7	−2.20
1327.0	−2.00		
1360.0	−1.11		

$$H(T)-H(0)=\sum_{k=0}^{3} c_k T^k. \tag{2.15}$$

The values of the coefficients are given in Table 2.14.

Table 2.15 gives the relative deviation of the experimental values of the enthalpy

$$\delta H=\frac{[H(T)-H(0)]_{\rm exp}-[H(T)-H(0)]_{\rm calc}}{[H(T)-H(0)]_{\rm calc}} \tag{2.16}$$

from those calculated on the basis of Eq. (2.15).

Equations (2.13)–(2.16) were used to calculate the caloric functions of solid and liquid lithium hydride at temperatures up to 1300 K. These caloric properties are presented in Table 5.1. According to these data, the heat of fusion turned out to be 21 340 J/mole.

Analysis of the experimental data on the enthalpy obtained for lithium hydride in this temperature region clearly shows that the heat capacity increases near the melting point. The heat of fusion is, of course, too high in those studies which ignore the increase in the heat capacity. In our analysis, we have taken into account this effect by making use of the experimental data near the melting point.

The errors of the data in Table 5.1 on lithium hydride should be determined on the basis of an analysis of the original studies from which the standard working equations (2.13)–(2.16) were obtained. It can be assumed from the estimates in Table 2.11 that the maximum relative error in the caloric functions of the solid-state lithium hydride in Table 5.1 is no greater than 2%. The maximum relative errors of the data on the enthalpy is also estimated to be 2% from liquid lithium hydride. The other caloric functions found by using a temperature-differentiated equation [Eq. (2.16)] have larger errors. Since the temperature dependence of enthalpy is relatively slight and rather simple, we estimate the maximum errors of the data on the heat capacity and the rest (except for the enthalpy) of the caloric functions of liquid lithium hydride to be 3%.

The errors of the data on the caloric properties (principally the heat capacity and enthalpy) increase near the solid-phase melting point (beginning roughly at temperatures of 850–900 K). It is difficult, however, to quantitatively estimate the increase in the value of this error.

1.3. Calculation of the caloric functions of the isotopic modifications of solid-state lithium hydride

Analysis of the experimental studies of the caloric properties of lithium hydride and of its solid-state isotopic compositions carried out in Sec. 1.1 of this chapter shows that lithium hydride has been studied experimentally essentially over the entire temperature range from 0 K to the melting point. The data on lithium deuteride are available only up to room temperature. The caloric properties of lithium tritide have not been studied experimentally.

Calculation and theoretical methods[35,39,85,88] have been used to determine the thermodynamic properties of the isotopic modifications of lithium hydride in the crystalline state for those cases in which there are no experimental data available.

The interparticle interaction in a crystal is basically caused by the interaction of the outer electrons of the atoms and remains essentially constant in the case of an isotopic substitution. The potential curves for the interaction of particles remain constant, and the interaction parameters, in particular, the depth of the potential well and the equilibrium interparticle distance, are the same in the harmonic approximation for the LiH, LiD, and LiT isotopes [the volume of the unit cell of a crystal in the harmonic approximation $\tilde{v} = \tilde{l}^3/4$, for example, is independent of the isotopic composition of lithium hydride (l is the lattice constant)]. At the same time, isotopic replacement has a substantial effect on the vibrational frequencies of a crystal, which are associated with the mass of the atoms that constitute a given crystal. The replacement of an atom by its isotope in a LiH-type crystal (especially in the case of an isotopic replacement of hydrogen) therefore has an appreciable effect on such properties as density and heat capacity.

The temperature dependence of the heat capacity at low temperatures has been studied theoretically sufficiently well for crystalline substances with a simple structure. At $T \ll \Theta_D(0)$, where $\Theta_D(0)$ is the characteristic Debye temperature for low temperatures of the crystal, only the long-wave oscillations—the low-energy acoustic phonons—are excited ($\hbar\omega \sim kT$, where $\hbar$, ω, and k are Planck's constant, the crystal-lattice oscillation frequency, and

Boltzmann's constant, respectively). In this case the heat capacity (per cell) is given by (see Ref. 30, for example)

$$c_V = 6k(4\pi^4/5)[T/\Theta_D(0)]^3. \tag{2.17}$$

The limiting Debye frequency ω_D, and hence the Debye temperature $\Theta_D(0) = \hbar\omega_D/k$, can be expressed in terms of the velocity of sound or the elastic constants of the crystal (see Refs. 39 and 208, for example).

In the case of cubic crystals, Eq. (2.17) generally describes satisfactorily the behavior of the heat capacity at temperatures between 0 and 30–50 K. In the case of lithium hydride, however, this relation cannot be used for specific calculations because of the presence of the heat-capacity peak in this temperature region. At the same time, Yates *et al.*[263] have shown experimentally that the heat capacity of ^{7}LiH differs negligibly from that of ^{7}LiD at temperatures up to 50 K (and even slightly higher). This difference becomes noticeable at $T \geqslant 100$ K (see Fig. 2.1). At these temperatures, both the long-wave and short-wave lattice vibrations occur in the crystal and the heat capacity at a constant volume, c_V, is affected by both the optical and acoustic branches (see Refs. 8 and 31, for example).

There are two methods in which the behavior of the heat capacity can be analyzed at these temperatures. The first method involves the use of the Debye approximation for the sound waves and the Einstein approximation for the optical waves. This method was used by Busey and Bevan[114] to analyze the experimental data on the heat capacity of lithium hydride and lithium deuteride. The second method, which yields good results at high temperatures, is based on the use of the Debye or Einstein approximation for the total range of c_V. The second method, widely used in solid-state physics theory, is more useful (where it applies) than the first, since it requires only the characteristic temperature. However, its applicability is limited to relatively high temperatures.

In the first method, the heat capacity of a crystal can be represented as a sum of two terms—acoustic term and optical term:

$$c_V = c_{\text{ac}} + c_{\text{opt}}. \tag{2.18}$$

The contribution to the heat capacity of crystalline lithium hydride from acoustic oscillations can be represented as a Debye approximation,

$$c_{\text{ac}} = 3D\left(\frac{\Theta_{\text{ac}}}{T}\right) = 9k\left(\frac{T}{\Theta_{\text{ac}}}\right)^3 \int_0^{\Theta_{\text{ac}}/T} \frac{x^4 e^x}{(e^x - 1)^2}\, dx. \tag{2.19}$$

All acoustic oscillations, including short-wave oscillations, become excited at intermediate and high temperatures. The frequencies of long-wave acoustic oscillations are proportional to $1/\sqrt{M+m}$, where M is the mass of the heavy atom and m is the mass of the light atom, and the frequencies of short-wave acoustic oscillations depend solely on the mass of the heavy atom in the unit cell and are proportional to $1/\sqrt{M}$ (see Ref. 23, for example). At high temperatures, we can accordingly assume that $\Theta_{\text{ac}} \sim 1/\sqrt{M}$, since the state density is high for short-wave oscillations. At $T \geqslant 100$ K, Θ_{ac} is, according to the data of Busey and Bevan, essentially constant (it is the same for ^{7}LiH and ^{7}LiD) and has the value of ~ 600 K. It thus seems reasonable that this value of Θ_{ac} also applies to the ^{7}LiT crystal.

For the isotopic modifications of ^{6}LiH(D,T), we find

$$\frac{\Theta_{ac}\,{}^6\mathrm{LiH(D,T)}}{\Theta_{ac}\,{}^7\mathrm{LiH(D,T)}} = \sqrt{\frac{M_{^7\mathrm{Li}}}{M_{^6\mathrm{Li}}}}, \tag{2.20}$$

where $\Theta_{ac,^6\mathrm{LiH(D,T)}} \approx 648$ K.

A good description of the optical contribution to the heat capacity is given by the Einstein function, especially when the mass of the heavy atom differs markedly from the mass of the light atom in the cell, since the long-wave and short-wave frequencies differ negligibly in this case [see Eqs. (2.24) and (2.25) below] and the lattice vibrations are nearly monochromatic:

$$c_{\mathrm{opt}} = E(\Theta_{LO}/T) + 2E(\Theta_{TO}/T), \tag{2.21}$$

where

$$\Theta_{LO} = \hbar\overline{\omega}_{LO}/k, \quad \Theta_{TO} = \hbar\overline{\omega}_{TO}/k; \tag{2.22}$$

$\overline{\omega}_{LO}$ and $\overline{\omega}_{TO}$ are the spectrum-averaged longitudinal and transverse optical frequencies of the crystal, respectively, and

$$E(\Theta/T) = E(x) = kx^2e^x/(e^x - 1)^2. \tag{2.23}$$

Using the data on the phonon spectra of ^{7}LiH and ^{7}LiD from Refs. 158 and 243, Busey and Bevan found the average optical frequencies of these two compounds and the characteristic temperatures Θ_{LO} and Θ_{TO} (Table 2.16).

The phonon spectra of ^{7}LiT and ^{6}LiH(D,T) crystals have not been studied. The following approach can be used to determine the typical optical temperatures of these compounds. In the limiting cases of long-wave and short-wave oscillations, the difference between the optical frequencies of a LiH crystal, in which the masses of the atoms differ markedly, is only slight, since for the long-wave oscillations, we have

$$\omega_{\mathrm{opt}} \sim \sqrt{(1/M) + (1/m)} = \sqrt{1/\mu}, \tag{2.24}$$

and for the short-wave oscillations, we have

$$\omega_{\mathrm{opt}} \sim \sqrt{1/m}. \tag{2.25}$$

Proceeding from these circumstances, we can find the average optical oscillation frequency of one isotopic modification of lithium hydride with respect to that of the other modification. Averaging of this sort can be carried out in many different ways. A test based on the data on phonon spectrum of ^{7}LiH and ^{7}LiD showed that the best approximations can be found by using the following relation for the averaging (of the longitudinal and transverse oscillations):

$$\frac{\overline{\omega}_{Mm_1}}{\overline{\omega}_{Mm_2}} = \frac{1}{2}\left(\sqrt{\frac{\mu_2}{\mu_1}} + \sqrt{\frac{m_2}{m_1}}\right) = \frac{1}{2}\sqrt{\frac{m_2}{m_1}}\left(1 + \frac{\sqrt{M+m_1}}{\sqrt{M+m_2}}\right). \tag{2.26}$$

For ^{7}LiH and ^{7}LiD, Eq. (2.26) in this case yields $\overline{\omega}_{^7\mathrm{LiH}}/\overline{\omega}_{^7\mathrm{LiD}} = 1.3735$, and an accurate analysis of the spectra carried out by Busey and Bevan on the basis of the experimental values of Refs. 158 and 243 gives $\overline{\omega}_{^7\mathrm{LiH}}/\overline{\omega}_{^7\mathrm{LiD}} = 1.3775$. The results are in satisfactory agreement with each other.

Equation (2.26) was used to determine the spectrum-averaged optical frequencies for ^{7}LiT and ^{6}LiH(D,T) on the basis of the data for ^{7}LiH and to

Table 2.16. Characteristic temperatures of the isotopic modifications of lithium hydride.

Temperature, K	^{7}LiH	^{7}LiD	^{7}LiT	^{6}LiH	^{6}LiD	^{6}LiT
Θ_{LO}	1496*	1086*	912	1505	1099	924
Θ_{TO}	1140*	809*	693	1145	834	705
Θ_{∞}	1170*	878	755	1182	894	774
Θ_E	906	680	585	916	692	600

*The values were obtained by Busey and Bevan (Ref. 114).

Table 2.17. Heat capacity c_V of ^{7}LiH, ^{7}LiD, and ^{7}LiT crystals in the harmonic approximation (a superposition of the acoustic and optical oscillations), cal/mole·deg.

		c_{opt}			$c_V = c_{ac} + c_{opt}$		
T, K	c_{ac}	^{7}LiH	^{7}LiD	^{7}LiT	^{7}LiH	^{7}LiD	^{7}LiT
100	1.58	0.01	0.08	0.20	1.59	1.66	1.78
150	3.00	0.12	0.61	1.02	3.12	3.61	4.02
200	3.95	0.50	1.44	2.03	4.45	5.39	5.98
250	4.54	1.06	2.27	2.90	5.61	6.81	7.44
300	4.92	1.69	2.98	3.56	6.61	7.90	8.48
350	5.18	2.28	3.54	4.06	7.46	8.72	9.24
400	5.34	2.80	3.97	4.42	8.14	9.31	9.76
450	5.46	3.24	4.31	4.70	8.70	9.77	10.17
500	5.55	3.61	4.57	4.91	9.17	10.12	10.46
550	5.62	3.92	4.78	5.07	9.54	10.40	10.69
600	5.67	4.18	4.95	5.20	9.85	10.62	10.87
650	5.72	4.40	5.08	5.31	10.11	10.80	11.03
700	5.75	4.58	5.19	5.39	10.33	10.94	11.14
750	5.78	4.73	5.28	5.47	10.50	11.05	11.24
800	5.80	4.86	5.36	5.52	10.66	11.16	11.32
850	5.81	4.97	5.42	5.58	10.78	11.23	11.39
900	5.83	5.07	5.49	5.63	10.90	11.32	11.46
950	5.84	5.15	5.54	5.66	10.99	11.38	11.50

determine the corresponding characteristic temperatures (Table 2.16).

The Debye functions $D(x)$ and the Einstein functions $E(x)$ have been tabulated in a number of publications (see Ref. 174, for example), so that at certain characteristic temperatures the heat capacity c_V can easily be found in the harmonic approximation of the isotopic modifications of lithium hydride in the crystalline state by using Eqs. (2.18), (2.19), and (2.21). The results of calculation of c_V in the harmonic approximation of the isotopic compositions of lithium hydride are given in Tables 2.17 and 2.18.

Figure 2.6 compares the effect on the heat capacities of the isotopic modifications ^{7}LiD and ^{7}LiT relative to the heat capacity of ^{7}LiH. We see that as the temperature is raised, the difference in the heat capacities increases, reaches a maximum at room temperature, and then decreases. It should be noted (see Tables 2.17 and 2.18) that the difference in the heat capacities of lithium hydride resulting from the isotopic replacement of lithium is an order of magnitude smaller than the difference in the heat capacities of lithium hydride caused by the isotopic replacement of hydrogen. This behavior can be

Table 2.18. Heat capacity c_V of the ^{6}LiH, ^{6}LiD, and ^{6}LiT crystals in the harmonic approximation (a superposition of the acoustic and optical oscillations), cal/mole·deg.

T,K	C_{ac}	C_{opt}			$C_V = C_{ac} + C_{opt}$		
		^{6}LiH	^{6}LiD	^{6}LiT	^{6}LiH	^{6}LiD	^{6}LiT
100	1.35	0.01	0.07	0.19	1.36	1.42	1.54
150	2.72	0.12	0.55	0.97	2.84	3.27	3.69
200	3.71	0.49	1.35	1.97	4.20	5.06	5.68
250	4.36	1.05	2.18	2.84	5.41	6.54	7.20
300	4.77	1.67	2.88	3.51	6.44	7.65	8.28
350	5.05	2.26	3.45	4.01	7.31	8.50	9.05
400	5.25	2.78	3.90	4.38	8.03	9.15	9.63
450	5.38	3.22	4.24	4.67	8.60	9.62	10.05
500	5.49	3.59	4.52	4.88	9.08	10.01	10.37
550	5.57	3.91	4.73	5.05	9.48	10.30	10.62
600	5.63	4.17	4.90	5.18	9.80	10.53	10.81
650	5.67	4.38	5.04	5.29	10.05	10.71	10.96
700	5.72	4.57	5.16	5.38	10.29	10.88	11.10
750	5.75	4.72	5.26	5.45	10.47	11.01	11.20
800	5.77	4.85	5.33	5.51	10.62	11.10	11.28
850	5.79	4.96	5.40	5.56	10.75	11.19	11.35
900	5.81	5.06	5.47	5.62	10.87	11.28	11.43
950	5.82	5.14	5.52	5.65	10.96	11.34	11.47

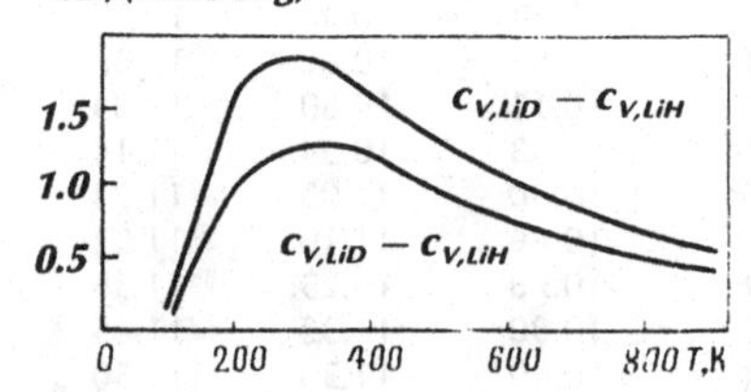

Figure 2.6. Comparison of the heat capacities c_V of the isotopic modifications of ^{7}LiH (D,T).

explained on the basis of Eqs. (2.24) and (2.26), which show that the optical oscillation frequency is virtually unaffected by the isotopic replacement of a heavy atom. For this reason, we have analyzed jointly the caloric data obtained for lithium hydride with natural lithium and with its isotope ^{7}Li.

Calculation of the heat capacity of a LiH crystal on the basis of a superposition of the acoustic and optical lattice vibrations makes it possible to determine the heat capacity even in the temperature region between 0 and 100 K (with the exception of a narrow band roughly between 5 and 15 K, in which a heat-capacity peak has been observed experimentally), where the characteristic acoustic temperature changes considerably (roughly from 880 K at $T = 0$ to 600 K at $T = 100$ K). Since the acoustic temperature and hence the acoustic component of the heat capacity differ negligibly in the case of the ^{7}LiH(D,T)-type isotopic modifications, we can write the following relation for calculating the heat capacity in this region:

$$c_{V,^7\mathrm{LiT}} = c_{V,^7\mathrm{LiH}} + (c_{\mathrm{opt},^7\mathrm{LiT}} - c_{\mathrm{opt},^7\mathrm{LiH}}). \tag{2.27}$$

For specific calculations, the values of $c_{V,^7\mathrm{LiH}}$ may be taken from the

Table 2.19. Heat capacity c_V of the isotopic compositions of lithium hydride in the Debye approximation, cal/mole deg.

T,K	^{7}LiH	^{7}LiD	^{7}LiT	^{7}LiH	^{7}LiD	^{7}LiT
300	6.18	8.10	8.89	6.09	7.90	8.69
350	7.21	8.89	9.56	7.12	8.72	9.46
400	8.10	9.47	10.02	8.00	9.38	9.93
450	8.70	9.93	10.36	8.68	9.84	10.36
500	9.18	10.28	10.68	9.18	10.19	10.60
550	9.66	10.53	10.83	9.56	10.44	10.82
600	9.93	10.76	11.04	9.80	10.68	10.96
650	10.19	10.90	11.17	10.18	10.83	11.11
700	10.36	11.04	11.23	10.36	10.97	11.23
750	10.60	11.10	11.34	10.52	11.11	11.34
800	10.76	11.23	11.40	10.68	11.20	11.41
850	10.83	11.30	11.45	10.83	11.29	11.43
900	10.97	11.37	11.50	10.96	11.35	11.50
950	11.10	11.43	11.55	11.04	11.40	11.54

experimental data of Ref. 263. Since the replacement of the ^{7}Li isotope by the ^{6}Li isotope has, as was indicated above, virtually no effect on the heat capacity of a LiH crystal, the heat capacity of ^{6}LiH(D,T) compounds may be assumed equal to within the experimental error to the heat capacity of ^{7}LiH(D,T) over the temperature interval under consideration.

In addition to the method considered above for calculating the heat capacity of a ^{7}LiH crystal in the harmonic approximation, we can also use at higher temperatures, with the help of the superposition of acoustic and optical oscillations, the Debye approximation with a characteristic Debye temperature Θ_∞ which can be expressed in terms of the mean-square oscillation frequency (see Ref. 31, for example):

$$\Theta_\infty = \frac{\hbar}{k}\sqrt{\frac{5}{3}\overline{\omega}^2} = \frac{\hbar}{k}\sqrt{\frac{5}{2}\frac{\tilde{l}}{\tilde{\kappa}}\frac{1}{\mu}}. \quad (2.28)$$

Here $\tilde{l}$ and $\tilde{\kappa}$ are, respectively, the lattice constant and crystal compressibility in the harmonic approximation, and μ is the reduced mass of the crystal cell (in grams).

Strictly speaking, the heat capacity in this case is determined by the Debye function

$$c_V = 6D(\Theta_\infty/T). \quad (2.29)$$

Table 2.16 gives the Debye temperatures Θ_∞ of various isotopic modifications of lithium hydride. The method of determining the values of $\tilde{l}$ and $\tilde{\kappa}$ necessary for this calculation is shown below (see Sec. 2.2).

Table 2.19 lists the heat capacities c_V of the isotopic compositions of lithium hydride in the harmonic approximation, obtained from relation (2.29). A comparison of these data with the data of Tables 2.17 and 2.18 shows that the heat capacity of a LiH crystal is described satisfactorily in the Debye approximation, beginning with the temperature of about 400 K. The difference in the heat capacities obtained by the two methods discussed above is no greater than 1% over the temperature interval 400–950 K.

At high temperatures, the Einstein approximation with the characteristic Einstein temperature

$$\Theta_E = \frac{\hbar}{k}\sqrt{\overline{\omega^2}} = \sqrt{\frac{3}{5}}\,\Theta_\infty \qquad (2.30)$$

can also be used to calculate the heat capacity of a LiH crystal in the harmonic approximation. The values of Θ_E are listed in Table 2.16.

The estimates carried out in Ref. 35 show that the contribution of anharmonicity to the heat capacity c_V accounts for no more than 1%–2%. The constant-volume heat capacity c_V of the isotopic modifications of crystalline lithium hydride can thus be calculated in the entire temperature interval of the compound in the solid state. Although the method of superposition of the acoustic and optical components is slightly more difficult than the Debye approximation (with Θ_∞) or the Einstein approximation (with Θ_E), since it requires the knowledge of three characteristic temperatures, Θ_{ac}, Θ_{LO}, and Θ_{TO}, it can be used over a broader temperature interval, and hence is preferable in calculating the heat capacity c_V.

The conversion from the constant-volume heat capacity c_V to the constant-pressure heat capacity c_P can be brought about by using a standard thermodynamic equation

$$c_P - c_V = 9\alpha^2 v T/\kappa_{iso}, \qquad (2.31)$$

where α is the coefficient of linear thermal expansion of the crystal, v is the molar volume, and κ_{iso} is the isothermal compressibility coefficient.

The coefficient α in the quasi-harmonic approximation can be determined by using the known Mie–Grüneisen equation (see Ref. 18, for example)

$$\alpha = \gamma c_V \kappa_{iso}/3v. \qquad (2.32)$$

Since $c_P/c_V = \kappa_{iso}/\kappa_{ad}$, where κ_{ad} is the adiabatic compressibility coefficient, we can find the following expression from Eqs. (2.31) and (2.32):

$$(c_P/c_V) - 1 = \gamma^2 c_V T (c_P/c_V)(\kappa_{ad}/v). \qquad (2.33)$$

On the basis of an analysis of the elastic properties of a crystal, Mel'nikov[36] found a simple and useful relation which relates the adiabatic compressibility coefficient to the compressibility coefficient in the harmonic approximation (see Sec. 2.3):

$$\kappa_{ad} = \tilde{\kappa}[1 + \gamma(\Delta v/\tilde{v})], \qquad (2.34)$$

where $\tilde{\kappa}$ is the compressibility coefficient in the harmonic approximation, γ is the Grüneisen constant, and $\Delta v = v - \tilde{v}$ is the deviation of the molar volume of the crystal from its value in the harmonic approximation. In the harmonic approximation, the compressibility coefficient is the same for all isotopes. The molar volume of a crystal can be represented in the form

$$v \equiv \tilde{v} + \Delta v \equiv \tilde{v}[1 + (\Delta v/\tilde{v})]. \qquad (2.35)$$

It follows from Eqs. (2.34) and (2.35) that

$$\frac{\kappa_{ad}}{v} = \frac{\tilde{\kappa}}{\tilde{v}}\,\frac{[1 + \gamma(\Delta v/\tilde{v})]}{[1 + (\Delta v/\tilde{v})]}. \qquad (2.36)$$

Table 2.20. Comparison of the experimental and calculated data on the heat capacity c_P solid lithium hydride, J/mole·deg.

T,K	$c_{P,\mathrm{exp}}$	$c_{P,(2.37)}$	$c_{P,(2.39)}$	T,K	$c_{P,\mathrm{exp}}$	$c_{P,(2.37)}$	$c_{P,(2.39)}$
100	6.37	6.67	6.67	550	44.00	44.04	44.58
150	13.38	13.16	13.16	600	46.00	46.06	46.81
200	19.03	18.92	18.93	650	48.07	47.91	48.93
250	23.96	24.07	24.09	700	50.31	49.62	50.97
300	29.15	28.67	28.71	750	52.79	51.10	52.86
350	33.50	32.73	32.82	800	55.56	52.60	54.86
400	36.90	36.16	36.31	850	58.64	53.91	56.77
450	39.63	39.13	39.37	900	62.05	55.28	58.93
500	41.93	41.79	42.16	950	65.74	56.50	61.08

Since $\Delta v/\tilde{v}$ is small over the entire temperature range, and since the Grüneisen constant for lithium hydride is ~ 1.2, according to the data of Jex,[160] Eq. (2.36) shows that κ_{ad}/v can be replaced by $\tilde{\kappa}/\tilde{v}$ in Eq. (2.33).

The compressibility coefficient for the isotopic modifications of lithium hydride is $\tilde{\kappa} = 2.84 \times 10^{-11}$ m^2/N (see Sec. 2.3). The volume of the unit cell $\tilde{v}$ was found (from the lattice constant $\tilde{l}$ in the harmonic approximation; see Sec. 2.2) to be 15.9365×10^{-24} cm^3 or $\tilde{v} = 9.597$ cm^3/mole scaled to the molar volume. Substituting these values into Eq. (2.33), we find the dependence of c_P (J/mole deg) on c_V (J/mole deg) and on T (K):

$$c_P = \frac{c_V}{1 - 0.4261 \times 10^{-5} c_V T}. \tag{2.37}$$

A comparison of the values of c_P found from this relation with the experimental data for lithium hydride showed that the agreement is very good up to temperatures on the order of 750–800 K. If the difference between κ_{iso} and κ_{ad} is ignored on the right-hand side of Eq. (2.31), we obtain the equation

$$\frac{c_P}{c_V} - 1 = \gamma^2 c_V T \left(\frac{c_P}{c_V}\right)^2 \frac{\kappa_{\mathrm{ad}}}{v}, \tag{2.38}$$

which, after the replacement of κ_{ad}/v by $\tilde{\kappa}/\tilde{v}$, gives a satisfactory agreement with experiment to higher temperatures.

Substituting into relation (2.38) the values of γ, $\tilde{\kappa}$, and $\tilde{v}$ given above, we find for c_P (J/mole deg):

$$c_P = c_V \frac{1 - \sqrt{1 - 1.7044 \times 10^{-5} c_V T}}{0.8522 \times 10^{-5} c_V T}. \tag{2.39}$$

The experimental and calculated [according to Eqs. (2.37) and (2.39)] data on the heat capacity c_P of solid lithium hydride are compared in Table 2.20. Table 2.21 is a comparison of the experimental and calculated [from Eq. (2.39)] data on the enthalpy and entropy of solid lithium hydride.* Both tables clearly show that the calculated data are in satisfactory agreement with the

*In the case of entropy, *experimental data* is, strictly speaking, referred to only conventionally, since the entropy of a substance cannot be measured directly.

Table 2.21. Comparison of the experimental and calculated data on the enthalpy and entropy of solid lithium hydride.

T,K	$[H(T)-H(0)]_{exp}$ J/mole	$[H(T)-H(0)]_{calc.}$ J/mole	$S(T)_{exp}$ J/(mole K)	$S(T)_{calc}$ J/(mole K)
100	172	178	2.28	2.20
150	669	670	6.11	6.13
200	1 468	1 488	10.87	11.12
250	2 560	2 567	15.64	15.90
300	3 880	3 891	20.44	20.70
350	5 451	5 430	25.27	25.43
400	7 214	7 156	29.97	30.03
450	9 130	9 048	34.47	34.47
500	11 170	11 085	38.77	38.76
550	13 319	13 251	42.86	42.88
600	15 569	15 535	46.77	46.85
650	17 920	17 926	50.53	50.67
700	20 378	20 417	54.17	54.35
750	22 955	23 005	57.70	57.85
800	25 662	25 689	61.21	61.38
850	28 516	28 470	64.66	64.74
900	31 532	31 351	68.10	68.03
950	34 725	34 339	71.55	71.26

Table 2.22. Comparison of the experimental and calculated data on the caloric properties of lithium hydride near the melting point.

T,K	Δc_p, $\frac{J}{mole\cdot K}$	$\frac{\Delta c_p}{c_p}$, %	ΔH $\frac{J}{mole}$	$\frac{\Delta H}{H}$, %	ΔS, $\frac{J}{mole\ K}$	$\frac{\Delta S}{S}$, %	ΔG, $\frac{J}{mole}$	$\frac{\Delta G}{G}$, %
800	0.9	1.7	− 2.7	− 0.1	− 0.17	− 0.3	− 0.13	− 0.4
850	2.0	3.5	46	0.2	− 0.08	− 0.1	0.14	− 0.4
900	3.4	5.8	181	0.6	0.07	0.1	− 0.13	− 0.4
950	4.5	8.0	386	1.1	0.29	0.4	− 0.11	− 0.3
965	5.4	8.8	463	1.3	0.38	0.5	− 0.10	− 0.3

experimental data. As the melting point is approached, the discrepancy between the experimental and calculated data increases.

Table 2.22 gives the difference between the caloric functions of lithium hydride obtained experimentally and those found from Eq. (2.39), and also the relationship between this difference and the calculated value. We see from the Table that the error in the calculated heat capacity just before the melting point is appreciable beginning at 850 K. In the case of enthalpy, this effect becomes noticeable only at 950 K. Entropy behaves in a similar manner. For the thermodynamic potential, the difference between the data obtained experimentally and the data calculated on the basis of the relations proposed above is insignificant near the melting point. At high temperatures, the difference between the experiment and calculations based on Eqs. (2.37) and (2.39) is attributable principally to the fact that these equations ignore the

Table 2.23. Difference between the heat capacities (J/mole deg) of $^7LiD(T)$ and ^7LiH.

T,K	$\Delta c_{P,\,^7LiD-^7LiH}$	$\Delta c_{P,\,^7LiT-^7LiH}$	T,K	$\Delta c_{P,\,^7LiD-^7LiH}$	$\Delta c_{P,\,^7LiT-^7LiH}$
100	0.21	0.80	600	4.06	5.40
150	2.09	3.84	650	3.74	5.00
200	4.07	6.64	700	3.39	4.52
250	5.31	8.13	750	3.14	4.24
300	5.83	8.49	800	2.93	3.88
350	5.85	8.29	850	2.73	3.70
400	5.56	7.74	900	2.59	3.48
450	5.23	7.15	950	2.48	3.27
500	4.75	6.47	967	2.40	—
550	4.41	5.93	968	—	3.16

anharmonicity and other complex effects which are clearly seen as the temperature of the crystals approaches the melting point. The heat capacity is the property that is most sensitive to these effects.

The relation for reliable calculation of the caloric properties of solid lithium deuteride and solid lithium tritide in the temperature interval up to the melting point can be obtained from the available experimental data on lithium hydride, and the isotopic effect can be determined from Eq. (2.37). The following expression for the two isotopic compositions can be obtained from this equation:

$$\Delta c_P = c_{P,2} - c_{P,1}$$

$$= \frac{c_{V,2} - c_{V,1}}{1 - 0.4261\times10^{-5}T\,(c_{V,1} + c_{V,2}) + 0.1816\times10^{-10}T^{\,2}c_{V,1}c_{V,2}}, \tag{2.40}$$

where subscript 1 denotes lithium hydride.

Clearly, the use of Eq. (2.40) makes it possible to take essentially full account of the isotopic effect in the heat capacity c_P of LiD and LiT. Using the same equation for solid lithium deuteride at temperatures not included in the experiments and for lithium tritide, which has not been studied experimentally, we find the following standard working equations:

$$c_{P,^7LiD} = c_{P,^7LiH} + \Delta c_{P,^7LiD-^7LiH}; \tag{2.41}$$

$$c_{P,^7LiT} = c_{P,^7LiH} + \Delta c_{P,^7LiT-^7LiH}, \tag{2.42}$$

where $c_{P,^7LiH}$ is taken from the data of Table 5.1 which were obtained experimentally. The difference between the heat capacities of lithium deuteride and lithium hydride, $\Delta c_{P,^7LiT-^7LiH}$, and between lithium tritide and lithium hydride, $\Delta c_{P,^7LiT-^7LiH}$, found from Eq. (2.40), are listed in Table 2.23.

To determine other caloric functions, we used the equation

$$c_P = \sum_{n=1}^{4} b_n(T-100)^n + b_0 \tag{2.43}$$

to approximate the heat capacity c_P [J/(mole deg)] found from Eqs. (2.41) and (2.42) for lithium tritide over the temperature interval 100–298.15 K, and we used the equation

Table 2.24. Coefficients in Eqs. (2.43) and (2.44).

n	b_n	m	c_m	
			^{7}LiD	^{7}LiT
0	7.17	0	34.6	37.5
1	2.184×10^{-1}	1	$1.094\,79\times10^{-1}$	$9.849\,838\times10^{-2}$
2	$-2.356\,651\times10^{-4}$	2	$-3.919\,604\times10^{-4}$	$-3.478\,047\times10^{-4}$
3	$-1.560\,018\times10^{-6}$	3	$8.997\,200\times10^{-7}$	$7.732\,907\times10^{-7}$
4	$5.466\,726\times10^{-9}$	4	$-9.188\,680\times10^{-10}$	$-7.375\,124\times10^{-10}$
		5	$3.866\,132\times10^{-13}$	$2.900\,111\times10^{-13}$

Table 2.25. Comparison of the heat capacities c_P and c_V of lithium hydride, cal/mole K.

T,K	c_P	c_V	T,K	c_P	c_V
100	1.59	1.59	600	11.17	9.85
200	4.52	4.45	700	12.15	10.33
300	6.86	6.61	800	13.07	10.66
400	8.67	8.14	900	14.02	10.90
500	10.06	9.17			

$$c_P = c_0 + \sum_{m=1}^{5} c_m (T-298.15)^m \qquad (2.44)$$

to approximate the heat capacity for lithium deuteride and lithium tritide over the temperature interval from 298.15 K to the melting point. The coefficients in the approximating equations are listed in Table 2.24, and the caloric functions for lithium deuteride and lithium tritide found from these equations are listed in Tables 5.2 and 5.3.

The heat capacity and other caloric properties of lithium deuteride determined up to a temperature of 300 K are based on the data of Ref. 263. At temperatures of 50 and 100 K, the caloric properties of lithium tritide were determined from the characteristic temperature of ^{7}LiT found above and from the tabulated thermodynamic functions,[174] since the difference between c_P and c_V in this temperature region is negligible (the heat capacities for ^{7}LiH are listed in Table 2.25). The maximum relative error of the caloric functions of lithium deuteride at temperatures up to 300 K, which are listed in Table 5.2, and which were determined directly from the approximated experimental data of Ref. 263, is estimated to be 2%, as in the case of lithium hydride. The errors in the theoretical calculations of the properties of solid lithium deuteride at the remaining temperatures, and of lithium tritide at temperatures corresponding to its solid state can be determined from the following considerations.

The agreement between the calculated and experimental data on the heat capacity and enthalpy of solid lithium hydride is satisfactory (see Tables 2.20 and 2.21), the discrepancy being within $\pm 1\%$ (except at temperatures near the melting point). The discrepancy between the entropy and the thermodynamic potential is even smaller, within ± 0.2–0.4%. The errors of the calculated data on the caloric properties of lithium deuteride and lithium tritide found from Eqs. (2.41) and (2.42) consist of the errors of the data on lithium hydride and the error in determining Δc_P. Judging strictly from the discrepancy between the calculated and experimental values of the heat capacity c_P (see Table 2.20), the error in determining Δc_P should be 0.2%–2.0%. The lower error corresponds to that obtained at room temperature and slightly higher temperatures.

As was noted above, the isotopic effect in the heat capacity of a ^{7}LiH crystal peaks at 15%–20% near room temperature. At the same time, the isotopic effect amounts to only 3%–4% at high temperatures, where the error in determining Δc_P is maximum. This comparison shows that the error in determining Δc_P contributes no more than 0.1% to the total relative error in the heat capacity c_P of lithium deuteride and lithium tritide at room temperature and at high temperatures.

From the analysis carried out above, we conclude that the relative errors of the data on the caloric properties of solid-phase lithium deuteride and lithium tritide, which were obtained by means of theoretical calculations and which are listed in Tables 5.2 and 5.3, are roughly within the same range as in the case of lithium hydride.

1.4. Determination of the caloric properties of lithium deuteride and lithium tritide in the liquid phase

The caloric properties of liquid lithium deuteride and liquid lithium tritide have not been studied experimentally. The theory of liquids, on the other hand, cannot yet be used for reliable calculation of the caloric properties of substances in the liquid state, especially complex systems such as liquid hydride. Accurate data on the caloric properties of liquid ^{7}LiD and ^{7}LiT can be obtained in this case by using our value (see Chap. 3, Sec. 2.3) of the equilibrium constant for the heterogeneous reaction of the type

$$\mathrm{LiH}_l \rightleftarrows \mathrm{Li}_l + \tfrac{1}{2}\,\mathrm{H}_{2\ \mathrm{gas}} \tag{2.45}$$

for Li–LiD(LiT) systems and also the corresponding standard thermodynamic relations

$$RT\ln K = [\Phi^0(T) - H^0(0)]_2 - [\Phi^0(T) - H^0(0)]_1 - \tfrac{1}{2}\,[\Phi^0(T) - H^0(0)]_3 + \Delta_f H^0(0); \tag{2.46}$$

$$RT^2\, d\ln K/dT = [H^0(T) - H^0(0)]_1 + \tfrac{1}{2}\,[H^0(T) - H^0(0)]_3 - [H^0(T) - H^0(0)]_2 - \Delta_f H^0(0) = \Delta H^0(T). \tag{2.47}$$

Here $\Delta H^0(T)$ and K are, respectively, the reaction energy and the equilibrium constant of heterogeneous reaction (2.45); $\Delta_f H^0(0)$ is the heat of formation of solid lithium hydride (solid lithium deuteride, solid lithium tritide) from solid Li and H_2 gas (D_2, T_2 gas) at a temperature of 0 K; and the indices 1, 2, and 3 correspond to liquid lithium, isotopic modifications of liquid lithium hydride, and hydrogen gas isotopes, respectively.

The difference between the heats of formation for ^{7}LiH and ^{7}LiD is relatively small. According to the data of Refs. 131 and 132, which are analyzed in

Table 2.26. Comparison of the calculated and experimental data on the caloric properties of liquid lithium hydride.

	$\Phi(T)-H(0)$, J/mole		$H(T)-H(0)$, J/mole		$c_P(T)$, J/(mole K)	
T, K	Calculation	Experiment	Calculation	Experiment	Calculation	Experiment
1000	– 37 960	– 37 690	59 680	59 160	43.9	58.3
1100	– 47 940	– 47 650	66 180	64 630	44.1	52.0
1200	– 58 310	– 58 080	68 490	69 720	44.2	50.7
1300	– 69 060	– 68 940	72 930	74 940	44.6	54.5

Sec. 1.3, at a temperature of 298 K, the heats of formation are

$$\Delta_f H^0(^7\text{LiH},298) = -21\,663 \pm 52 \text{ cal/mole}; \quad (2.48)$$

$$\Delta_f H^0(^7\text{LiD},298) = -21\,784 \pm 21 \text{ cal/mole}. \quad (2.49)$$

Using the data on the enthalpy of lithium,[68,136] on the enthalpy of lithium hydride and lithium deuteride from Tables 5.1 and 5.2, and on the enthalpy of hydrogen and deuterium[15] to scale the values in Eqs. (2.48) and (2.49) to 0 K, we find

$$\Delta_f H^0(^7\text{LiH},0) = -20\,460 \text{ cal/mole}; \quad (2.50)$$

$$\Delta_f H^0(^7\text{LiD},0) = -20\,735 \text{ cal/mole}. \quad (2.51)$$

The heat of formation of lithium tritide was found previously (see Sec. 1.3):

$$\Delta_f H^0(^7\text{LiT},0) = -20\,822 \text{ cal/mole}. \quad (2.52)$$

The difference between the heats of formation of ^{7}LiH and ^{7}LiD is relatively small at room temperature [see Eqs. (2.48) and (2.49)]. The difference between the heats of formation of ^{7}LiD and ^{7}LiT must be even smaller. Ignoring this difference and assuming that $\Delta_f H^0$ (^{7}LiD,298) = $\Delta_f H^0$ (^{7}LiT,298), we find that scaling the heat of formation of LiT to a temperature of 0 K by using the same sources and the data of Table 5.3 gives a value approximately equal to that in Eq. (2.52).

Before calculating the caloric properties of liquid lithium deuteride and liquid lithium tritide according to Eqs. (2.46) and (2.47), we used this method to calculate the properties of liquid lithium hydride and compared the results with experiment. Confining ourselves to this approach in determining the heat capacity, we must differentiate the logarithm of the equilibrium constant K twice with respect to temperature. Table 2.26 compares the calculated values of the thermodynamic potential, the enthalpy, and the heat capacity of liquid lithium hydride with the data obtained on the basis of a caloric experiment. The agreement is satisfactory for the thermodynamic potential and the enthalpy. The agreement for the heat capacity is, of course, worse. In these calculations, the caloric properties of lithium were taken from the sources indicated above, the equilibrium constant was taken from Table 3.22 (see Chap. 3, Sec. 2.3), and $\Delta_f H^0$ (^{7}LiH,0) was determined from Eq. (2.50).

The average discrepancy between the data on the thermodynamic potential, which were obtained experimentally and from Eq. (2.46), is 0.5%. The average discrepancy in the enthalpy, where the first derivative of ln K with respect to temperature must be used in Eq. (2.47), is 2.5%. As for the heat capacity, we see from Table 2.26 that an accurate value of $d^2 \ln K/dT^2$ requires that the temperature dependence of the equilibrium constant be determined more accurately. Another approach should therefore be used to calculate the heat capacity. Since the second derivative of $\ln K$ determines the temperature evolution of the thermal effect of reaction (2.45), we can assume in calculating the heat capacity of liquid lithium deuteride and liquid lithium tritide that the temperature evolution of the thermal effect of this reaction is identical for all three isotopic modifications (^{7}LiH, ^{7}LiD, and ^{7}LiT) in the relevant temperature range ($T_m = 1300$ K). In this case, Eq. (2.47) implies

$$c_P^0(^7\text{LiD}) = c_P^0(^7\text{LiH}) - \tfrac{1}{2}\, c_P^0(\text{H}_2) + \tfrac{1}{2}\, c_P^0(\text{D}_2). \qquad (2.53)$$

The error arising from an inaccuracy in determining the temperature dependence (see Chap. 3, Sec. 2.3) of the equilibrium constant of ^{7}Li–^{7}LiH(D,T) systems can be reduced by calculating the enthalpy of liquid lithium deuteride from the equation

$$[H^0(T) - H^0(0)]_{^7\text{LiD}} = [H^0(T) - H^0(0)]_{^7\text{LiH}} + \Delta H_{(^7\text{LiD} - ^7\text{LiH})}, \qquad (2.54)$$

where [see Eq. (2.47)]

$$\Delta H_{(^7\text{LiD} - ^7\text{LiH})} = \tfrac{1}{2}[H^0(T) - H^0(0)]_{\text{D}_2} - \tfrac{1}{2}[H^0(T) - H^0(0)]_{\text{H}_2}$$
$$- \Delta_f H^0(^7\text{LiD},0) + \Delta_f H^0(^7\text{LiH},0) - RT^2 \frac{d \ln(K_{\text{Li}-\text{LiD}}/K_{\text{Li}-\text{LiH}})}{dT}. \qquad (2.55)$$

The enthalpy of liquid lithium tritide can be calculated by using a similar equation.

The thermodynamic potential can be calculated from Eq. (2.46):

$$[\Phi^0(T) - H^0(0)]_{^7\text{LiD}} = [\Phi^0(T) - H^0(0)]_{^7\text{Li}} + \tfrac{1}{2}[\Phi^0(T) - H^0(0)]_{\text{D}_2}$$
$$- \Delta_f H^0(^7\text{LiD},0) + RT \ln K. \qquad (2.56)$$

The expression for the thermodynamic potential of ^{7}LiT is similar. The entropy of liquid ^{7}LiD and ^{7}LiT can be determined from Eqs. (2.54) and (2.56).

The caloric properties of liquid ^{7}LiD and liquid ^{7}LiT, calculated from the equations indicated above, are listed in Tables 5.2 and 5.3. In the calculations, the equilibrium constant K was taken from Table 3.22 for the Li–LiD system and was calculated from Eq. (3.77) for the Li–LiT system (see Chap. 3, Sec. 2.3). At the temperature 0 K the heat of formation was calculated from Eqs. (2.50)–(2.52). The caloric properties of lithium hydride were calculated from the relations in Sec. 1.2, and the caloric functions of hydrogen isotopes in the ideal gas state were taken from Ref. 15.

It is important to note that the caloric data on isotopic modifications of lithium hydride in the solid and liquid phases obtained by the methods suggested above give consistent results. The values of the Planck function (of the reduced thermodynamic potential) at the melting point, which are determined in this section, are essentially in agreement with those found in the preceding section (Table 2.27). The average values of the reduced thermodynamic potential at the melting point are given in Tables 5.1–5.3.

The errors of the data on the caloric properties of liquid lithium deuteride and liquid lithium tritide listed in Tables 5.2 and 5.3 can be estimated in the following way. Equation (2.54) implies the errors of the data on the enthalpy of lithium deuteride are equal to the sum of the errors of the data on the enthalpy of lithium hydride and the error in determining the term $\Delta H_{(^7LiD-^7LiH)}$, which is found to be $\sim 15\%$ of the enthalpy of lithium hydride (for lithium tritide this term is about 20% of the enthalpy of lithium hydride). The difference in $\Delta H_{(^7LiD-^7LiH)}$ for the derivative of $RT^2\, d \ln (K_{Li-LiD}/K_{Li-LiH})/dT$, which corresponds to $\sim 30\%$ of $\Delta H_{(^7LiD-^7LiH)}$, has been determined least accurately [see Eq. (2.55)]. It is difficult to accurately determine the error of this derivative. If, on the other hand, this error is arbitrarily assumed to be 20%, then this component will contribute only $\sim 6\%$ to the total error in $\Delta H_{(^7LiD-^7LiH)}$, adding 0.9% to the error of the enthalpy of lithium hydride. [A 20% error seems to be significantly higher than the actual error, since we have used, as will be shown in Chap. 3, Sec. 2.3, two separate methods and also the data on the partial pressure of hydrogen and deuterium[102,234] to determine the equilibrium constants of reaction (2.45) for the Li–LiH and Li–LiD systems. The results obtained by us turned out to be in fairly good agreement; see Eqs. (3.56) and (3.57) and Table 3.20.] The remaining components contribute much less to the error in determining $\Delta H_{(^7LiD-^7LiH)}$). It thus follows that the maximum relative errors of these data on the enthalpy of liquid lithium deuteride amount to $\sim 3\%$, of which 2% corresponds to the errors of the data on the enthalpy of lithium hydride.

This method can also be used to estimate the error of the enthalpy of liquid lithium tritide, although the temperature dependence of the equilibrium constant of reaction (2.45) for lithium tritide has been obtained by a slightly different method (see Chap. 3, Sec. 2.3).

The errors of the data (given in Tables 5.2 and 5.3) on the heat capacity of liquid lithium deuteride and liquid lithium tritide can be estimated on the basis of similar considerations which can be summarized as follows. According to Eq. (2.53), the error contained in c_P^0 (^{7}LiD, ^{7}LiT) is comprised of the error of c_P^0 (^{7}LiH) (the principal part) and an additional part which contains an error resulting from the assumption that the temperature evolution of the thermal effect of reaction (2.45) is identical for ^{7}LiH, ^{7}LiD, and ^{7}LiT.

Although it is difficult to determine the last error, it is clear that since the heat capacity of lithium deuteride and lithium tritide in the liquid phase differs only slightly (by 2%–3%) from the heat capacity of lithium hydride, the contribution of the additional part of the error to the total error in determining c_P^0 (^{7}LiD) or c_P^0 (^{7}LiT) cannot be large. We conclude, therefore, that the maximum relative errors of the data on the heat capacity of liquid lithium deuteride and liquid lithium tritide may be assumed the same as those for liquid lithium hydride, i.e., 3%. The errors of the rest of the caloric functions in the liquid phase, listed in Tables 5.2 and 5.3, also have the same value.

In conclusion we wish to emphasize that the agreement between the values of the thermodynamic potential at the melting points of ^{7}LiH, ^{7}LiD, and ^{7}LiT, found by independent methods for the solid and liquid phases (see Table 2.27), shows that analysis of the experimental data on the caloric properties of lithium hydride and the suggested methods of calculating the caloric functions of lithium deuteride and lithium tritide in the solid and liquid phases yield sufficiently accurate results.

Table 2.27. Comparison of Planck's functions (the reduced thermodynamic potentials) of the isotopic modifications of lithium hydride at the melting point found for the solid and liquid phases, J/mole deg.

Hydride	$\left(S - \frac{H(T) - H(0)}{T}\right)_s$	$\left(S - \frac{H(T) - H(0)}{T}\right)_l$
^{7}LiH	35.51	35.59
^{7}LiD	40.61	40.55
^{7}LiT	43.47	43.69

2. Thermal properties

2.1. Review and analysis of experimental studies of the thermal properties—the lattice constant, density, and thermal expansion—in the solid phase

The lattice constant l of LiH was determined experimentally primarily by x-ray diffraction. It was also determined by neutron diffraction and by measuring the density by flotation based on a standard formula for a NaCl-type fcc lattice (see Ref. 23, for example):

$$\rho = 4(M + m)/N_A l^3, \tag{2.57}$$

where ρ is the crystal density; M and m are the masses of the heavy and light ions, respectively; and N_A is Avogadro's number.

At high temperatures, the lattice constants found from measurement of the density are more accurate, because lithium hydride, by virtue of low electron density of lithium and hydrogen ions, is transparent to hard x-ray emission. As the temperature is raised, the diffraction lines become very faint and broaden, sharply reducing the accuracy in determining the lattice constant. X-ray techniques are thus not very effective at high temperatures. At low temperatures (down to room temperature), however, most researchers prefer x-ray diffraction because of its accuracy and because this method makes it possible to determine the crystal structure. The results of x-ray diffraction and neutron diffraction measurements of the lattice constant of lithium hydride and the results of calculations on the basis of the density measurements are in good agreement with each other.

Table 2.28 gives the measurement results of the lattice constant of the isotopic modifications of lithium hydride at room temperature. All the results, with the exception of those of Brückner *et al.*,[113] were obtained by x-ray diffraction. Brückner *et al.*[113] used neutron diffraction. Of the authors cited in Table 2.28, only Smith and Leider, Zalkin and Silvera, and Anderson *et al.* carried out measurements of the lattice constants over a relatively broad temperature interval.

Smith and Leider[214] used x-ray diffraction to measure the lattice constants of several substances, including lithium hydride and lithium deuteride

Table 2.28. Lattice constants l of ^7LiH and ^7LiD crystals at room temperature.

References	l,Å	
	^7LiH	^7LiD
Zintl and Harder (Ref. 262)	4.085 ± 0.001	4.065 ± 0.001
Tronstad and Wegeland (Ref. 228)	4.085 ± 0.001	—
Ahmeld (Refs. 94, 95)	4.085 ± 0.02	—
Starizky and Walker (Ref. 221)	4.0834 ± 0.0005	4.0684 ± 0.0005
Kogan and Omarov (Ref. 24)	4.084 ± 0.001	4.070 ± 0.001
Brückner *et al.* (Ref. 113)	4.083 ± 0.001	4.069 ± 0.001
Zalkin and Silvera (Ref. 41, 190)	4.083 ± 0.001	4.069 ± 0.001
Smith and Leider (Ref. 214)	4.0856	4.0708
Haussühl and Skorczyk (Ref. 138)	4.084	4.069
Anderson *et al.* (Ref. 96*)	4.0829	4.0693
Zimmerman (Ref. 261)	4.0831 ± 0.0004	—

*The lattice constant of ^7LiT, which is 4.0633 Å, was also measured in this study.

Table 2.29. Lattice constants of the isotopic modifications of lithium hydride from the experimental data of Smith and Leider (Ref. 214).

T,K	^7LiH	T,K	^7LiH	T,K	^7LiD	T,K	^6LiD
16.9	4.0648	16.2	4.0660	15.7	4.0492	12.1	4.0492
80	4.0656	80	4.0664	81	4.0500	80	4.0496
154	4.0688	154	4.0692	152	4.0536	152	4.0536
189	4.0720	193	4.0730	193	4.0576	195	4.0568
235	4.0768	230	4.0780	233	4.0616	235	4.0612
281	4.0830	273	4.0822	272	4.0668	275	4.0668
		298	4.0856	299	4.0708	298	4.0704

isotopes ^7LiH, ^6LiD, and ^7LiD over the temperature interval 12–300 K. Their experimental results are presented in Table 2.29. The last significant figure for the lattice constant is approximate. These authors assume that the discrepancy between the results of the two sets of measurements with ^7LiH at low temperatures extends beyond the limits of the experimental error and suggest that the reason for this discrepancy is traceable to the difference in the compositions of the impurities contained in the samples, since the uncertainty in the data of the spectral analysis, in their view, is large enough. The authors use the same arguments to explain a slight discrepancy between their results and the room-temperature data found in the literature.

Zalkin and Silvera (see Refs. 41 and 190) carried out x-ray diffraction measurements of the lattice constant of ^nLiH over the temperature interval 298–800 K and found the following relation, taken from Ref. 190:

$$l = 4.083[1 + 4.2\times10^{-5}(T-25) + 1.9\times10^{-8}(T-25)^2], \qquad (2.58)$$

where l is the lattice constant (Å), and T is the temperature (°C).

The experimental data of Zalkin and Silvera are given in Ref. 41:

T,°C....	25	200	400	525
l, Å....	4.083	4.116	4.159	4.188

Table 2.30. Lattice constants *l* (Å) of the isotopic modifications of lithium hydride from the experimental data of Anderson *et al.* (Ref. 96) as a function of the temperature.

Crystal	– 190 °C	25 °C	140 °C	240 °C
^{7}LiH	4.0657	4.0829	4.1005	4.1224
^{7}LiD	4.0477	4.0693	4.0893	4.1119
^{7}LiT	4.0403	4.0633	—	—
^{6}LiH	4.0666	4.0851	4.1013	4.1218
^{6}LiD	4.0499	4.0708	4.0888	4.1110

Anderson *et al.*[96] carried out one of the more recent measurements of the thermal properties of lithium hydride and its isotopic modifications. They determined the lattice constants of the crystals of various isotopic modifications of LiH by x-ray diffraction at temperatures of – 190 °C and 25 °C. They measured the density by the flotation method at temperature of 25, 140, and 240 °C.

In the x-ray measurements, the temperature of the sample was measured with a Chromel–Alumel thermocouple within $\pm 1°$. In the flotation measurements, the sample was placed in a quartz test tube containing a liquid (thoroughly dried and degased benzene, octane, heptane, etc.), which was inserted into a thermostat. The temperature was measured with a platinum resistance thermometer within $\pm 0.05°$, which resulted in a 0.005% error in the density. The thermometer's stability was checked daily against the melting point of ice. The authors present their results in the form of a table of lattice constants l (Table 2.30), noting that at a constant temperature the lattice constant of the isotopic modifications that have been studied satisfies the expression

$$l = A + B/\sqrt{\mu}, \tag{2.59}$$

where A and B are coefficients which do not depend on the isotopic modification, and μ is the reduced mass of the compound ($1/\mu = 1/M + 1/m$, where M is the atomic mass of the lithium isotope and m is the atomic mass of the hydrogen isotope). The value of B depends on the temperature:

T,°C...	– 190	25	140	240
B, Å...	0.0665	0.0523	0.0464	0.0407

As was noted by Anderson *et al.*,[96] replacement of one isotope by another in the chemical compound changes, according to the data of Ubbelohde,[63] the rotational and vibrational frequencies of the molecules without substantially changing the shape of the potential-energy curve $U(r)$. In the case of a lighter isotope, the frequency is higher and, hence, the interatomic distance r increases in the crystal because of the anharmonicity. As a result, we find the relationship among the lattice constants of the isotopic modifications of lithium hydride to be $l_{^7\mathrm{LiH}} > l_{^7\mathrm{LiD}} > l_{^7\mathrm{LiT}}$. Anderson *et al.*[96] have used lithium hydride samples containing both pure lithium isotopes (^{7}Li and ^{6}Li) and natural lithium ($M = 6.941$) to study the lattice constants and densities. They presented no measurement results for natural lithium.

X-ray diffraction studies by Zimmerman[261] have shown that Vegard's rule,[232] according to which the lattice constant of a mixed crystal depends linearly on the isotopic content, is satisfied in the case of lithium hydride. In other words, we have

$$l = a_1 l_1 + a_2 l_2, \tag{2.60}$$

where l_1 and l_2 are the lattice constants of pure isotopic compositions such as ^{7}LiH and ^{6}LiH, and a_1 and a_2 are, respectively, the atomic concentrations of the isotopes in the mixed crystal. For pure ^{7}LiH, Zimmerman found the value $l = 4.0831 \pm 0.0004$ Å at room temperature. Estimates based on Eq. (2.60) have shown that the difference between the lattice constants of lithium hydride resulting from the replacement of ^{7}Li by a natural lithium is no greater than 0.02%; i.e., it is within the error limits of the experiment.

The following general conclusions can be drawn from the analysis of all studies:

(1) The isotopic effect is very strong when D or T is substituted for H and is very weak upon the replacement of lithium isotopes.

(2) The lattice constant of a LiH-type crystal with a light isotope is larger than the lattice constant of a crystal with a heavy isotope.

(3) The maximum relative change in the lattice constant resulting from the exchange of one isotope for another occurs at low temperatures.

(4) The isotopic effect decreases with increasing temperature.

The density of lithium hydride was first studied in 1920. Moers[194] used a pycnometer (1.5 cm^3 in volume) to study the density of solid lithium hydride (Table 2.31). The sample in the pycnometer was drenched with hexane which was dried through distillation and by sodium. The temperature was measured within $\pm 0.1°$. The experiment lasted only about 10 minutes. The average density was found to be $\rho = 0.816 \pm 0.004$ g/cm^3 on the basis of the results of two measurements at a temperature of 21.45 °C.

Bode[104] also used the pycnometric method to measure the density of lithium hydride. Particular attention was given to the preparation of the pycnometric liquid, which in this case was hexane that was subjected to preliminary purification by sulfuric acid and then dried by sodium. After distillation, the hexane was dried again by calcium hydride and lithium hydride. The volume of the pycnometer, which was calibrated against mercury, was ~ 1 cm^3. During the filling of the pycnometer, the time of contact with air of lithium hydride used in the experiment was reduced to a minimum. The quantity of hydride was determined from the volume of hydrogen and from the quantity of LiOH found after the hydrolysis. The average density was 0.705 ± 0.005 g/cm^3 at room temperature.

Although Moers estimated the error of his results to be no more than 1%, this figure appears to be much too low. This gross underestimation of error also seems to apply to the data of Bode. A common feature of both these studies is the extremely small content of the sample studied (a fraction of a gram), which always increases the error in determining the mass or volume of a sample. The influence of external effects on the sample, such as that of gas

Table 2.31. Principal experimental studies in which the density ρ and the thermal expansion coefficient α of solid-phase lithium hydride and lithium deuteride were measured.

Temperature (K)	Characteristic	Method	References
294	ρ	Pycnometric	Moers (Ref. 194)
298	ρ	Pycnometric	Bode (Ref. 104)
298	ρ	Gas densitometric	Gibb and Messer (Ref. 127)
298	ρ	—	Brodsky and Burstein (Ref. 111)
293	ρ	Flotation	Haussühl and Skorczyk (Ref. 138)
293	ρ,α	Flotation	Baikov *et al.* (Ref. 5)
300–850	α	Dilatometric	Welch (Refs. 250, 251)
300–920	α	Dilatometric	Fieldhaus *et al.* (see Ref. 251)
78–300	α	x-ray measurements of the lattice constants	Kogan and Omarov (Ref. 24)
77–300	α	Dilatometric	Brückner *et al.* (Ref. 113)
80–298	α	x-ray measurements of the lattice constants	Smith and Leider (Ref. 214)
83–513	α	Flotation, x-ray measurements of the lattice constants	Anderson *et al.* (Ref. 96)
20–300	α	Dilatometric	Eash (Ref. 121)
273–573	α	x-ray diffraction analysis	Tannenbaum and Ellinger (Ref. 224)

Table 2.32. Density of lithium hydride at room temperature from the data of Ref. 111.

Sample	Isotopic composition of lithium, %	Density, g/cm^3
^{7}LiH	92.5	0.7754
^{7}LiH	99.9	0.7829
^{6}LiH	95	0.6891

during transfer processes, that of pycnometric liquid, etc., also increases the error.

In their review,[127] Gibb and Messer mentioned that the density of ^{n}LiH was measured in their laboratory at room temperature with a gas densitometer in which helium gas was used as pycnometric liquid. The volume of the samples of a known mass was determined from the volume of the displaced gas at constant pressure and temperature. The density of lithium hydride was 0.788 ± 0.002 g/cm^3.

Brodsky and Burstein[111] studied experimentally the spectral characteristics and other characteristics of a lithium hydride crystal grown from natural lithium and its isotopes in a reflected infrared spectrum. In their rather thorough study, Brodsky and Burstein present the results of measurements (Table 2.32) of the density of the initial samples carried out at room temperature (unfortunately, they do not indicate the method used in the measurements).

Haussühl and Skorczyk[138] have studied the elastic properties of the isotopic modifications of crystalline lithium hydride. They carried out control measurements of the density of the samples at a temperature of 293 K and found the density of ^{n}LiH to be 0.776 g/cm^3 and the density of ^{n}LiD to be 0.893 g/cm^3.

Table 2.33. Experimental results of Baikov *et al.* (Ref. 5) for the measurement of the volume expansion coefficient β and the density ρ.

Sample	Temperature range (°C)	β, 10^{-6} (deg^{-1})	ρ 20 °C (g/cm^3)
^{n}LiH	8–69	104.2 ± 0.3	0.7773 ± 0.0002
^{n}LiD	13–69	121.1 ± 0.4	0.8841 ± 0.0002

Baikov *et al.*[5] published measurement results of the density and thermal expansion of lithium hydride and lithium deuteride at temperatures between 8 and 70 °C. The measurements were carried out by the flotation method. As the flotation liquid they used a mixture of cyclohexane and benzene (for lithium hydride) and a mixture of benzene with toluene and ethylene bromide (for lithium deuteride), which were dehydrated beforehand by metallic sodium, distilled in a vacuum, and dehydrated again by lithium hydride in hermetically sealed containers. The expansion coefficient of the flotation liquid was determined in a separate experiment. Unfortunately, the authors do not give the initial density of the liquid.

For the experiment, lithium hydride (deuteride) was obtained by a direct action of hydrogen (deuterium) gas on metallic lithium of natural isotopic composition at 700 °C. The lithium hydride crystals to be studied were relatively small (2–4 mm). The authors noted that the values of the density ρ were reproduced with a scatter of $\pm 5 \times 10^{-5}$ g/cm^3; the thermal cubic expansion coefficient β was reproduced with a scatter of $\pm 3 \times 10^{-7}$ deg^{-1}; and the temperature was thermostatically controlled within $\pm 0.002°$. The measurement results are given in Table 2.33.

Bijvoet and Karssen[99,100] used the measured values of the lattice constant to calculate the density of lithium hydride at room temperature. The value found by them was 0.76 ± 0.01 g/cm^3. Zintl and Harder[262] suggested that this value contains an error stemming from the fact that the measurements were carried out in air. There are several other studies in which Eq. (2.57) was used to calculate the density of lithium hydride on the basis of the results of x-ray structural analysis of the crystal lattice. Chretien *et al.*[117] found the density to be 0.774 ± 0.001 g/cm^3 on the basis of calculations using the data of Zintl and Harder. Gibb and Messer[127] found the density of lithium hydride to be 0.7738 ± 0.0004 g/cm^3. Their calculation was based on crystallographic data of an unknown origin. Staritzky and Walker[221] found the densities of ^{n}LiH and ^{n}LiD to be 0.775 and 0.881 g/cm^3, respectively, on the basis of their x-ray measurements of the lattice constant of these compounds. In all these studies, the calculations of the density based on x-ray structural analysis data correspond to room temperature.

Many studies have dealt with the task of finding the average and true thermal expansion coefficients of lithium hydride and lithium deuteride. Welch found the average linear thermal expansion coefficient $\bar{\alpha}$ of ^{n}LiH to be $34.8 \pm 0.5 \times 10^{-6}$ deg^{-1} over the temperature interval 30–580 °C (the samples studied were compressed rods; he carried out a total of 22 measurements, using three samples). Using these data, Welch calculated the density of solid

Table 2.34. Average coefficient of thermal expansion of lithium hydride from the data of Fieldhouse *et al.* (see Ref. 251).

T,°C	$\bar{\alpha}$,10^{-6} deg^{-1}	T,°C	$\bar{\alpha}$,10^{-6} deg^{-1}	T,°C	$\bar{\alpha}$,10^{-6} deg^{-1}
93	31.5	315	45.0	538	52.4
149	33.5	371	47.5	593	51.5
204	37.6	426	50.0	648	49.5
260	41.9	482	52.4		

Table 2.35. Density ρ of lithium hydride from the data of Messer (Ref. 190).

T,°C	ρ,g/cm^3	T,°C	ρ,g/cm^3
25	0.775	600	0.705
200	0.757	688_s	0.69
400	0.733	688_l	0.58 ± 0.03
525	0.718		

lithium hydride at a temperature of 580 °C. The value obtained by him was 0.703 g/cm^3. Welch gave no indication of the origin of the initial density, without which the calculation would have been impossible.

Welch[251] presented data on the average linear thermal expansion of lithium hydride obtained at temperatures between 26.8 °C and those given in Table 2.34 by Fieldhouse, Hidgi, and Lang using samples pressed into rods. The lower values of a obtained at high temperatures were attributed by Welch to the possible sintering of the sample.

Messer[190] assumed the density to be 0.775 g/cm^3 at room temperature and used the thermal expansion coefficients, given in Table 2.34, to calculate the density of solid lithium hydride up to the melting point (688 °C) (Table 2.35). To determine the density of liquid lithium hydride at a temperature of 688 °C (Table 2.36), Messer calculated the change in the volume as a function of the structure of the crystal at the melting point, using the following line of reasoning. If LiH had a purely ionic structure, it would expand by ~26% under fusion, consistent with the results obtained by Schinke and Sauerwald[210] for LiF, LiCl, NaF, and NaCl (25%–29%). The bond covalence of the structure of the salts AgBr and AgCl accounts for a 8%–9% change in the volume at the melting point. The change in volume, $\Delta V/V$, during melting as a function of covalence was expressed graphically by the Pauling method.[45] The value of $\Delta V/V$ for ^{n}LiH was thus found to be within 16%–20%. The mean value of 18% was used in the calculations. In their experiment, Gibb and Messer[127] assumed a 16% solidification-induced change in the volume of lithium hydride, which was found by Kisson. They did not disclose the method used to obtain this value.

Brückner *et al.*[113] measured thermal expansion by the dilatometric method. The results obtained by them are given in a plot form. They found the following numerical values of the true linear thermal expansion coefficients at a temperature of 20 °C: $(35.2 \pm 0.5)\times 10^{-6}$ and $(36.1 \pm 0.8)\times 10^{-6}$ deg^{-1} for

Table 2.36. Average linear thermal expansion coefficients $\overline{\alpha}$, 10^{-6} deg^{-1}, of the isotopic modifications of lithium hydride from the data of Anderson *et al.* (Ref. 96).

Temperature interval, K	^6LiH	^6LiD	^7LiH	^7LiD	^7LiT
83–298	21.0 ± 0.3	24.0 ± 1.0	19.8 ± 0.4	24.8 ± 0.4	26.4 ± 0.5
298– 41	34.4 ± 0.8	38.4 ± 1.8	37.4 ± 0.4	42.9 ± 1.0	—
413–513	50.0 ± 1.0	54.3 ± 1.0	53.3 ± 0.6	55.0 ± 1.4	—

Table 2.37. Average linear thermal expansion coefficient $\overline{\alpha}$, 10^{-6} deg^{-1}.

Temperature interval, K	^nLiH	^nLiD	References
77–300	21	25	Kogan and Omarov (Ref. 24)
77–300	21.4	24.7	Brückner *et al.* (Ref. 113)
80–298	21.4	22.6	Smith and Leider (Ref. 214)
83–298	19.8	24.8	Anderson *et al.* (Ref. 96)

lithium hydride and $(41.2 \pm 1.0) \times 10^{-6}$ deg^{-1} for lithium deuteride. They also found the average expansion coefficient to be $\bar{a} = 21.4 \times 10^{-6}$deg^{-1} for ^nLiH and $\bar{a} = 24.7 \times 10^{-6}$ deg^{-1} for ^nLiD in the temperature range 77–300 K. We conclude from this plot that at a temperature of 80 K the true linear thermal expansion is approximately 6×10^{-6} deg^{-1} for ^nLiH.

Working in the same temperature interval, Kogan and Omarov[24] found by measuring the lattice constant at temperatures of 78 and 300 K that the average linear thermal expansion of lithium hydride is 21×10^{-6} deg^{-1} and that of lithium deuteride is 25×10^{-6} deg^{-1} in this temperature interval.

By measuring the lattice constants of ^nLiH and ^nLiD, Smith and Leider[214] found that the average linear thermal expansion coefficients are 21.4×10^{-6} and 22.6×10^{-6} deg^{-1}, respectively, in the temperature interval 80–298 K.

Of particular interest is the study of Anderson *et al.*[96] cited above, in which they determined the average thermal expansion coefficient $\bar{a}$ over a broad temperature interval for five isotopic modifications (Table 2.36).

It can be seen from Table 2.36 that in all the modifications, $\bar{a}$ increases with temperature. The increase, however, is greater for heavier isotopes than for light isotopes. As a rule, $\bar{a}_{^7LiT} > \bar{a}_{^7LiD} > \bar{a}_{^6LiD}$ in each temperature interval, and a weak inverse isotopic effect occurs only at low temperatures (83–298 K); i.e., $\bar{a}$ is slightly larger for ^6LiH than for ^7LiH.

The data on the average linear thermal expansion of lithium hydride and lithium deuteride obtained in different studies are nearly the same over the temperature interval 77–300 K (Table 2.37).

Finally, it should be mentioned that the isotopic effect manifests itself most strongly in the density. Since the mass of the unit cell increases and its volume decreases in a heavier isotope, the density of a heavy isotope of crystalline hydride is, according to Eq. (2.57), markedly higher than the density of a light isotope. Experimental studies of the density of solid lithium hydride and other isotopic compounds confirm this conclusion.

The lattice constants of the isotopic modifications of lithium hydride measured by Smith and Leider[214] by x-ray diffraction over the temperature range 15–300 K, and the thermal expansion of ^{6}LiD measured by Eash[121] with a quartz dilatometer over the temperature range 20–300 K*, show that the linear expansion of a LiH crystal is negligible at temperatures between 0 K and about 80 K and that its lattice constant at a temperature of 0 K is the same as it is at 80 K.

It can generally be concluded that, so far, not enough experimental data on the thermal properties of solid lithium hydride and its isotopic modifications have been published. This is particularly true for the temperature interval from room temperature to the melting point. Lithium hydride has been studied experimentally up to a temperature of 800 K, lithium deuteride up to 515 K, and lithium tritide up to room temperature.

2.2. Determination of the thermal properties of lithium hydride and of its solid-phase isotopic modifications: Theory and calculation

In Sec. 1.1., we showed that lithium hydride in all solid-state isotopic combinations is an ionic crystal with a NaCl-type lattice. The general physical characteristics of these crystals have been well studied. In particular, the density of a solid crystal having this structure can be expressed by a known relation [see Eq. (2.57)]:

$$\rho = 4(M+m)/N_A l^3,$$

where (for a LiH crystal) M is the mass of the lithium atom, m is the mass of the hydrogen atom or of its isotope, l is the lattice constant, and N_A is the Avogadro number. Theoretical calculation of the density of lithium hydride in the crystalline state thus reduces to determining the lattice constant (see Ref. 38).

The behavior of ionic crystals with a simple structure is described sufficiently well by the Mie-Grüneisen equation of state:

$$P + \frac{du}{dv} = \gamma \frac{\epsilon_{\text{vib}}}{\tilde{v}}. \tag{2.61}$$

Here P is the pressure, $\tilde{v}$ is the volume of the unit cell, u and ϵ_{vib} are, respectively, the potential energy and the vibrational energy of the cell, and γ is the Grüneisen constant. Ignoring the pressure, we obtain the following expressions from this equation in the lowest-order approximation based on the deviations from the harmonic equilibrium (see Ref. 31):

$$(v - \tilde{v})/\tilde{v} = \gamma \tilde{\kappa} \tilde{\epsilon}_{\text{vib}}(T)/\tilde{v} \tag{2.62}$$

* Here the results of measurements carried out at temperatures between 20 and 300 K are given only as a plot and the linear thermal expansion coefficient, which is equal to 37×10^{-6} K^{-1}, is found only at room temperature.

and

$$(l - \tilde{l})/\tilde{l} = \gamma\tilde{\kappa}\tilde{\varepsilon}_{\text{vib}}(T)/3\tilde{v}, \tag{2.63}$$

where $\tilde{l}$ is the lattice constant, $\tilde{v} = \tilde{l}^3/4$ is the volume of the unit cell, $\tilde{\kappa} = 1/\tilde{v}\,(d^2u/dv^2)_{v=\tilde{v}}$ is the compressibility, and $\tilde{\varepsilon}_{\text{vib}}$ is the vibrational energy (in a harmonic approximation).

At $T = 0$, the energy of the lattice constant can be determined from Eq. (2.63):

$$[l(0) - \tilde{l}]/\tilde{l} = \gamma\tilde{\kappa}\tilde{\varepsilon}_{\text{vib}}(0)/3\tilde{v}. \tag{2.64}$$

According to Debye's harmonic theory,

$$\tilde{\varepsilon}_{\text{vib}}(0) = \frac{9}{4}\hbar\sqrt{\frac{5}{2}\frac{\tilde{l}}{\tilde{\kappa}}\frac{1}{\mu}}, \tag{2.65}$$

where $\hbar$ is Planck's constant, and $\mu = 1/(1/M + 1/m)$ is the reduced mass (in grams). We can infer from this expression that

$$\frac{l(0) - \tilde{l}}{\tilde{l}} = \frac{3}{4}\frac{\hbar\gamma\tilde{\kappa}}{\tilde{v}}\sqrt{\frac{5}{2}\frac{\tilde{l}}{\tilde{\kappa}}\frac{1}{\mu}} \tag{2.66}$$

or that

$$l(0) = \tilde{l}\left(1 + \frac{3}{4}\frac{\gamma\hbar\kappa}{\tilde{v}}\sqrt{\frac{5}{2}\frac{\tilde{l}}{\kappa}\frac{1}{\mu}}\right). \tag{2.67}$$

From these equations, we can determine $\tilde{l}$ by using the experimental data of Anderson *et al.*[96] on the lattice constants of ^{7}LiH, ^{7}LiD, and ^{7}LiT crystals obtained at low temperatures. As was noted in Sec. 2.1, Anderson *et al.*[96] found that the lattice constants for the isotopic compositions of lithium hydride satisfy Eq. (2.59).

Comparing Eqs. (2.59) and (2.67) and ignoring thermal expansion in the temperature range 0–83 K, i.e., assuming that $l(0) = l(83)$, we find

$$A(83) = \tilde{l} \text{ and } B = 1.1858\frac{\gamma\hbar}{\tilde{l}}\sqrt{\frac{\tilde{\kappa}}{\tilde{l}}}. \tag{2.68}$$

Using these relations and the data of Ref. 96, we find $\tilde{l}$(Å) to be 3.9949 for ^{7}LiH, 3.9945 for ^{7}LiD, 3.9945 for ^{7}LiT, and 3.9950 for ^{6}LiH. A good agreement of the results of this calculation confirms the theory's conclusion that the lattice constant of the crystal is independent of the isotopic composition in the harmonic approximation. The average value of the lattice constant is $\tilde{l} = 3.9947$ Å.

It is clear from Eq. (2.66) that at $T = 0$ the isotopic effect in crystals of the type considered by us can be expressed in the form (γ is essentially independent of the isotopic composition; see the discussion below)

$$[l_1(0) - \tilde{l}]/[l_2(0) - \tilde{l}] = \sqrt{\mu_2/\mu_1}\,. \tag{2.69}$$

Differentiating Eq. (2.61) with respect to temperature at a constant volume, we find the Mie–Grüneisen equation for linear thermal expansion α of the crystal:

$$\alpha = \frac{1}{l}\frac{dl}{dT} = \gamma \frac{\kappa_{\mathrm{ad}} c_P}{3v}. \tag{2.70}$$

In Sec. 1.3, it was assumed that the ratio κ_{ad}/v can be replaced by $\tilde{\kappa}/\tilde{v}$ in the harmonic approximation [see Eq. (2.36)]. From Eq. (2.70) we then find

$$\alpha = \gamma \frac{\tilde{\kappa} c_P}{3\tilde{v}} \tag{2.71}$$

and

$$l(T) = l(0)\exp\left(\gamma \frac{\tilde{\kappa}}{3\tilde{v}} [H(T) - H(0)]\right). \tag{2.72}$$

The Grüneisen constant for the LiH crystal was determined from the thermodynamic data on the basis of Eq. (2.70) and from the pressure derivatives of the elastic constants or the vibrational energy (see Refs. 125, 160, and 184). This question was studied most thoroughly by Jex,[160] who used a model-based ionic potential whose parameters were selected on the basis of the best agreement between the calculated phonon spectra and the experimental data.[159,243] Calculations have shown that beginning at a temperature of ~50 K, the Grüneisen constant varies only slightly with temperature and isotopic composition, being approximately 1.2 both for ^{7}LiH and ^{7}LiD. In the harmonic approximation, the compressibility factor $\tilde{\kappa}$, which is the same for all isotopic compositions, was determined from Eq. (2.34):

$$\tilde{\kappa} = \kappa_{\mathrm{ad}} \left(1 + \gamma \frac{\Delta v}{\tilde{v}}\right). \tag{2.73}$$

Substituting into this equation the value of κ_{ad} at zero temperature, taken from Ref. 225, $v(0)$ taken from Ref. 96, and $\tilde{v}$ determined from $\tilde{l}$, which was obtained above, we find $\tilde{\kappa} = 2.84 \times 10^{-12}$ cm^2/dyn.

Since the volume of the unit cell of LiH is $\tilde{v} = \tilde{l}^3/4$ in the harmonic approximation, we find the following working relations after substituting in Eqs. (2.71) and (2.72) the values found above:

$$\alpha = 1.1837 \times 10^{-6} c_P \tag{2.74}$$

and

$$l(T) = l(0)\exp\{1.1837 \times 10^{-6}[H(T) - H(0)]\}. \tag{2.75}$$

In these equations α is the thermal expansion coefficient (K^{-1}), c_P is the heat capacity (J/mole K), and H is the enthalpy, (J/mole).

Table 2.38 can be used to compare the lattice constants of the isotopic modifications of lithium hydride determined experimentally with those calculated from Eq. (2.75). Figure 2.7 is a graphic comparison with experiment. Both Table 2.38 and Fig. 2.7 show a good agreement between the calculated and experimental data. Note that as the melting point is approached, the difference in the lattice parameters of the isotopic modifications of lithium hydride and hence the difference in the molar volume $v = N_A l^3/4$ decrease.*

In the calculation of the lattice constant, the enthalpy of the isotopic compositions of lithium hydride was taken from Tables 5.1–5.3 with the help

*London[179] attempted to develop a theory in which a correlation would be established between the isotopic replacement and the specific volume of monatomic and diatomic crystals.

Table 2.38. Comparison of the experimental and calculated values of the lattice constant *l* of the isotopic modifications of lithium hydride.

Isotopic modification	Temperature K	*l*, Å Experiment	Calculation	References (experimental data)
^7LiH	154	4.069	4.0692	Smith and Leider (Ref. 214)
	298	4.085 ± 0.001	4.0842	Zintl and Harder (Ref. 262)
	298	4.085 ± 0.001	4.0842	Tronstad and Wergeland (Ref. 228)
	298	4.085 ± 0.002	4.0842	Ahmed (Refs. 94, 95)
	298	4.0834 ± 0.0005	4.0842	Starizky and Walker (Ref. 221)
	298	4.083	4.0842	Zalkin and Silvera (see Ref. 41)
	298	4.083 ± 0.001	4.0842	Brückner *et al.* (Ref. 113)
	298	4.0856	4.0842	Smith and Leider (Ref. 214)
	298	4.084	4.0842	Haussühl and Skorczyk (Ref. 138)
	298	4.0829	4.0842	Anderson *et al.* (Ref. 96)
	300	4.084 ± 0.001	4.0844	Kogan and Omarov (Ref. 24)
	413	4.1005	4.1027	Anderson *et al.* (Ref. 96)
	473	4.116	4.1144	Zalkin and Silvera (see Ref. 41)
	513	4.1224	4.1221	Anderson *et al.* (Ref. 96).
	673	4.159	4.1584	Zalkin and Silvera (see Ref. 41)
	798	4.188	4.1907	*ibid.*
^7LiD	150	4.0536	4.0514	Smith and Leider (Ref. 214)
	298	4.965 ± 0.001	4.0695	Zintl and Harder (Ref. 262)
	298	4.0684 ± 0.0005	4.0695	Starizky and Walker (Ref. 221)
^7LiD	298	4.070 ± 0.001	4.0695	Kogan and Omarov (Ref. 24)
	298	4.069 ± 0.001	4.0695	Brückner *et al.* (Ref. 113)
	298	4.069	4.0695	Haussühl and Skorczyk (Ref. 138)
	298	4.0693	4.0695	Anderson *et al.* (Ref. 96)
	299	4.0708	4.0697	Smith and Leider (Ref. 214)
	413	4.0893	4.0913	Anderson *et al.* (Ref. 96)
	513	4.1119	4.1133	*ibid.*
^7LiT	298	4.0633	4.0637	*ibid.*
^6LiH	298	4.0851	4.0843	*ibid.*
	413	4.1013	4.1025	*ibid.*
	513	4.1218	4.1217	*ibid.*
^6LiD	150	4.052	4.0530	Smith and Leider (Ref. 214)
	298	4.070	4.0704	*ibid.*
	298	4.0708	4.0704	Anderson *et al.* (Ref. 96)
	413	4.0888	4.0916	*ibid.*
	513	4.1110	4.1133	*ibid.*

of a linear interpolation. The agreement between the experimental and calculated data has made it possible to calculate the lattice constant and the density of the isotopic modifications of lithium hydride in the uninvestigated region up to the melting point by using Eqs. (2.57) and (2.75) and the data on the enthalpy from Tables 5.1–5.3. The calculated values of the lattice constant are given in Table 2.39, and the values of the density are listed in Table 5.4 and shown in Fig. 2.8. Also shown in this figure are results of other authors. We see, on the one hand, a quite satisfactory agreement between the

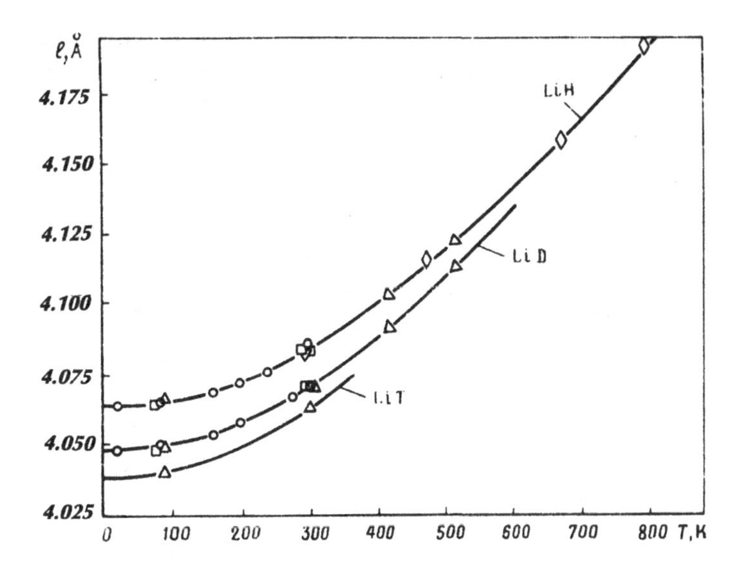

Figure 2.7. Temperature dependences of the lattice constants of LiH, LiD, and LiT according to the data of different authors. ○—Smith and Leider; △—Kogan and Omarov; □—Anderson *et al.*; ◇—Zalkin and Silvera; — —results of our calculations.

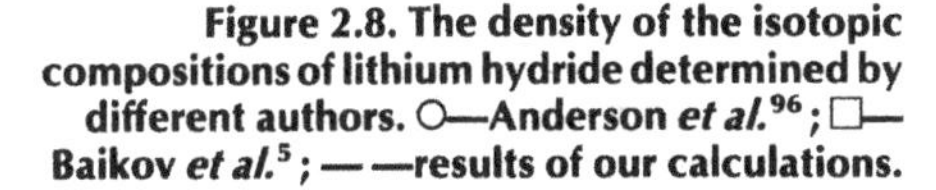
Figure 2.8. The density of the isotopic compositions of lithium hydride determined by different authors. ○—Anderson *et al.*[96]; □—Baikov *et al.*[5]; — —results of our calculations.

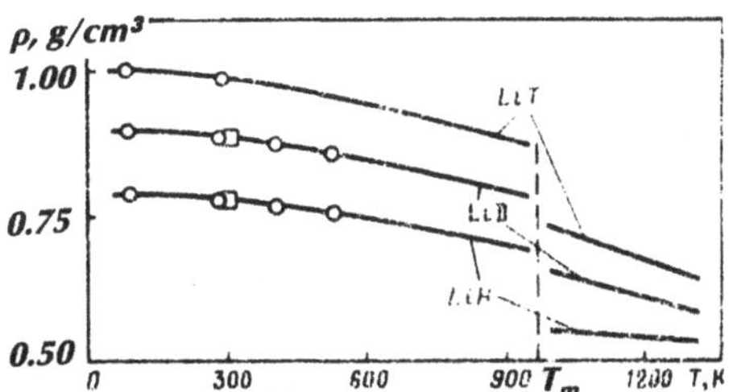

calculated data and the data obtained experimentally and, on the other, a very strong isotopic effect. In the calculations, the following values of the mass (amu) of the hydrogen isotopes, taken from the data of Refs. 61 and 62, were used: $m_H = 1.007\,82$, $m_D = 2.014\,10$, and $m_T = 3.016\,04$. According to Eq. (2.69), the lattice constant of hydride containing natural lithium (92.5% ^{7}Li and 7.5% ^{6}Li, $M = 6.941$ amu) is essentially the same as that of hydride with ^{7}Li.

At high temperatures, the difference in the values of the lattice constants of the isotopic compositions of lithium hydride decreases substantially. The accuracy of the data in Table 2.39 is dependent principally on the accuracy of $l(0)$.

The thermal expansion coefficient of the isotopic compositions of lithium hydride in the crystalline state can be calculated from Eq. (2.74). Table 5.5 gives the linear thermal expansion coefficients for ^{n}LiH, ^{n}LiD, and ^{n}LiT crystals. The heat capacity c_P was taken from Tables 5.1–5.3 in the tabulation. Figure 2.9 can be used to compare the linear thermal expansion coeffi-

Table 2.39. Lattice constants of the isotopic modifications of lithium hydride, *l*.

T,K	*l*, Å		
	^{7}LiH	^{7}LiD	^{7}LiT
100	4.0665	4.0486	4.0412
150	4.0689	4.0512	4.0441
200	4.0729	4.0557	4.0493
250	4.0780	4.0620	4.0563
298.15	4.0842	4.0695	4.0643
300	4.0844	4.0698	4.0647
350	4.0920	4.0788	4.0743
400	4.1006	4.0887	4.0847
450	4.1099	4.0993	4.0960
500	4.1198	4.1104	4.1074
550	4.1303	4.1220	4.1193
600	4.1413	4.1340	4.1317
650	4.1529	4.1465	4.1445
700	4.1650	4.1594	4.1577
750	4.1777	4.1730	4.1715
800	4.1911	4.1871	4.1859
850	4.2053	4.2020	4.2010
900	4.2203	4.2177	4.2169
950	4.2363	4.2343	4.2337
Melting point*	4.2413	4.2401	4.2400

**Note:* The melting points of ^{7}LiH, ^{7}LiD, and ^{7}LiT are 965, 967, and 968 K, respectively.

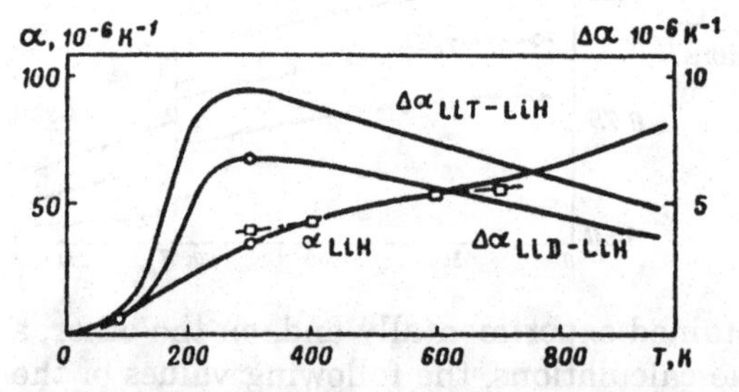

Figure 2.9. Comparison of the linear thermal expansion coefficients α of lithium hydride, lithium deuteride, and lithium tritide in the solid state. ○—Bruckner *et al.*[113]; □—Zalkin and Silvera (see Ref. 31); — —results of our calculations.

cients of the isotopic compositions of lithium hydride. The curve for the difference in the coefficients, $\Delta\alpha$, is similar to that in Fig. 2.6.

A slight discrepancy with the data of Zalkin and Silverman stems from the fact that the working equation for the temperature dependence of the lattice constant of the ^{n}LiH crystal [see Eq. (2.58)], which they used to determine the thermal expansion, was derived from only four experimental points and therefore cannot exactly reproduce the temperature derivative, especially at the limits of the temperature interval studied. Furthermore, we have used the x-ray method, which in the case of lithium hydride leads to an error at high temperatures, as was indicated in Sec. 2.1.

The isotopic effect is most pronounced in the thermal expansion in the intermediate temperature range. The difference decreases at low and high temperatures. At low temperatures, the thermal expansion of all crystalline solids is nearly zero. At high temperatures, when the vibrational levels of the

Table 2.40. Comparison of the data on the density of the isotopic modifications of crystalline lithium hydride obtained experimentally, ρ_{exp}, and calculated, ρ_{calc} (g/cm³).

	^{7}LiH			^{7}LiD			^{7}LiT		
T,K	ρ_{exp}	ρ_{calc}	$\delta\rho$,%	ρ_{exp}	ρ_{calc}	$\delta\rho$,%	ρ_{exp}	ρ_{calc}	$\delta\rho$,%
293	0.7773	0.7753	0.25	0.8841	0.8831	0.11	—	—	—
298	0.7814	0.7806	0.10	0.8884	0.8883	0.01	0.9915	0.9912	0.03
413	0.7714	0.7701	0.17	0.8754	0.8741	0.15	—	—	—
513	0.7591	0.7593	−0.01	0.8611	0.8602	0.10	—	—	—

crystal are virtually totally excited, the molecular heats of the crystals, c_V, tend toward the limiting value determined by the Debye theory or the Einstein theory, and the difference in the thermal expansions decreases, according to Eq. (2.74) [the fact that Eq. (2.74) contains the heat capacity c_P, instead of c_V, is of no consequence in this analysis].

The accuracy of the data on the density of the isotopic modifications of lithium hydride in the crystalline state, which are listed in Table 5.4 and which were calculated theoretically, can be reliably determined only by comparing the calculated and experimental results. The values of the density obtained directly, which are found in the literature, correspond to temperatures no higher than room temperature. Anderson *et al.*[96] have measured the densities of the isotopic modifications of lithium hydride at higher temperatures, but they cited only the lattice constants that were calculated on the basis of these measurements.

The density of the crystalline modifications of lithium hydride was measured by the pycnometer method, gas-densitometer method, and the flotation method. The most accurate results, however, were obtained by the flotation method, which was used in Refs. 5 and 96. In these studies, the experiments were set up in such a way as to suggest that they were accurate. In particular, the maximum relative error in the measurement of the density in Ref. 5 was within 0.03%.

If the procedure is reversed by going from the lattice constant, in accordance with the data of Ref. 96, to the density by making use of Eq. (2.57), and the value of ρ_{exp} obtained is then compared with our calculated data on the density of the isotopic modifications of lithium hydride, ρ_{calc}, then the agreement will be quite satisfactory. This comparison at temperatures of 298, 413, and 513 K is given in Table 2.40, along with a comparison of the results of our calculations with the experimental data of Ref. 5 on the isotopic modifications of lithium hydride with natural lithium obtained at a temperature 293 K; $\delta\rho = (\rho_{exp} - \rho_{calc})/\rho_{calc}$.

The discrepancy between the calculations and experiment is $\sim$0.15%, on the average. Accordingly, it can be assumed that the relative error in the calculated data on the density of the isotopic modifications of lithium hydride in the crystalline state (with allowance for the error of the initial experiment) is no greater than 0.2%.

2.3. Elastic properties of crystalline lithium hydride and of its isotopic modifications

The elastic properties of solid lithium hydride are of interest in themselves and also because they determine many other characteristics of this compound. Accordingly, the compressibility is taken into account in the working equations for the heat capacity c_P, for the thermal expansion coefficient α, and for other thermodynamic properties of a crystal. Furthermore, the elastic moduli c_{ij} determine the lattice scattering of x rays and neutrons, the infrared absorption and reflection, the Raman and Brillouin scattering, and the scattering and propagation of sonic and ultrasonic waves in a crystal. The data on the elastic constants and their pressure and temperature derivatives are the key to understanding the nature of the cohesive forces in a given substance.

The crystals with a cubic symmetry, such as LiH, have three elastic moduli c_{11}, c_{12}, and c_{44}. The hydrostatic-compression modulus B and the compressibility κ can be expressed in terms of these moduli in the following way:

$$B = (c_{11} + 2c_{12})/3, \quad \kappa = 1/B. \tag{2.76}$$

The elastic properties of isotopic polycrystals are determined by the longitudinal-compression modulus, Young's modulus, and the bulk modulus. These quantities can be measured by the static and dynamic methods. The static measurements yield isothermal elastic and bulk moduli, whereas the adiabatic moduli can be obtained by the dynamic methods. In the initial experiments, a static method was generally used to measure the isothermal compressibility in a high-pressure apparatus. The dynamic methods, which were employed later, have the advantage of allowing the use of smaller samples, of simplifying the temperature control and of improving the measurement accuracy, because these methods are based on the measurement of the resonant frequency or the propagation velocity of ultrasound, rather than a slight displacement.

The adiabatic compressibility κ_{ad} is related to the isothermal compressibility κ_{iso} by (see Ref. 30, for example)

$$\kappa_{iso} - \kappa_{ad} = 9Tv\alpha^2/c_P \tag{2.77}$$

or

$$\kappa_{iso}/\kappa_{ad} = c_P/c_V. \tag{2.78}$$

The difference between the compressibilities in Eq. (2.77) is proportional to T at high temperatures, to T^4 at low temperatures, and vanishes at $T = 0$; this difference is always smaller than the compressibilities themselves.

Hearmon[129] has shown that for solids this difference is small (on the order of 1% of the compressibility) and is usually ignored, since it is smaller than the error of the measurement results obtained by using the static methods. In the case of lithium hydride, for example, at room temperature we have $\kappa_{iso} - \kappa_{ad} \approx 0.036 \times 10^{-12}$ cm^2/dyn, which is ~1% of κ_{iso}.

Review and analysis of experimental studies

The isothermal compressibility κ_{iso} of lithium hydride and of some of its isoto-

Table 2.41. Density ρ and isothermal compressibility κ_{iso} of lithium hydride and of its isotopes at room temperature.

Isotopic modification	ρ,kg/cm^2	κ_{iso}, 10^{-12} cm^2/dyn	B_{iso}, 10^{11} dyn/cm^2	References
^{n}LiH	—	3.77	2.66	Weil and Lawson (Ref. 248)
^{n}LiH	—	4.38	2.23	Voronov *et al.* (Ref. 10)
^{7}LiH	0.782	2.88(2.98)	3.48(3.36)	Stephens and Lilly (Ref. 222)
^{7}LiD	0.893	2.85(2.94)	3.51(3.50)	*ibid.*
^{6}LiH	0.663	2.83(2.95)	3.53(3.38)	*ibid.*
^{6}LiD	0.750	2.76(2.85)	3.62(3.51)	*ibid.*

Note: κ_{iso} and B_{iso} were determined by using the Born-Mayer equation of state. The values enclosed in the parentheses were determined in the analysis of the experimental data on the basis of Murnahan's equation of state (see Ref. 222).

pic modifications at room temperature were measured by Weil and Lawson,[248] Voronova *et al.*,[10] and Stephens and Lilly.[222] Weil and Lawson[248] measured the compressibility κ_{iso} of ^{n}LiH up to 6 kbars with the Bridgman dilatometer. The sample to be studied was pressed from a powder under a pressure of 10 kbar. Voronov *et al.*[10] determined the compressibility of lithium hydride at pressures of up to 20 kbar by the "piston-displacement" method in a high-pressure apparatus. They obtained a linear dependence of the relative change in the volume on the pressure P (dyn/cm^2)

$$-\Delta V/V = 4.38\times10^{-13}P. \tag{2.79}$$

The values of $\Delta V/V$ were measured within 2%.

Stephens and Lilly[222] measured the compressibility of ^{7}LiH, ^{7}LiD, ^{6}LiH, and ^{6}LiD in the pressure range up to 40 kbar. The density of the compressed-powder samples was approximately equal to the theoretical density. The ^{7}LiH samples were 99.9% pure by mass, and the ^{6}LiH samples were 96.1% pure by mass. The error in the data was estimated to be 2%. At room temperature, the values of the compressibility κ_{iso}, according to the results of these studies, are given in Table 2.41.

Many studies dealing with the measurement of the elastic constants and the compressibility by the dynamic method have recently been published. We briefly review these studies and compare the results of the static and dynamic methods of measuring the compressibility.

The dynamic measurements were carried out with lithium hydride single crystals and polycrystalline samples fabricated from a powder that was pressed to a density close to the theoretical value. The longitudinal V_L and transverse V_T velocities of sound in polycrystalline samples were measured by the ultrasonic-echo method at frequencies of several tens of megahertz. The adiabatic modulus of uniform compression, V_{ad}, and hence the adiabatic compressibility κ_{ad} can be determined from the total velocity of sound V_b:

$$B_{ad} = \rho V_b^2; \quad V_b^2 = V_L^2 - \tfrac{4}{3} V_T^2. \tag{2.80}$$

In the experiments with single crystals, the longitudinal and transverse velocities of sound in the (110) direction were also measured by the ultrasonic-

Table 2.42. Measured values of the elastic constants of the isotopic modifications of single-crystal lithium hydride at room temperature.

Isotopic modification	ρ, g/cm²	$c_{11,ad}$ 10^{11} dyn/cm²	$c_{12,ad}$ 10^{11} dyn/cm²	$c_{44,ad}$ 10^{11} dyn/cm²	B_{ad} 10^{11} dyn/cm²	κ_{ad} 10^{-12} cm²/dyn	References
^{n}LiH	0.776	6.531	1.485	4.501	3.167	3.155	Haussühl and Skorczyk (Ref. 138)
^{7}LiH	0.775	6.68	1.54	4.59	3.250	3.08	Guinan and Cline (Ref. 130)
^{n}LiH	Theor.	6.633	1.476	4.604	3.195	3.14	Terras (Ref. 225)
^{7}LiH	0.783	6.720	1.493	4.637	3.235	3.10	Gerlich and Smith (Ref. 125)
^{n}LiH	0.776	6.36	1.50	4.24	3.13	3.20	Marsh (Ref. 184)
^{7}LiH	0.783	6.37	1.54	4.28	3.15	3.18	*ibid.*
^{7}LiH*	0.778	6.71	1.75	4.60	3.40	2.94	Laplaze *et al.* (Ref. 176)
^{7}LiH**	0.778	6.68	1.54	4.59	3.25	3.08	*ibid.*
^{n}LiD	0.893	6.626	1.462	4.553	3.183	3.14	Haussühl and Skorczyk (Ref. 138)
^{7}LiD	Theor.	6.70	1.397	4.725	3.165	3.16	Guinan and Cline (Ref. 130)
^{7}LiD	0.890	6.26	1.71	4.19	3.23	3.08	Marsh (Ref. 184)

*The results were obtained by the method of Brillouin scattering of hypersound.

**The results were obtained by the method of ultrasound propagation.

echo method or the Brillouin-scattering method. The elastic constants in this case can be determined from the measured velocities by using the equations

$$\rho V_L^2 = (c_{11} + c_{12} + 2c_{44})/2; \quad \rho(V_T^{\parallel})^2 = c_{44};$$

$$\rho(V_T^{\perp})^2 = \frac{c_{11} - c_{12}}{2}, \tag{2.81}$$

where V_L is the longitudinal velocity of sound in the (110) direction, and $V_T^{\parallel}$ and $V_T^{\perp}$ are the transverse velocities of sound in the (110) and $(1\bar{1}0)$ directions, respectively. The elastic moduli $c_{ij,\mathrm{ad}}$ in lithium hydride single crystals were measured by many researchers (see Table 2.42), three of whom—Haussühl and Skorczyk,[138] Guinan and Cline,[130] and Marsh[184] —measured the adiabatic elastic moduli of ^{7}LiH and ^{7}LiD, making it possible to study the isotopic effect in elastic properties.

To determine the adiabatic elastic moduli c_{11}, c_{12}, and c_{44}, Haussühl and Skorczyk[138] measured the propagation velocity of elastic waves in ^{n}LiH and ^{n}LiD crystals in the (100) and (110) directions using an improved method of Bergmann–Scheffer (this method and other methods of measuring the elastic characteristics of various crystals are described in a review by Huntington[64]). The ultrasonic frequency was 25 MHz.

Guinan and Cline[130] measured the elastic constants of ^{7}LiH and ^{7}LiD single crystals at a temperature of 298 K by determining the ultrasonic velocity at 10 MHz. In their measurements they used four pure crystals (two ^{7}LiH samples and two ^{7}LiD samples), whose acoustic surfaces were oriented at right angles to the (100) and (110) axes. The crystals had a high isotopic purity: ^{7}Li isotope was 99.99% pure, H was 100% pure, and D was 99.65% pure. After fabrication, the samples were held in either an H_2 or a D_2 atmosphere at a temperature of 823 K for 25 days. All measurements were carried out in pure argon. Three measurements were made for each sample (two transverse sound velocities and one longitudinal velocity). The elastic constants were determined by Gauss's averaging method from a set of six measurements.

Marsh[184] reported the measurement of the elastic moduli of ^{n}LiH, ^{7}LiH, and ^{7}LiD by the method indicated above. He measured the longitudinal and transverse velocities of sound in the (100) and (110) directions. Terras,[225] Gerlich and Smith,[125] and Laplaze *et al.*[176] measured the elastic constants of ^{7}LiH at room temperature by the ultrasonic-echo method. In the last study, the elastic properties of ^{7}LiH were also determined from the Brillouin scattering of hypersound at 60 GHz. The results of all these studies are summarized in Table 2.42.

Most of the studies (except that of Guinan and Cline[130]), including the measurements of polycrystalline samples (Table 2.43) and also the static measurements of the compressibility carried out by Stephens and Lilly[222] (Table 2.42), show that the compressibility of ^{7}LiD is approximately 0.6%–3% lower than that of ^{7}LiH. The elasto-isotropic properties of lithium hydride and of its isotopes were measured in polycrystalline powders pressed to single-crystal density (see Table 2.43).

Marsh[184] determined the longitudinal velocity V_L and transverse velocity V_T of sound by the ultrasonic-echo method and then determined the elasto-isotropic moduli for four isotopic modifications of lithium hydride. Ter-

Table 2.43. Elastic properties of polycrystalline lithium hydride and of its isotopic modifications at room temperature.

Hydride	V_L km/s	V_T, km/s	c_b, km/s	B_{ad} 10^{11} dyn/cm²	μ 10^{11} dyn/cm²	E, 10^{11} dyn/cm²	ν'	Measurement method	References
^7LiH	10.15	6.83	6.39	3.20	3.65	7.93	0.089	Averaging according to Voigt	Guinan and Cline[a]
	10.05	6.75	6.34	3.14	3.56	7.76	0.086	Ultrasound	Marsh[b]
	10.05	6.70	—	—	—	—	—	Shock waves	Braun and Danegan[c]
	—	—	—	3.25	3.50	7.60	0.08	Ultrasound	Terras and Moussin[d]
	—	—	—	3.25	3.78	8.18	0.082	Averaging according to Voigt	*ibid.*
^7LiD	9.57	6.49	5.96	3.17	3.75	8.07	0.076	Ultrasound	Guinan and Cline[a]
	9.56	6.43	6.02	3.24	3.69	8.03	0.077	Ultrasound	Marsh[b]
	9.55	6.35	—	—	—	—	—	Shock waves	Braun and Danegan[c]
^6LiH	10.67	7.18	6.72	3.14	3.59	7.81	0.086	Ultrasound	Marsh[b]
^6LiD	10.10	6.80	6.35	3.22	3.69	8.02	0.085	Ulrasound	Marsh[b]

[a] Ref. 130
[b] Ref. 184
[c] Ref. 130
[d] Ref. 226

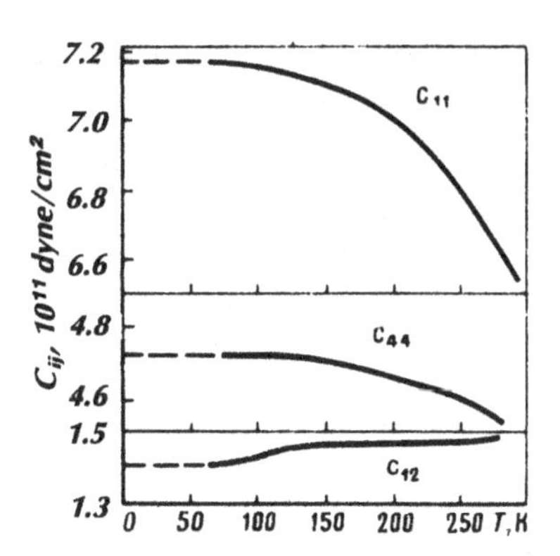

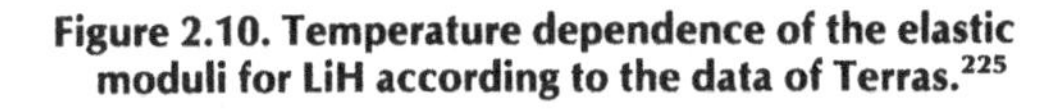
Figure 2.10. Temperature dependence of the elastic moduli for LiH according to the data of Terras.[225]

ras and Moussin[226] and Guinan and Cline[130] found the effective isotropic moduli as Voigt averages.[244] Braun and Danegan (see Ref. 130) measured the longitudinal and transverse velocities of sound by the shock-wave method. The elastic properties of polycrystalline lithium hydride and of its isotopic compositions at room temperature are given in Table 2.43 (ν' is the Poisson number, E is Young's modulus, and μ is the longitudinal-compression modulus). It can be seen from Table 2.43 that the values of the compressibility κ_{ad} and of the uniform-compression modulus B_{ad} obtained for polycrystalline powders by different authors are in reasonably good agreement with each other and with the values obtained from single crystals.

A comparison of the measurement results of κ_{iso} (see Table 2.42) and κ_{ad} (see Tables 2.43 and 2.44) shows that the experimental values of κ_{iso} obtained by Stephens and Lilly[222] are in best agreement with the values of κ_{ad} obtained by different authors. Judging from its description, the study of Stephens and Lilly[222] is the most comprehensive of all studies of the static compressibility. The samples used in the experiment were, according to the authors, initially pressure treated and had a density approximately equal to the theoretical density. Voronov *et al.*[10] and Weil and Lawson,[248] however, did not report the density of the samples. The large values of the compressibility obtained in these experiments indicate that the density of the samples presumably was not theoretical, which may account for the high values of κ_{iso}.

The temperature dependence of the adiabatic elastic moduli and, correspondingly, of the hydrostatic-compression modulus B_{ad} was measured by Haussühl and Skorczyk,[138] Terras,[225] and Gerlich and Smith.[125]

Terras[225] measured the elastic constants c_{11}, c_{44}, and c_{12} over the temperature interval 77–300 K. Figure 2.10 shows that c_{11} and c_{44} decrease with increasing temperature (c_{44} decreases slower than c_{11}), while c_{12} increases. At $T \gg 77$ K, no change in $c_{ij,ad}$ with temperature is observed. The following elastic constants found at $T \approx 77$ K were thought by Terras to be the values recorded at a temperature of 0 K: $c_{11} = 7.16 \times 10^{11}$ dyn/cm²; $c_{12} = 1.39 \times 10^{11}$ dyn/cm²; $c_{44} = 4.72 \times 10^{11}$ dyn/cm²; $B_{ad} = 3.31 \times 10^{11}$ dyn/cm²; and $\kappa_{ad} = 3.02 \times 10^{-12}$ cm²/dyn.

The thermoelastic constants found by Haussühl and Skorczyk[138] (at 273 K) and Gerlich and Smith[125] (at 298 K) are given in Table 2.44 ($\lambda_{ij} = d \ln c_{ij,ad}/dT$).

Table 2.44. Thermoelastic constants of lithium hydride and lithium deuteride (10^{-3} K^{-1}).

Isotopic composition	λ_{11}	λ_{12}	λ_{44}	$\frac{1}{B_{ad}}\frac{dB_{ad}}{dT}$	References
^{n}LiH	−0.560	0.620	−0.265	—	Haussühl and Skorczyk (Ref. 138)
^{7}LiH	−0.58	0.66	−0.33	−0.195	Gerlich and Smith (Ref. 125)
^{7}LiD	−0.615	0.670	−0.305	—	Haussühl and Skorczyk (Ref. 138)

These results suggest that the temperature dependence and isotopic dependence of the elastic moduli and of the compressibility need further experimental study. Measurements have been carried out for only LiH and LiD, while there are no measurements for LiT.

Determination of the elastic properties of a LiH-type crystal: Theory and calculations

The isotopic and temperature dependences of the compressibility of crystals with a simple structure, whose behavior satisfies the Mie–Grüneisen equation of state [see Eq. (2.61)]

$$P+\frac{du}{dv}=\gamma\frac{\epsilon_{\text{vib}}}{\tilde{v}},$$

have been studied theoretically by Mel'nikova[36,37] in the lowest-order approximation with respect to the anharmonicity. Differentiating this equation with respect to v, restricting ourselves to terms linear in $\Delta v/\tilde{v}=[v(T)-\tilde{v}]/\tilde{v}$ [$\tilde{v}$ is the cell volume in the harmonic approximation and $v(T)$ is the cell volume at a given temperature], and transforming from the derivatives of v to the derivatives of r (r is the distance between the nearest ions), we find the relation for the thermal variation of the compressibility (see Ref. 8)

$$\frac{1}{\kappa_{\text{iso}}}-\frac{1}{\tilde{\kappa}}=\frac{\Delta v}{v}\left(\frac{\tilde{r}}{3\tilde{\kappa}}\frac{d^3u/dr^3}{d^2u/dr^2}\bigg|_{r=\tilde{r}}\right)+\frac{\gamma^2}{\tilde{v}}e_{\text{vib}}-\gamma^2\frac{T}{tv}\tilde{c}_V. \tag{2.82}$$

The tilde denotes values in the harmonic approximation.

Since the exact form of the interionic potential $u(r)$ is generally not known, we must express the values in Eq. (2.82) in terms of experimentally measured parameters to link the different thermodynamic properties. The second derivative can be written in the form

$$\frac{d^2u}{dr^2}\bigg|_{\tilde{r}}=\frac{18\tilde{r}}{\tilde{\kappa}}. \tag{2.83}$$

The third derivative is related to thermal expansion in the lowest-order approximation in the anharmonicity (see Ref. 108):

$$\alpha=-\frac{\tilde{c}_V(d^3u/dr^3)_{r=\tilde{r}}}{2\tilde{r}(d^2u/dr^2)^2_{r=\tilde{r}}}. \tag{2.84}$$

In turn, in this approximation [see Eq. (2.62)]

$$\Delta v/\tilde{v} = \gamma\tilde{\kappa}\tilde{\epsilon}_{\text{vib}}(T)/\tilde{v}$$

and [see Eq. (2.71)]

$$\alpha = \gamma\tilde{c}_V\tilde{\kappa}/3\tilde{v}.$$

Using these relations, we can reduce Eq. (2.82) to

$$\frac{\kappa_{\text{iso}} - \tilde{\kappa}}{\tilde{\kappa}} = \gamma\frac{\Delta v}{\tilde{v}} + \gamma^2\frac{\tilde{\kappa}}{\tilde{v}}T\tilde{c}_V. \tag{2.85}$$

At the same time, the crystal theory (see Ref. 8, for example) shows that

$$\frac{1}{\kappa_{\text{iso}}} - \frac{1}{\kappa_{\text{ad}}} = -\gamma^2\frac{T\tilde{c}_V}{\tilde{v}}. \tag{2.86}$$

Substituting this equation into (2.85), we find the following expression for the adiabatic compressibility:

$$(\kappa_{\text{ad}} - \tilde{\kappa})/\tilde{\kappa} = \gamma\Delta v/\tilde{v}. \tag{2.87}$$

These equations make it possible to analyze the temperature dependence and isotopic dependence of the compressibility. For any two isotopic compositions,

$$\delta\kappa_{\text{ad}}/\tilde{\kappa} = \gamma\delta v/\tilde{v}, \tag{2.88}$$

where $\delta\kappa_{\text{ad}}$ and δv are, respectively, the differences in the compressibilities and molar volumes of the two isotopic compounds. Since in the approximation under consideration the molar volume of the light-isotope compound is always larger than that of the heavy-isotope compound, κ_{ad} of the light-isotope compound is larger than that of the heavy-isotope compound. The largest isotopic effect, observed at $T = 0$, decreases with increasing temperature. At room temperature, the isotopic effect in κ_{ad} is $\sim 2\%$ if H is replaced by D in lithium hydride. This value falls outside the limits of the measurement error of this quantity.

The isotopic and temperature dependences of κ_{iso} are slightly more complicated. From Eq. (2.85) we find

$$(\delta\chi_{\text{iso}}/\tilde{\kappa}) = \gamma(\delta v/\tilde{v}) + \gamma^2(T\tilde{\kappa}/\tilde{v})\delta\tilde{c}_V \tag{2.89}$$

for any two compositions, where $\delta\tilde{c}_V$ is the difference in the heat capacities of the two isotopic compositions. At $T = 0$, the heat capacity is zero and it increases with increasing temperature, reaching a maximum near room temperature and tending to zero at high temperatures; the sign of $\delta\tilde{c}_V$ is opposite to that of δv, since the heat capacity of a heavy-isotope compound is higher than that of a light-isotope compound. At $T = 0$ K the isotopic effect in κ_{iso} is therefore determined by the first term and is equal to $\delta\kappa_{\text{ad}}$. As the temperature is raised, $\delta\kappa_{\text{iso}}$ first decreases (the first term decreases in accordance with the modulus, and the second term, whose sign is opposite to that of the first, increases) and at a certain temperature its sign can theoretically change. In the limit of high temperatures, however, $\delta\kappa_{\text{iso}}$ and $\delta\kappa_{\text{ad}}$ tend to zero.

Equation (2.87) can be used to calculate κ_{ad} or $B_{\text{ad}} = 1/\kappa_{\text{ad}}$ at different temperatures. Since the volume of the crystal cell is $v = l^3/4$ for LiH, we have

$$\kappa_{\text{ad}} = \tilde{\kappa}\left(1 + 3\gamma\frac{l(T) - \tilde{l}}{\tilde{l}}\right); \quad B_{\text{ad}}(T) = \tilde{B}\left(1 - 3\gamma\frac{l(T) - \tilde{l}}{\tilde{l}}\right). \tag{2.90}$$

The value of $\tilde{\kappa}$ can be determined by using the experimental data on $l(T)$ and $\kappa_{ad}(T)$, for example, at room temperature. The most recent reliable experimental studies of the compressibility[125] and lattice constant[96] give the values $\tilde{\kappa} = 2.84\times 10^{-11}\ \mathrm{m^2/N}$ and $\tilde{B} = 3.52\times 10^{10}\ \mathrm{N/m^2}$.

As shown in Sec. 2.2 of this chapter, this value of the compressibility gives satisfactory results when used in calculations of the heat capacity and the linear thermal expansion of a LiH crystal. It should be pointed out, however, that the temperature derivatives of the compressibility or of the hydrostatic-compression modulus determined from Eq. (2.90) are not absolutely accurate. At room temperature, a calculation based on Eq. (2.90) yields $1/B_{ad}(dB_{ad}/dT) \approx -(0.14\text{–}0.17)\times 10^{-3}\ \mathrm{K^{-1}}$, whereas Gerlich and Smith's experiment yields $1/B_{ad}(dB_{ad}/dT)_P = -0.195\times 10^{-3}\ \mathrm{K^{-1}}$.

The lowest-order approximation with respect to the anharmonicity, which is used in the derivation of these relations, does not seem to adequately describe the temperature dependence of $B_{ad}(T)$ and $\kappa_{ad}(T)$ and hence $B_{iso}(T)$ and $\kappa_{iso}(T)$, but can be used to calculate the difference $c_P - c_V$, the lattice constant $l(T)$, and several other characteristics which do not require differentiation with respect to temperature of the elastic modulus. There is, however, a way in which the temperature dependence of the elastic moduli B_{iso} and B_{ad} can be determined more accurately (see Ref. 39). In this connection, we write the expression (see Ref. 32)

$$B_{iso} = v\left(\frac{\partial^2 F}{\partial v^2}\right)_T = v\frac{\partial^2 u}{\partial v^2} + v\left(\frac{\partial^2 F_{vib}}{\partial v^2}\right)_T . \tag{2.91}$$

In the quasiharmonic approximation

$$\left(\frac{\partial F_{vib}}{\partial v}\right)_T = -\gamma\frac{\epsilon_{vib}}{v}, \tag{2.92}$$

where ϵ_{vib} is the vibrational energy per cell, v is the cell volume, and γ is the Grüneisen constant. Assuming that γ is independent of the volume, the temperature, and the isotopic composition, we find from Eq. (2.92),

$$\left(\frac{\partial^2 F_{vib}}{\partial v^2}\right)_T = -\gamma\left[\frac{\partial}{\partial v}\left(\frac{\epsilon_{vib}}{v}\right)\right]_T = \gamma(\gamma+1)\frac{\epsilon_{vib}}{v^2} - \gamma_2\frac{Tc_V}{v^2}. \tag{2.93}$$

After substituting (2.93) into (2.91), we obtain

$$B_{iso} = v\left(\frac{\partial^2 u}{\partial v^2}\right)_T + \frac{1}{v}\gamma(\gamma+1)\epsilon_{vib} - \frac{\gamma^2}{v}Tc_V. \tag{2.94}$$

Writing Eq. (2.86) in quasiharmonic form

$$B_{iso} - B_{ad} = -\gamma^2\frac{Tc_V}{v} \tag{2.95}$$

and substituting it into (2.94) with the help of Eq. (2.61), we find

$$B_{ad} = v\left(\frac{\partial^2 u}{\partial v^2}\right)_T + (\gamma+1)\left(P + \frac{du}{dv}\right). \tag{2.96}$$

Differentiation of this equation yields

$$\left(\frac{\partial B_{ad}}{\partial T}\right)_P = \beta\left(v(\gamma+2)\frac{d^2u}{dv^2} + v^2\frac{d^3u}{dv^2}\right), \tag{2.97}$$

where β is the volume thermal expansion coefficient.

The expression in square brackets can be expanded in a series in small deviations $\Delta V/\tilde{V}$. Confining ourselves to the dominant terms in the expansion, we then find

$$\left(\frac{\partial B_{ad}}{\partial T}\right)_P = \beta\left(\tilde{v}(\gamma+2)\frac{d^2u}{dv^2}\bigg|_{\tilde{v}} + \tilde{v}^2\frac{d^3u}{dv^2}\bigg|_{\tilde{v}}\right) \equiv -\Gamma\beta. \quad (2.98)$$

In this expression the square brackets contain quantities which, in the harmonic approximation, are independent of the isotopic composition of the crystal. It thus follows that

$$\left(\frac{\partial B_{ad}}{\partial T}\right)_P^{^7\text{LiH}} \Bigg/ \left(\frac{\partial B_{ad}}{\partial T}\right)_T^{^7\text{LiD}} = \beta_{^7\text{LiH}}/\beta_{^7\text{LiD}}. \quad (2.99)$$

According to the data of Haussühl and Skorczyk,[138] who measured the coefficients $1/B_{ad}(\partial B_{ad}/\partial T)_P$ for ^{n}LiH and ^{n}LiD, the ratio $(\partial B_{ad}/\partial T)_{P,^n\text{LiH}}/(\partial B_{ad}/\partial T)_{P,^n\text{LiD}}$ is ~ 0.86 at a temperature of 273 K, whereas calculations based on our data (see Table 5.4 in Chap. 5) yield a value of ~ 0.84 for this ratio. This agreement is quite satisfactory.

Using Gerlich and Smith's data on $B_{ad}(T)$ at room temperature and the thermal expansion of the isotopic modifications of lithium hydride (see Table 5.5), we find, with the help of Eq. (2.98), the value of Γ which is 18.725×10^3 dyn/cm^2. Integrating Eq. (2.98) from 0 K to T, we find

$$B_{ad}(T) - B_{ad}(0) = \Gamma \ln\frac{v(T)}{v(0)} = -3\Gamma \ln\frac{l(T)}{l(0)}. \quad (2.100)$$

From this expression we can calculate $B_{ad}(0)$ with use of the known lattice constants of the isotopic modifications of lithium hydride (see Table 2.39) and the values of $B_{ad}(T)$ at room temperature, taken from the study of Gerlich and Smith. Calculations have shown that $B_{ad}(0)_{^7\text{LiH}} = 3.315\times10^{11}$ dyn/cm^2, whereas Terras's experiment yields* $B_{ad}(0)_{^7\text{LiH}} = 3.31\times10^{11}$ dyn/cm^2.

For those isotopic modifications of lithium hydride which have no data on B_{ad}, we can find $B_{ad}(0)$ by using Eq. (2.100) in the form

$$B_{ad}(0) - \tilde{B} = -3\Gamma \ln\left[l(0)/\tilde{l}\right]. \quad (2.101)$$

Using this equation for the ^{7}LiH crystal, for example, for which this equation has all the necessary information, and also using the value of $\tilde{l}$ found in Sec. 2.2, we find $\tilde{B} = 3.64\times10^{10}$ N/m^2. This value is slightly higher than that found previously from a linearized equation [Eq. (2.90)]: $\tilde{B} = 3.52\times10^{10}$ N/m^2.

Using the value $\tilde{B} = 3.64\times10^{10}$ N/m^2, which does not depend on the isotopic composition, and the data of Ref. 96 on the lattice constants $l(0)$ (see Sec. 2.2), we can find from Eq. (2.100) the value of $B_{ad}(0)$, and then, with the help of Eq. (2.101) and Table 2.39, we can find the adiabatic moduli of the hydrostatic compression and the adiabatic compressibility as a function of temperature for all isotopic compositions of lithium hydride. The results of these calculations are given in Table 2.45.

* The coefficients c_{ij} are calculated in the Appendix.

Table 2.45. Adiabatic moduli of hydrostatic compression B_{ad}, 10^{10} N/m², and adiabatic compressibility coefficients κ_{ad}, 10^{-11} m²/N, of the isotopic modifications of lithium hydride.

	^{7}LiH		^{7}LiD		^{7}LiT	
T,K	B_{ad}	κ_{ad}	B_{ad}	κ_{ad}	B_{ad}	κ_{ad}
0	3.31	3.02	3.40	2.94	3.45	2.90
100	3.30	3.03	3.39	2.95	3.44	2.91
200	3.28	3.05	3.37	2.97	3.40	2.94
298	3.24	3.09*	3.30	3.03	3.33	3.00
300	3.23	3.10	3.29	3.04	3.32	3.01
400	3.15	3.17	3.21	3.12	3.24	3.09
500	3.07	3.26	3.11	3.22	3.13	3.19
600	2.97	3.37	3.01	3.32	3.02	3.31
700	2.87	3.49	2.89	3.46	2.90	3.45
800	2.75	3.64	2.76	3.62	2.78	3.60
900	2.62	3.82	2.63	3.80	2.64	3.79
Melting point	2.53	3.96	2.53	3.96	2.54	3.94

*The value of κ_{ad} was taken from the experimental study of Gerlich and Smith.

2.4. The densities of lithium hydride, lithium deuteride, and lithium tritide in the liquid phase

In Ref. 250, Welch presents measured values of the density of molten lithium hydride obtained as part of a research program of General Electric's division of nuclear aircraft engines:

T,°C...	700	750	800	850	900	950
ρ,g/cm³...	0.550	0.546	0.540	0.532	0.518	0.492

The samples were 97.5% pure. In another study,[251] Welch presents strictly experimental values obtained for the same temperature interval:

T,°C...	710	730	830	945
ρ,g/cm³...	0.548	0.549	0.536	0.496

Unfortunately, Welch does not disclose the experimental procedure used in determining these values. Analysis of experimental data does, however, allow one to make certain assumptions. If, for example, Welch's density of lithium hydride varies by 0.73% over the temperature interval 700–750 °C, it will then vary by 5.28% in the temperature interval 900–950 °C. Such a temperature evolution of the density is typical of pycnometric measurements when the sample has a gas-filled cavity, which accounts for the generally lower measured values of the density.

At the Moscow Engineering–Physics Institute in 1959, Novikov, Akhmatova, and Solov'ev measured the density of liquid lithium hydride at temperatures between 680 and 1100 °C by the dilatometric method (see Ref. 70). These measurements by Novikov and co-workers, along with the measurements of other properties of lithium hydride—enthalpy, pressure of the dissociation products, thermal conductivity, viscosity—were the first measurements anywhere to be carried out for this substance at such high temperatures and were a fundamental contribution to the knowledge of the thermophysical properties of lithium hydride.

The closed volume of the dilatometer consisted of a working cylinder 30 mm in diameter and 100 mm high, which contained the molten sample, and a guiding tube 14 mm in diameter and 200 mm high, into which the contact needle was inserted. The apparatus itself was constructed from stainless steel.

The density ρ of the substance under study was calculated from the relation

$$\rho = M/V, \tag{2.102}$$

where M is the mass of the substance placed into the apparatus and V is its volume. The mass M of ~ 20 g was determined by weighing the working vessel before and after the insertion of the sample. At the temperature T, the volume occupied by the substance to be tested was determined from preliminary calibration of the volume of the working cylinder at room temperature (in a solution of sodium chloride in water) as a function of height. In the experiment, the level of liquid hydride was determined by making electrical contact between the needle and the surface of the liquid. The temperature was measured by a Chromel–Alumel thermocouple which was attached to the case that was welded to the bottom of the working cylinder. The contact needle was insulated from the housing by a Teflon bushing. The thermal expansion of the apparatus was taken into account in the calculation of the density. The substance to be tested was loaded into the apparatus in an argon atmosphere. The chemical composition of the lithium hydride under study was not specified.

The authors estimate the maximum relative experimental error to be 2.5%. Error components such as the meniscus effect and the adhesion of a droplet to the contact needle were ignored. Three sets of measurements were performed. The authors used an interpolation equation for ρ (g/cm^3) as a function of temperature T (°C) to describe the results of experimental measurements (Table 2.46):

$$\rho = 0.612[1 - 2.81 \times 10^{-4}(T - 680)]. \tag{2.103}$$

Yakimovich *et al.*[86,212] experimentally measured the density of liquid lithium hydride up to a temperature of 1300 K by the method of maximum pressure in a gas bubble with a single capillary. An important characteristic of this method at high temperatures is that it is virtually insensitive to the sample's mass loss due to evaporation during the experiment and to the thermal expansion of the structural material, since the linear dimensions in the standard working formula used for calculating the density have the power of 1. The experimental apparatus is shown schematically in Fig. 2.11. A high-pressure furnace (1) was incorporated into this apparatus. The working section in the form of a squirrel cage constructed from tungsten rods is located inside a single-phase heater. This section consists of a 65-mm-high crucible (2) with 8-mm wall thickness and 40-mm inside diameter. The crucible contains a thin-walled, 36-mm-i.d. beaker (3), which holds the melt under study and which stabilizes the composition of the lithium hydride. The gap between the external surface of the beaker and the internal surface of the crucible is filled

Table 2.46. Experimental data on the density of liquid lithium hydride obtained by Novikov, Akhmatova, and Solov'ev.

T,°C	ρ,g/cm³	T,°C	ρ,g/cm³	T,°C	ρ,g/cm³
Set I		874	0.578	708	0.606
714	0.612	884	0.577	733	0.601
716	0.611	896	0.576	783	0.594
763	0.593	909	0.575	818	0.588
804	0.583	926	0.573	854	0.584
849	0.583	959	0.566	888	0.578
867	0.580	972	0.563	896	0.575
903	0.572	975	0.559	911	0.571
927	0.567	990	0.558	925	0.567
950	0.563	995	0.556	967	0.564
964	0.559	1011	0.552	986	0.562
984	0.556	1023	0.551	997	0.557
989	0.555	1036	0.550	1017	0.555
Set II		1044	0.547	1039	0.552
704	0.610	1073	0.545	1049	0.549
723	0.606	1087	0.543	1079	0.545
761	0.601	Set III		1091	0.544
770	0.599	751	0.600	1103	0.543
817	0.586	748	0.600	1110	0.540
860	0.581	738	0.601		
865	0.580	730	0.602		

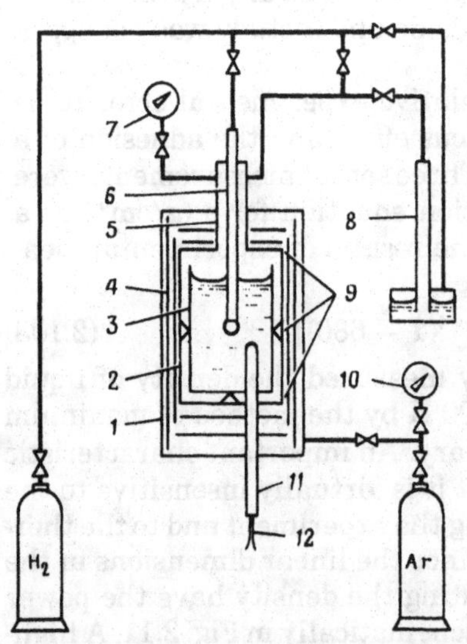

Figure 2.11. Schematic of the experimental apparatus for measuring the density and surface tension of lithium hydride by the method of maximum pressure in the gas bubble.

with hydrogen. In the absence of this auxiliary device, the upper layers of liquid lithium hydride would be enriched with lithium as a result of hydrogen yield from the sample through the crucible walls. To provide uniform heating along the height of the crucible, the crucible was surrounded by a system of radial and end-face screens (4) and was directly connected to the flange of the furnace through a connecting tube (5). A capillary (6), whose motion in the vertical direction was controlled, was placed axially inside this tube. The argon pressure in the furnace and the hydrogen pressure in the crucible were determined from the readings of spring-loaded pressure gauges (7) and (10). A cathetrometer with a resolution of 10^{-3} mm was used, with the help of a cup-shaped differential pressure gauge (8), to measure the pressure drop of hydrogen inside and outside the capillary. Dibutyl phthalate was the pressure fluid.

The temperature of the working section was monitored by three Chromel–Alumel thermocouples (9) which were placed along the height of the external surface of the crucible. An additional thermocouple (12) inside the tube (11), which was welded to the bottom of the crucible, was used in the set of experiments II–IV. The choice of the material for thermocouples was determined by the stability of the material's characteristics in the reducing (hydrogen) medium. Before the experiment, all thermocouples were tested against a standard type PP-1 (second-class) platinorhodium–platinum thermocouple. The difference in the readings of platinorhodium–platinum and Chromel–Alumel thermocouples was within the error limit of the latter, which was 2 K at a temperature of 1300 K, for example.

Lithium hydride was obtained directly in the crucible by filling it beforehand with pure lithium in a special device and then introducing hydrogen into it in the apparatus itself at a temperature of 970 K. Hydrogenation was appreciably exothermic, as indicated by the sharp increase in the thermocouple readings. A constant hydrogen pressure in the crucible indicated that equilibrium was attained and that hydrogenation was complete. The apparatus was then operated in the desired temperature regime, after which the maximum pressure in the gas bubble was measured at different immersion levels of the capillary, beginning with the melt surface. The time required for the bubble to form in the experiments was in the range of 1 to 6 min. There was no evidence of the effect of time on the measurement results.

Three different methods were used to produce gauge pressure in the capillary. The first method was based on the fact that the gas pressure in the crucible decreased at a rate that depended on the temperature because of the passage of hydrogen through the walls of the crucible. In this spontaneous process, the pressure drop in the capillary-crucible system reached levels on the order of a 80-mm water column in 8.5 min at a temperature of 1000 K and in 3 min at 1200 K. The second and third methods of producing gauge pressure in the capillary involved either introducing hydrogen into the capillary or removing it from the external cavity.

A special experiment was performed to determine the section in the furnace corresponding to the lowest temperature gradient along the height of the crucible, which in this case was ~ 1 deg/mm. In addition, the temperature gradient was determined directly in the lithium hydride melt using a special crucible with two bushings welded to the bottom. These bushings contained thermocouples spaced 12 mm apart vertically. In most of the measurements, the temperature drop varied between 2 and 10 K, depending on the measurement series. At two points, the temperature drop greatly exceeded this value (see Table 2.49).

The effect of possible dissolution of hydrogen in dibutyl phthalate (pressure fluid) on its density was studied. Hydrogen was introduced into a transparent cell filled with dibutyl phthalate, and the change in the height of the

Table 2.47. Impurity content (mass %) in the original lithium used in the preparation of the lithium hydride sample.

Element	Content	Element	Content
Na	0.13	Mn	0.001
K	0.005	Fe	0.003
Ca	0.019	Si	0.0065
Mg	0.02	N	0.051
Al	0.003		

dibutyl phthalate column was measured as a function of hydrogen pressure. As the pressure was changed from vacuum to 10 bar, the column height changed by 0.05%, which was of no concern for the basic experiment. Four sets of measurements of the density of lithium hydride were carried out using different capillaries. The inside diameters of the capillaries, determined with an optical comparator, were 1.004 mm in the first set of measurements, 1.402 mm in the second set, and 1.424 mm in the third and fourth sets of measurements. The crucible and the capillaries were constructed from 1X18H9T stainless steel.

After the experiment, the thermocouples were again checked against a standard type PP-1 (second-class) platinorhodium-platinum thermocouple. These readings showed an increase of 7 K at a temperature of 1200 K. A correction was made for this value after determining the temperature to which the experiments were scaled.

The chemical composition of the lithium used in the experiments is shown in Table 2.47. The total volume concentration of the impurities in hydrogen was $\sim 10^{-3}\%$, of which $10^{-7}\%$ corresponded to oxygen.

The original relation for calculating the density is the equation that follows from the analysis of the equilibrium at the pole of the gas bubble (see Refs. 72, 73, and 213);

$$P = (2\sigma / r_{\text{eff}}) - (\sigma_0 / R) + \rho g h. \tag{2.104}$$

Here P is the maximum pressure drop in the gas bubble at equilibrium; σ and σ_0 are the surface tensions in the bulk of the melt and at its surface, respectively; r_{eff} is the effective radius of the capillary; R is the major radius of curvature of the melt surface at the pole; ρ is the density of the melt under study; g is the acceleration due to gravity; and h is the depth to which the capillary is immersed, with allowance for the vertical dimension of the gas bubble.

The value of P can be determined from the drop in the level of the pressure-fluid column—dibutyl phthalate—in the differential pressure gauge (if the density of a similar gas column is disregarded):

$$P = \Delta H K_1 \rho_1 g, \tag{2.105}$$

where ΔH is the drop in the pressure-fluid level in the transparent elbow of the differential pressure gauge; K_1 is a constant of the differential pressure gauge ($K_1 = 1 + d_0^2 / D_0^2$, where d_0 and D_0 are the inside diameters of the transparent tube and of the extended cup, respectively; $K_1 = 1.0036$); and ρ_1 is the density of dibutyl phthalate ($\rho_1 = 1.0411\ \text{g/cm}^3$).

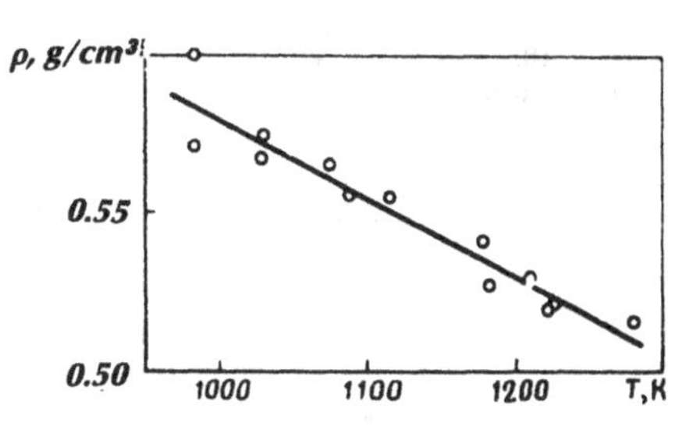

Figure 2.12. The density of liquid lithium hydride according to the experimental data of Yakimovich *et al.*[86]

Differentiating Eq. (2.104) with respect to the axial displacement of the capillary provides the working relation

$$\rho = \frac{\rho_1 K_1 K_2}{1+\alpha T}\frac{d\Delta H}{d\Delta x} - \Delta\rho = \rho' - \Delta\rho, \tag{2.106}$$

where K_2 is a coefficient which takes into account the change in the level of the melt in the crucible upon the introduction of a capillary into it, a is the coefficient of thermal expansion of the material from which the capillary and crucible are fabricated, T is the melt temperature, Δx is the displacement of the capillary reckoned from the end of the capillary that is outside the furnace, and $\Delta\rho$ is the correction for the nonuniform heating of the melt along the height. Several gas bubbles with corresponding values of the maximum pressure were detected at each vertical position of the capillary.

All initial data were approximated by a linear equation

$$\Delta H = a_0 + a_1 \Delta x, \tag{2.107}$$

where a_0 and a_1 are the parameters of the equation which are determined by the method of least squares. Table 2.48 gives the coefficients of the approximating equation (2.107). In all sets of measurements, the mean-square deviation of the experimental points from Eq. (2.107) is no greater than $\pm 0.1\%$. In the analysis of the results of the initial measurements, the derivative $d\Delta H/d\Delta x$ in the principal working equation [Eq. (2.106)] is the coefficient a_1 in Eq. (2.107), and the error of this derivative is the mean-square error of this coefficient, δa_1, which is $\sim 0.5\%$.

The experimental data on the density of liquid lithium hydride are given in Table 2.48 and in Fig. 2.12. Also given in Table 2.48 is the hydrogen pressure above the melt measured in the experiments; dT/dh is the temperature gradient along the height of the melt.

The experimental values of the density of liquid lithium hydride were approximated by the method of least squares using the linear equation

$$\rho = b_0 + b_1 T. \tag{2.108}$$

The results of the first set of measurements at temperatures of 1178.1 and 1280.4 K were not analyzed because of the large temperature gradients in these melts. Analysis showed that, within the scatter of the experimental data, the hydrogen pressure has no effect on the measured density.

The coefficients of the approximating equation [Eq. (2.108)] are $b_0 = 787$ kg/m^3 and $b_1 = -0.209$ kg/m^3 K. The mean-square deviation of the experimental values for Eq. (2.108) is 0.9% in the principal temperature interval. A

Table 2.48. Results of analysis of the experimental data on the density of liquid lithium hydride (Refs. 86 and 212).

T,K	a_0,mm	a_1	dT/dh, deg/cm	ρ',kg/m³	$\Delta\rho$,kg/m³	ρ,kg/m³
			Set I, $P_{H_2} = 10^6$ Pa			
981.6	—	0.6171	11	612.2	41.1	571.1
1178.1	72.679	0.6057	20	598.4	52.0	546.4
1280.4	70.000	0.5889	35	580.1	52.5	527.6
981.6	79.274	0.6529	13	647.7	48.6	599.1
			Set II, $P_{H_2} = 10^6$ Pa			
1226.3	—	0.5319	5.5	534.6	6.4	528.2
			Set III, $P_{H_2} = 10^6$ Pa			
1087.9	53.545	0.5760	10	581.2	25.3	555.9
1221.2	50.753	0.5427	10	546.2	12.6	533.5
1182.1	51.355	1	10	552.8	18.0	534.8
1115.9	53.154	0.5762	10	580.8	23.5	557.3
1073.8	54.246	0.5860	10	591.3	26.0	565.3
1026.7	56.025	0.5870	10	592.9	26.6	566.3
			Set IV, $P_{H_2} = 2\times10^5$ Pa			
1030.7	56.044	0.5740	2	579.7	5.0	574.7
1210.2	52.694	0.5318	2	535.0	0.1	534.9

Note: The dash in the table indicates that a_0 has not been determined experimentally.

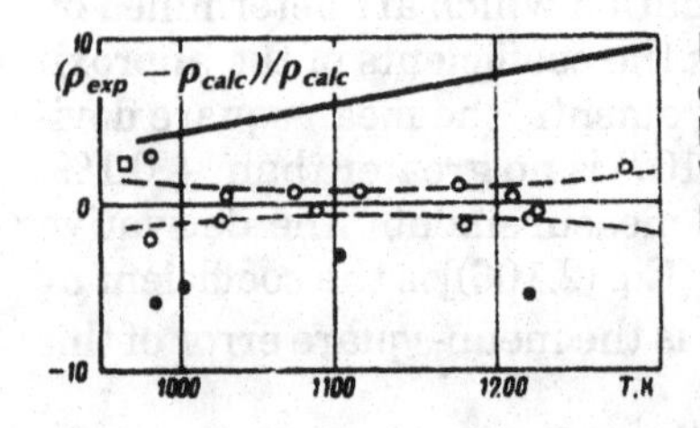

Figure 2.13. The discrepancy observed between the experimental data on the density, ρ_e, of liquid lithium hydride and ρ_c, calculated on the basis of Eq. (2.108). ●—Welch[250]; ○—Yakimovich *et al.*[86]; □—Skuratov[54]; — — Novikov *et al.* (see Ref. 70).

slightly larger deviation is observed near the melting point. The mean-square deviation increases to 1.4% if two experimental points at $T = 981.6$ K are taken into account.

It can be seen from Fig. 2.13 that there are definite discrepancies in the measurement results of the density of liquid lithium hydride obtained by different authors. Possible reasons for these discrepancies in the data of Novikov and co-authors were considered in a book by Shpil'rain and Yakimovich.[70] A subsequent analysis showed that the principal part of the error of these data, which is attributable to the meniscus of the melt in the crucible, is slightly smaller than that obtained by Shpil'rain and Yakimovich.[70] A more accurate estimate shows that this error source accounts for approximately 4% higher experimental values of the density of liquid lithium hydride. The explanation for the possible causes of error in Welch's[250,251] data was given in

Table 2.49. Comparison of the change in the molar volume, δv of different substances during melting caused by the isotopic substitution.

Compound	v_s,cm³/mole	v_l,cm³/mole	$\delta v = \frac{v_l - v_s}{v_s}$, %
H_2O	19.643	18.0207	− 8.3
D_2O	19.670	18.1235	− 7.9
nH_2	23.25	26.10	12.3
nD_2	20.48	23.14	13.0

the beginning of this section. As for the data of Skuratov *et al.*,[54] we present here only the experimental value of the density of liquid lithium hydride at the melting point, without any additional description.

Equation (2.108) can thus be used to calculate the density of liquid lithium hydride up to temperatures of about 1300 K. The calculation results are given in Table 5.4.

The densities of liquid lithium deuteride and liquid lithium tritide, which have not been studied experimentally, can be determined from the following considerations. In general, experiments have shown that isotopic substitutions in a compound result in only slight differences in the relative change in the molar volume during melting. The data of Ref. 52, given in Table 2.49, support this conclusion. Accordingly, relative changes in the molar volume during melting of lithium deuteride and lithium tritide can be assumed to be the same as that for lithium hydride. Additionally, according to the data of Ref. 52, the thermal expansion of liquids is essentially unaffected by isotopic substitution in the case of many hydrocarbons and water. Using the results for the densities and molar volumes of solid lithium hydride, solid lithium deuteride, and solid lithium tritide determined in Sec. 2.2 and assuming that the thermal expansion of liquid LiD and liquid LiT is the same as that of liquid LiH, we find the following working relations for the temperature dependence of the densities of liquid lithium deuteride, $\rho_{\mathrm{LiD},l}$, and liquid lithium tritide, $\rho_{\mathrm{LiT},l}$:

$$\rho_{\mathrm{LiD},l} = \rho^0_{\mathrm{LiD},s}\,\frac{1-\beta(T-T_m)}{1+\delta v}; \tag{2.109}$$

$$\rho_{\mathrm{LiT},l} = \rho^0_{\mathrm{LiT},s}\,\frac{1-\beta(T-T_m)}{1+\delta v}, \tag{2.110}$$

where $\delta v = (v^0_{\mathrm{LiH},l} - v^0_{\mathrm{LiH},s})/v^0_{\mathrm{LiH},s}$ is the change in the molar volume of lithium hydride at the melting point; $\rho^0_{\mathrm{LiD},s}$ and $\rho^0_{\mathrm{LiT},s}$ are the densities of solid lithium deuteride and solid lithium tritide at the melting point; β is the volume thermal expansion coefficient, which is calculated on the basis of an equation derived from Eq. (2.108) for the density of liquid lithium hydride (kg/m³)

$$\rho_{\mathrm{LiH},l} = 585[1-\beta(T-T_m)] \tag{2.111}$$

and which is 35.8×10^{-5} K^{-1}; and T_m is the melting point of the hydride under consideration. The melting point of each compound was selected from the recommendations in Sec. 1.4. In accordance with our data, the value of δv is 18.4%.

As a result, we obtain equations for the densities ρ (kg/m^3) of liquid lithium deuteride and liquid lithium tritide

$$\rho_{\mathrm{LiD},l} = 659[1 - \beta(T - T_m)] = 866 - 0.236\,T; \qquad (2.112)$$

$$\rho_{\mathrm{LiT},l} = 733[1 - \beta(T - T_m)] = 986 - 0.262\,T. \qquad (2.113)$$

The calculated densities of liquid lithium hydride, liquid lithium deuteride, and liquid lithium tritide are given in Table 5.4 and in Fig. 2.8. As in the case of the density in the solid state, we see a considerable isotopic effect. The density of lithium tritide differs from the density of lithium hydride by as much as 25%. This difference stems principally from the difference in the molecular masses, since the molar volumes of these compounds in the liquid state are essentially the same.

The error in the data on the density of liquid lithium hydride, which were obtained on the basis of experimental studies and which are given in Table 5.4, was estimated at the 0.95 confidence level. This error turned out to be 2% over the entire temperature range of the experiment. Analysis of its structure showed that the systematic and random components are 1.5% and 1.4%, respectively. The main part of the systematic error is the error in determining the slope angle of the straight line $\Delta H = f(\Delta x)$ during the motion of the capillary [see Eq. (2.107)].

The error in the calculated values of the density of liquid deuteride and tritide is difficult to determine directly, principally because of the assumptions in this calculation that the relative change in the volume due to melting and the thermal expansion in the liquid phase are equal for LiH, LiD, and LiT, although these assumptions are reasonable for the isotopic modifications and have been confirmed on the basis of other substances.

We wish to emphasize again the circumstance mentioned above that, as the melting point is approached, the lattice constants of LiH, LiD, and LiT become approximately equal to each other (see Table 2.39). This means that the molar volumes of these compounds in the solid state [see Eq. (2.57)]

$$v_s = N_A l^3/4 = (M + m)/\rho \qquad (2.114)$$

are essentially equal at the melting point:

Hydride...	^{n}LiH	^{n}LiD	^{n}LiT
v_s^0, cm^3/mole...	11.49	11.48	11.48

In view of this circumstance, the assumptions made above support the argument that the molar volumes of the isotopic modifications of lithium hydride in the liquid phase are virtually the same and remain so upon a change in temperature. This behavior, not obvious at first glance, has been confirmed experimentally for other substances. Rabinovich[52] presents data on the molar volumes of H_2O, D_2O, and T_2O and on the ordinary and deuterium substituted benzenes, alcohols, and other substances which show that the molar volumes of the isotopic modifications of liquids differ very little (by only 0.2%–0.5%) over a broad temperature interval.

In sum, this analysis confirms that the methods developed above for cal-

Table 2.50. Results of experimental study of the surface tension of lithium hydride (Ref. 82).

T,K	ΔH_0,mm	σ,N/m	T,K	ΔH_0,mm	σ,N/m
	Set I, $P_{H_2} = 10^6$ Pa			Set III, $P_{H_2} = 10^6$ Pa	
1196.2	72.6789	0.1917	1098.9	53.5451	0.1997
1300.5	70.0003	0.2081	1232.3	50.7531	0.1899
995.6	79.2741	0.2081	1192.2	51.3553	0.1920
963.5	82.1660	0.2137	1127.0	53.1539	0.1984
			1084.9	54.2461	0.2023
	Set II, $P_{H_2} = 10^6$ Pa		1037.7	56.0249	0.2087
			998.6	56.3832	0.2099
				Set IV, $P_{H_2} = 2\times10^5$ Pa	
1004.6	59.5687	0.2118	1032.7	56.0438	0.2088
1052.8	58.3517	0.2077	1212.2	52.6944	0.1971
			1124.0	53.9200	0.2013

culating the density of the isotopic modifications of lithium hydride both in the solid phase and the liquid phase are correct. These methods collectively lead to an agreement in the molar volumes of ^{n}LiH, ^{n}LiD, and ^{n}LiT in the liquid phase. Further, this analysis shows that the error of the data on the densities of liquid lithium deuteride and liquid lithium tritide in Table 5.4 is nearly equal to the error for liquid lithium hydride determined above and is no more than 2.5%–3.0%.

2.5. The surface tension of lithium hydride

The surface tension of liquid lithium hydride was measured only by Shpil'rain *et al.*[82] The measurements were carried out by the method of maximum pressure in a gas bubble with use of the same experimental apparatus that was developed for measuring the density of liquid lithium hydride (see Sec. 2.4). In connection with the study of surface tension, we wish to emphasize that the method of maximum pressure in a gas bubble has fewer flaws and is a simpler and more advanced analytical method than other widely used methods such as the use of a hanging or lying droplet. In particular, it should be noted that a surface regeneration which occurs each time a new bubble forms reduces the uncertainty due to the adsorption of surface-active substances, which is perhaps one of the more serious drawbacks of other methods of determining surface tension.

The drop in the level, ΔH, of a differential pressure gauge was measured experimentally at a given temperature and constant hydrogen pressure above the melt as a function of the immersion depth Δx of the capillary. In Sec. 2.4 of this chapter it was shown that this functional dependence is linear [see Eq. (2.107)].

The equation derived from Eqs. (2.104) and (2.105), which were used for the case in which the capillary comes in contact with the surface of the melt, i.e., when $\Delta x = 0$, was used as the working relation for measuring the surface tension:

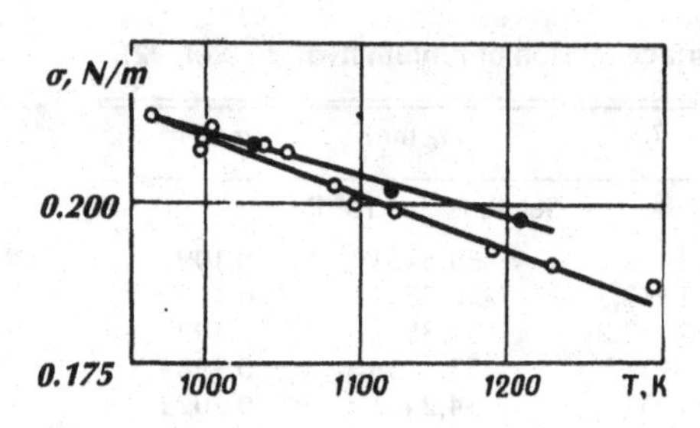

Figure 2.14. Temperature dependence of the surface tension, σ, of lithium hydride. ●— $P_{H_2} = 2x10^5$ Pa; ○—$P_{H_2} = 10^6$ Pa.

$$\sigma = \frac{K_1 g \rho_1 \Delta H_0}{(2/r_{eff}) - (1/R)}, \tag{2.115}$$

where $\Delta H_0 = \Delta H_{\Delta x = 0} = a_0$ [see Eq. (2.107)].

The values of r_{eff} and R were determined by a method used in Refs. 72 and 73. The surface tension necessary for calculating the capillary constant of lithium hydride $a = (2\sigma/\rho g)^{1/2}$ in first approximation was determined from Eq. (2.115) after replacing in it r_{eff} and R, respectively, by the inside radius r_0 of the capillary and the difference between the inside radius R_0 of the crucible and the outside radius r of the capillary, which was inserted into it (i.e., $R_0 - r$), with allowance for thermal expansion.

The experimental data on ΔH_0 and surface tension of liquid lithium are presented in Table 2.50 and in Fig. 2.14. Table 2.50 also gives the hydrogen pressure above the melt. The experimental error at the 0.95 confidence level is 1% over the entire experimental temperature interval. The systematic and random components of this error are each equal to 0.7%.

The experimental data from each set of measurements were analyzed in accordance with the linear law by the method of least squares:

$$\sigma = b_0 + b_1 T. \tag{2.116}$$

A statistical analysis of the measurement results based on the procedure used in Ref. 2 showed that there is a systematic discrepancy between the data of measurement sets I–III at a hydrogen pressure of 10 atm and the data of measurement set IV at a hydrogen pressure of 2 atm. The experimental data obtained at a hydrogen pressure of 2 atm above the melt were therefore analyzed separately from those obtained at a hydrogen pressure of 10 atm. The surface tension of liquid lithium hydride, in contrast with its density, depends on the hydrogen pressure above the melt and hence on the composition of the liquid phase. A decrease in the hydrogen pressure, i.e., an increase in the concentration of lithium in the melt, increases the surface tension. This effect diminishes as the temperature is lowered.

This behavior of σ is explained by taking into account (see Tables 5.7–5.9) that the difference in lithium concentrations on the isotherm decreases (in the region where the mole fraction of lithium hydride is close to unity) with decreasing temperature if the pressure changes uniformly. Consequently, with decreasing temperature of liquid lithium hydride, the composition of the liquid phase becomes less dependent on the pressure, and hence the difference in the surface tension of the melt is reduced.

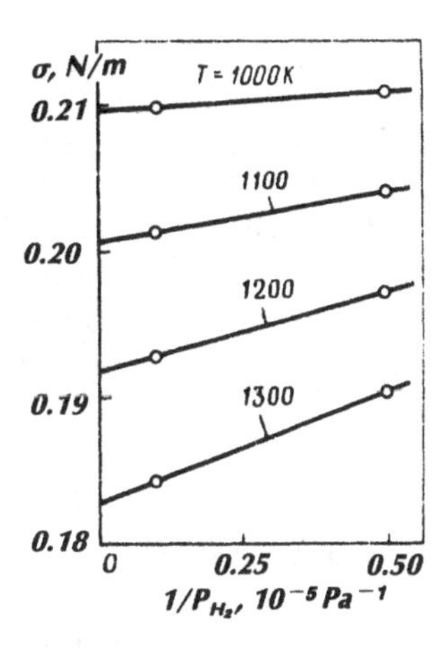

Figure 2.15. Determination of the surface tension of stoichiometrically pure lithium hydride.

The temperature dependence of the surface tension of liquid lithium hydride $\sigma = f(T)$ at constant hydrogen pressure above the melt is such (see Fig. 2.14) that the isobars nearly converge at the melting point. For this reason, the point at a temperature of 963.5 K was included in the analysis of the experimental results both for the isobar 2×10^5 Pa and for the isobar at 10^6 Pa. The coefficients of the approximating equation [Eq. (2.116)] at hydrogen pressures above the melt of 2×10^5 Pa and 10^6 Pa are

	2×10^5 Pa	10^6 Pa
b_0, 10^{-3} N/m...	278.9	296.0
b_1, N/m K)...	-68.1×10^{-6}	-85.9×10^{-6}

The mean-square deviation of the experimental points relative to Eq. (2.116) is 0.5% for the isobar 2×10^5 Pa and 0.8% for the isobar at 10^6 Pa.

After the completion of the experimental measurements of the surface tension, Shpil'rain *et al.*[82] estimated the contact angle of wetting of the inside wall of the crucible by lithium hydride. Upon opening the cooled crucible containing lithium hydride, its inside wall was found to be nearly totally wetted. This observation was confirmed by the results of a preliminary γ-ray bombardment of the cooled cell containing the solidified LiH sample after the completion of the experiment. Such an analysis is, of course, purely approximate, because the sample is in the solid rather than the liquid state. It can be used, however, to estimate that the contact angle for lithium-hydride wetting of the 1X18H9T stainless-steel wall of the cell is nearly zero. Accordingly, on the basis of this experiment the following two equations were obtained for calculating the surface tension σ (10^{-3} N/m) of lithium hydride:

$$\sigma_{P_{H_2}=10^6\,\mathrm{Pa}} = 296.0 - 85.9\times10^{-3}T \tag{2.117}$$

for a hydrogen pressure of 10^6 Pa and

$$\sigma_{P_{H_2}=2\times10^5\,\mathrm{Pa}} = 278.9 - 68.1\times10^{-3}T \tag{2.118}$$

for a hydrogen pressure of 2×10^5 Pa.

Different hydrogen pressures in the vapor phase lead to different equilibrium compositions of the liquid phase if the temperature remains constant. With increasing pressure, the lithium metal impurity decreases in the liquid hydride. The surface tension of lithium hydride, calculated from Eqs. (2.117) and (2.118), is shown in Fig. 2.15 as a linear function of $1/P_{H_2}$ at various temperatures. The surface tension of a hypothetically pure lithium hydride can be determined by extrapolating the isotherm to $1/P_{H_2} = 0$. The working equation found for this case is (σ, 10^{-3} N/m)

$$\sigma = 299.0 - 89.5 \times 10^{-3} T. \tag{2.119}$$

The linear dependence $\sigma = f(1/P_{H_2})$ used in the plots of Fig. 2.15 contributes an uncertainty in the extrapolation to the abscissa origin, because only two pressures have been obtained experimentally. A relatively small extrapolation region justifies the assumption, however, that the error arising from the extrapolation is not large. The extrapolation error decreases with decreasing isotherm slope (with decreasing temperature). The table for the surface tension of lithium hydride (Table 5.6) was compiled on the basis of Eqs. (2.117)–(2.119).

Estimates of the surface tension of lithium deuteride and lithium tritide with the help of a well-known expression for the parachor

$$P = \sigma^{1/4} \frac{M}{\rho' - \rho''} \approx \sigma^{1/4} \frac{M}{\rho'} = \sigma^{1/4} v \tag{2.120}$$

(where M is the molecular mass; ρ' and ρ'' are the densities of the liquid and vapor, respectively; and v is the molar volume) show that the surface tension of these compounds is the same as that of lithium hydride, since their molar volumes in the liquid state are, as was noted in Sec. 2.4 of this chapter, virtually the same as that of lithium hydride.

Chapter 3
Thermodynamic properties of Li-LiH (LiD, LiT) systems

1. The phase diagram of the Li-LiH system: Eutectic and monotectic properties

The phase diagram of the Li-LiH system is complex. Many investigators have studied its individual parts by different methods such as plotting thermograms, analysis of the electrical resistance as a function of temperature and composition of the solution, and measurement of the partial pressure of hydrogen or of its isotopes when the condensed phase is in equilibrium with the vapor. These studies have identified several systematic features in the phase diagram of the Li-LiH system.

The temperature-composition diagram for the systems under consideration is shown schematically in Fig. 3.1 (x is the mole fraction of lithium hydride in the condensed phase). The α phase (on the left-hand side), which is enriched with lithium, is arbitrarily distinguished from the β phase (on the right-hand side), which is enriched with hydride. The subscript l or s denotes the liquid or the solid phase, respectively. This system has eutectic properties at point c as well as monotectic properties (the line dfg).

Above the monotectic temperature, the liquid phase has a broad immiscibility zone. Experimental data show that the boundaries of the immiscibility zone (the curve dkf) are essentially the same for the Li-LiH and Li-LiD systems. At the same time, the experiment shows that the position of point c with respect to temperature and concentration is slightly different in these systems.

The characteristic points in the diagram for a Li-LiH system are given in Table 3.1 (the concentrations at points f and g are rough estimates).

Figure 3.2 shows the behavior of the isotherms in the pressure-composition diagram of the Li-LiH system above the temperature of the monotectic. The concentration interval inside the immiscibility zone (the dashed curve), in which the vapor pressure remains constant, is called a "two-dimensional" region, and the vapor pressure in this region is called a two-dimensional pressure.

It is interesting to note the behavior of the isotherms as the condensed-phase composition approaches the stoichiometric composition of lithium hydride. The fact that P goes to infinity as $x \rightarrow 1$ means that the stoichiometric composition of lithium hydride is actually an unattainable hypothetical

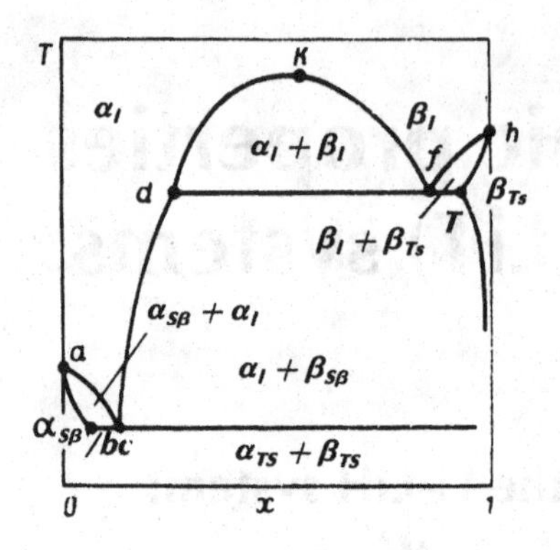

Figure 3.1. The temperature-composition phase diagram of the Li-LiH system (schematic); *k* is the critical solution point.

Table 3.1. Characteristic points of the temperature-composition diagram for the Li-LiH system.

Coordinates	*a*	*b*	*c*	*d*	*k*	*f*	*g*	*h*
T,K	453.64	453.56	453.56	961	1217	961	961	965
x	0	0.2×10^{-4}	1.6×10^{-4}	0.32	0.57	0.99	0.995	

state. This isotherm behavior applies not only to the Li-LiH system but also to many other systems containing metal hydrides.

The curve *acdfh* (see Fig. 3.1) is the solidification line of the Li-LiH system (the liquidus line). This curve was studied in detail for the first time by Messer *et al.*[40] In analyzing the behavior of the Li-LiH system when the condensed phase is in equilibrium with the vapor, they found that upon the removal of hydrogen from the vapor phase, the solidification temperature of the melt falls to the monotectic temperature of 685 ± 1 °C, remains constant at the molar concentrations of LiH between 98% and 26%, and then falls to 624 °C at 13% LiH. This part of the curve is described by Messer *et al.* in terms of the functional dependence

$$\log[x/(1+x)] = -(3381/T) + 2.835, \tag{3.1}$$

where x is the mole fraction of LiH in the solution. The heat of solution of solid lithium hydride in liquid lithium was found to be 64.72 kJ/mole on the basis of this equation.

The first data on the liquidus line in Li-LiH and Li-LiD systems below 624 °C were published by a group of British scientists headed by Adams and Hubberstey.[90,91,147–149,207*] The solubility was determined from the change in the electrical resistance during the dissolution of H_2 or D_2 in lithium and from the precipitation at the time of saturation.

*An analysis of these studies and also those of Ref. 240 on the determination of the liquidus line in the temperature interval between the eutectic and monotectic temperatures and a consistent analysis of the results of these studies were carried out by G. S. Aslanyan.

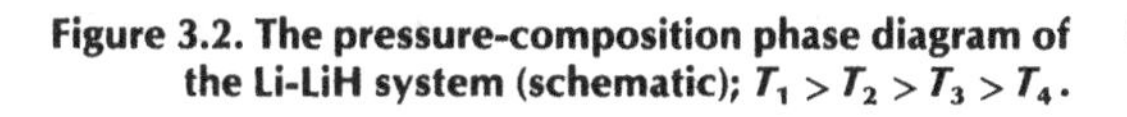

Figure 3.2. The pressure-composition phase diagram of the Li-LiH system (schematic); $T_1 > T_2 > T_3 > T_4$.

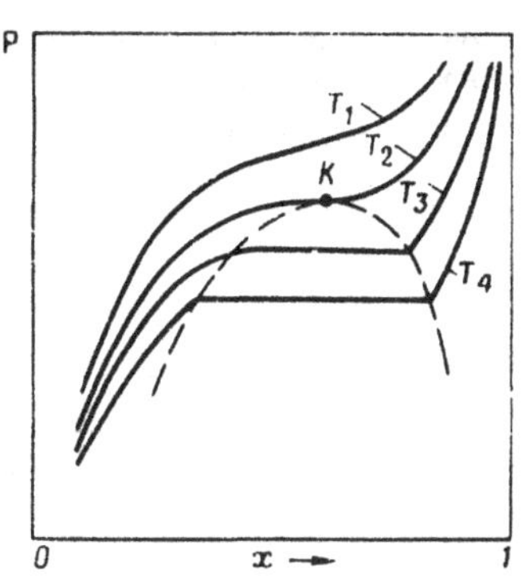

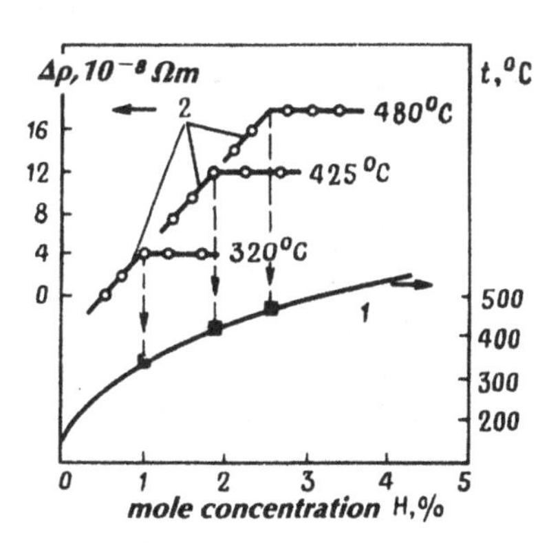

Figure 3.3. The curve for the solubility (curve 1) of the Li-LiH system plotted according to the data on the change in the resistivity of the solution (curve 2).

This method is based on a strong increase in the resistance with increasing concentration of H_2 (D_2) in the solution and on the appearance of a discontinuity in the $\bar{\rho} - x$ diagram ($\bar{\rho}$ is the resistivity, and x is the composition of the solution) at the saturation point of the solution (the point the hydride precipitates out).

The typical isotherms in Fig. 3.3 show that the resistivity of the solution increases linearly with increasing hydrogen concentration up to the saturation line and then remains nearly constant. The change in resistivity is clearly defined, which accounts for the high sensitivity in determining the solubility. The experimental data on the solubilities of LiH and LiD in lithium obtained in the studies under consideration are presented in Table 3.2. These results were approximated by the authors of the cited studies in the form of the equations

$$\log x_{\mathrm{H}} = 3.523 - \frac{2308}{T} \, (523\ \mathrm{K} < T < 775\ \mathrm{K}); \tag{3.2}$$

$$\log x_{\mathrm{D}} = 4.321 - \frac{2873}{T} \, (523\ \mathrm{K} < T < 724\ \mathrm{K}), \tag{3.3}$$

Table 3.2. Experimental data on the solubilities of hydrogen and deuterium in lithium (x is the atomic concentration of H or D, %).

T,°C	x_H	T,°C	x_D
221	0.037	551	5.680
227	0.063	276	0.122
257	0.140	305	0.229
275	0.212	328	0.330
296	0.301	340	0.527
326	0.462	350	0.493
344	0.596	365	0.691
361	0.753	375	0.773
376	0.892	389	0.990
395	1.148	397	0.990
397	1.195	402	1.095
411	1.420	417	1.393
441	1.930	451	2.350
499	3.480	456	2.631
525	4.540	500	3.900

where x is the atomic concentration of H or D in the solution (%).

These equations yield the values 44.2 and 54.8 kJ/mole for the partial molar enthalpies and 29.2 and 44.2 J/(mole deg) for the partial molar entropies of LiH and LiD, respectively. A comparison of the hydrogen and deuterium solubilities led the authors to conclude that at low temperatures, deuterium is less soluble than hydrogen, but at $T = 435$ °C and $x = 1.83\%$, the situation changes.

The solubility of lithium deuteride in liquid lithium between the eutectic and monotectic temperatures was also determined experimentally by Veleckis *et al.*[240] The experimental method used by them consisted of three stages: (1) preparation of the equilibrium solution of Li-LiD, (2) removal of the filtered solution into metallic tubes, and (3) analysis of the samples of the solution to determine whether they contain deuterium.

The deuterium content of the sample was determined by thermal-decomposition. Each sealed tube containing the sample was inserted into a container made from silica and then heated to 840 °C for 10 days in order to completely remove the dissolved deuterium. The evolved gas was collected into a container of specified volume, and the deuterium content in it was determined by mass spectrometry. The data obtained by this method are given in Table 3.3.

Analysis of the available experimental data on the solubility of hydrogen and deuterium in the temperature interval between the eutectic and monotectic temperatures, respectively, in Li-LiH and Li-LiD systems, and the additional evaluation of these data have shown that the experimental data are best described by an equation of the type

$$\ln x = A + B[(1/T) - (1/T_M)] + C\ln(T/T_M), \tag{3.4}$$

where x is the mole fraction of LiH or LiD in a solution with lithium and T_M is the temperature of the monotectic.

Table 3.3. Solubility of lithium deuteride in the Li-LiD system from the data of Ref. 240 (x is the molar concentration of LiD, %).

T,°C	x	T,°C	x
198.9	0.0514	351.0	0.633
221.1	0.0768	375.1	0.866
246.3	0.129	397.2	1.14
271.5	0.181	397.2	1.17
303.6	0.322	451.4	2.08
323.3	0.427	498.0	3.32

Table 3.4. Results of analysis of the experimental data on the solubility of hydrogen and deuterium in the Li-LiH and Li-LiD systems. Analysis of the data was based on Eq. (3.4).

System	A	B,K^{-1}	C	T_M,K	σ,%
Li-LiH	− 1.517	− 3998	3.004	961	1.9
Li-LiD	− 1.517	− 1197	6.83	963	1.9

The coefficients in Eq. (3.4) and the mean-square deviations of the experimental data, σ, are summarized in Table 3.4.

A decrease in the solidification temperature of lithium due to the dissolution of hydrogen and deuterium in it (curve ac in Fig. 3.1) was studied in Refs. 92, 148, and 149. The measurements were carried out by the thermal-analysis method (thermography).

Small portions (1 cm^3 under normal conditions) of soluble gas were added to pure lithium (30 g of 99.98% pure lithium). The solution was cooled slowly at the rate of 10 deg/h, which made it possible to reliably determine the solidification temperature. Continuous mixing of the solution kept it homogeneous. The melting point of pure lithium was assumed to be 180.49 ± 0.02 °C. With increasing atomic concentration of hydrogen or deuterium, the depression of the solidification temperature, ΔT, increased to the maximum value of 0.082 °C at 0.016% H and to 0.075 °C at 0.013% D, which corresponded to the eutectic points of the Li-LiH and Li-LiD systems (Fig. 3.4).

The solidus lines (curve ab in Fig. 3.1) were described under the assumption that the solid and liquid phases in Ref. 92 are ideal phases:

$$x_l - x_c = \frac{(T_m - T)\Delta H_m}{T_m TR}. \tag{3.5}$$

Here x_l and x_c, are the atomic concentrations of hydrogen or deuterium on the liquidus and solidus curves; T_m and ΔH are, respectively, the temperature and the heat of fusion of pure lithium; and R is the universal gas constant. The concentration at point b (see Fig. 3.1) for the Li-H system ($x_b = 0.002\%$) and for the Li-D system ($x_b = 0.001\%$) was determined from Eq. (3.5).

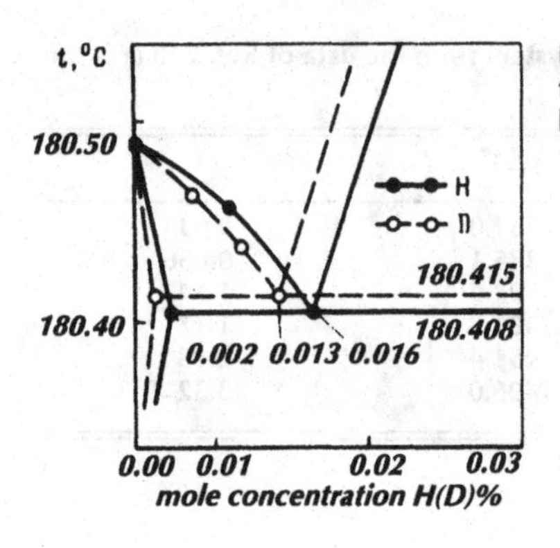

Figure 3.4. The decrease in the solidification temperature of lithium as a result of dissolution of hydrogen and deuterium.

The temperature of the monotectic T_M of the Li-LiH system was determined for the first time by Messer *et al.*[40] by the thermographic method. The value $T_M = 685 \pm 1$ °C obtained by them, however, turned out to be too low, as was indicated in the subsequent experiments, e.g., by Smith and Webb.[218] Analysis of the behavior of the pressure isotherm in the Li-LiH system, when the condensed phase is in equilibrium with the vapor, as a function of the composition of the condensed phase, shows that in the temperature region of the monotectic the isotherm plotted as $P = f(x)$ is horizontal (see Fig. 3.2) in the region df (see Fig. 3.1). The temperature of the monotectic can therefore be determined from the point of intersection of the functions $P = f(x)$ measured above and below the temperature of the monotectic.

At the temperature of the monotectic, there are four phases in a thermodynamic equilibrium: the vapor phase, a liquid phase with the composition corresponding to point d (see Fig. 3.1), a liquid phase with the composition at point f, and the solid phase with composition at point g.

In the temperature region slightly above T_M, the vapor phase and the two liquid phases are in equilibrium, whereas in the temperature region slightly below T_M, the vapor phase, the liquid phases and the solid phase are in equilibrium. The heat and entropy of dissociation in the reaction

$$\mathrm{LiH_{cond}} \rightleftarrows \mathrm{Li_{cond}} + \tfrac{1}{2}\,\mathrm{H_{2,gas}}, \tag{3.6}$$

which determines the pressure of the vapor phase, at $T > T_M$ and $T < T_M$ will therefore differ from the heat and entropy of fusion of lithium hydride. This accounts for the discontinuity in the function $P = f(T)$ at the monotectic temperature.

A consistent analysis* of Smith and Webb's[218] data for the Li-LiH (LiD) systems near temperatures of ± 100 K relative to T_M in the form of the relation

$$\ln P = A + B/T \tag{3.7}$$

*This analysis was performed by G. S. Aslanyan in collaboration with A. N. Drozdov.

yields (the coefficients A and B are assumed to be constant because of the narrow temperature interval) the following values: $T_M = 961 \pm 2$ K for the Li-LiH system and $T_M = 963 \pm 2$ K for the Li-LiD system.

For the Li-LiT system, the analysis, which took into account the experimental data of Refs. 144 and 236 (these studies are analyzed in detail in Sec. 2.1) for the entire temperature interval studied (because of their paucity, all data were taken into account), yielded the value $T_M = 964 \pm 3$ K.

Smith and Webb[218] and Veleckis[236] also determined the temperatures of the monotectic in Li-LiH (LiD, LiT) systems by analyzing their data on the vapor pressure in the two-dimensional region over a broad temperature range. However, the values obtained by them (the data of Ref. 218 are enclosed in parentheses), $T_M = 694$ °C (699.6 °C) for Li-LiH, $T_M = 690$ °C (687.7 °C) for Li-LiD, and $T_M = 688$ °C for Li-LiT, do not seem to be legitimate, principally because of their isotopic sequence.

The data on the melting point of the isotopic compounds found in the literature show that the heavier isotopic modifications have a higher melting point. Here are several melting points at 10^5 Pa, taken from the data presented in Refs. 52 and 63:

Substance	H_2	D_2	CH_4	CD_2	H_2O	D_2O
T, K	13.95	18.65	90.64	98.78	273.15	276.95

Although at the monotectic point the system is, strictly speaking, a solution rather than a pure substance, the isotopic dependence of the melting point (of the monotectic) must obviously reflect this trend.

The use by the authors of Refs. 218 and 236 of a linear approximation $\ln P = f(1/T)$ [see Eq. (3.7)] over a broad temperature range clearly is a possible source of error in determining the monotectic temperature. The temperature dependence of the heat and entropy of dissociation of LiH (LiD, LiT) in this case could affect the calculation results.

2. Thermodynamic properties of Li-LiH (LiD, LiT) systems when the liquid phase is in equilibrium with the vapor

2.1. A review of the data in the literature on the pressures of the dissociation products when the Li-LiH (LiD, LiT) systems are in phase equilibrium

In Li-LiH systems, equilibrium between the condensed phase and the vapor is characterized by the presence of the heterogeneous dissociation reaction (3.6) and by several vapor-phase chemical reactions which produce in the vapor a mixture of the components LiH, Li, H_2, Li_2, Li_2H, etc. The lithium vapor which is produced during the dissociation condenses, becoming mixed with the existing condensed phase. The study of the pressure and composition of the vapor of such a system as functions of the temperature and composition of the liquid solution is therefore a rather complex problem.

Analysis of the phase diagram of the Li-LiH system in Sec. 3.1 showed that in a certain temperature region there is an immiscibility zone, in which the vapor pressure and, in particular, the partial hydrogen pressure remain constant as the concentration of the components in the condensed phase changes. At the same time, the hydrogen pressure increases sharply beyond the immiscibility zone as the concentration of the components in the condensed phase approaches the stoichiometric composition of LiH.

The following conventional line of reasoning can be used to explain this phenomenon. Let us assume that the closed system in the liquid phase originally contained a pure lithium hydride of stoichiometric composition which evaporates in the first stage of the process. If there were no chemical reaction in the vapor phase, the amount of evaporated lithium hydride would be such that the vapor pressure would correspond to the saturation vapor pressure of LiH at the temperature T. In the vapor phase, however, there is a dissociation reaction

$$LiH \rightleftarrows Li + \tfrac{1}{2} H_2. \tag{3.8}$$

As a result, the number of LiH molecules in the vapor decreases, causing the partial pressure of the LiH vapor to drop (if it is assumed that the degree to which the vapor deviates from the ideal vapor before the dissociation is the same as the degree to which it deviates after the dissociation). At the same time, the partial pressure of one of the products of the dissociation reaction—atomic lithium—is higher than the saturation pressure of lithium at the temperature T under consideration (if all the produced lithium remains in the vapor state). Since we are dealing with a closed system, the excess lithium must condense. If there were no mixing of lithium with condensed lithium hydride, a new portion of lithium hydride would have to be converted to the vapor phase in order to restore the initial vapor pressure of LiH, which would be contingent in this case solely upon the temperature being kept constant. The evaporation and decomposition would evidently continue until all condensed lithium hydride would be converted to vapor and only pure lithium would remain in the condensed phase.

This situation, in fact, does not exist, which suggests that condensed lithium hydride in a pure form cannot coexist in equilibrium with the vapor. Coexistence can occur only if a certain amount of lithium dissolves in the lithium hydride. The partial vapor pressure of LiH drops to a value at which the condensed Li-LiH phase is in equilibrium with the vapor of a corresponding composition.

Since a solution of lithium in lithium hydride forms in the liquid phase, the partial pressure P_1 of lithium is proportional to the concentration of lithium in the liquid phase if the liquid is in a phase equilibrium with the vapor:

$$P_1 = P_1^0 \gamma_1 (1 - x), \tag{3.9}$$

where P_1^0 is the saturation pressure of pure lithium, γ_1 is the activity coefficient of lithium in a binary Li-LiH solution, and x is the mole fraction of lithium hydride in the solution.

Table 3.5. Principal studies in which the vapor pressure was measured at phase equilibrium in the Li-LiH (LiD, LiT) systems.

System	Temperature, K	Limiting pressure	References
Li-LiH	953	27 mmHg	Guntz (Refs. 134, 135)
Li-LiH	780– 930	8 mmHg	Hurd and Moore (Ref. 150)
Li-LiH	973–1077	1 atm	Heuman and Salmon (Ref. 144)
Li-LiD			
Li-LiT			
Li-LiH	973	28 mmHg	Messer, Damon, *et al.* (Ref. 40)
Li-LiH	973–1373	28.5 atm	Novikov, Solov'ev, *et al.* (see Ref. 70)
Li-LiH	888–1061	175 mmHg	Welch (Refs. 250, 251)
Li-LiH	873–1178	1 atm	Blander, Maroni, Veleckis, *et al.* (Refs. 102, 103, 181–183, 233–239, 241, and 242)
Li-LiD			
Li-LiT			
Li-LiH	973–1273	1 atm	Smith *et al.* (Refs. 215–217, 219)
Li-LiD			
Li-LiH	733–1043	15×10^3 Pa	Smith and Webb (Ref. 218)
Li-LiD			
Li-LiD	773–1073	100 mmHg	McCracken and Goodal (Refs. 128, 185)
Li-LiH	623–1013	50 mmHg	Katsuta *et al.* (Ref. 168)
Li-LiH	700– 800	137 Pa	Ihle and Wu (Refs. 153–155, 258)
Li-LiD	773– 973	0.6×10^{-3}	
Li-LiH	890–1650	20 atm	Shpil'rain, Yakimovich, *et al.* (Refs. 76–78)

The equilibrium constant of dissociation reaction (3.8),

$$K_p = P_1P_3^{1/2}/P_2, \tag{3.10}$$

is a function of only the temperature (here P_2 is the partial pressure of lithium hydride, and P_3 is the partial pressure of hydrogen). Since at $T=$ const, the partial pressure of lithium $P_1 \to 0$ as $x \to 1$, and since the pressure of lithium hydride, P_2, cannot tend to zero in this case, the partial hydrogen pressure must tend to infinity in this process. Although these arguments are based on the assumption that the vapor phase is an ideal phase, a similar regularity observed experimentally confirms these conclusions and indicates that at thermodynamic equilibrium the purity of a lithium hydride sample (i.e., the admixture with lithium) is proportional to the equilibrium hydrogen pressure.

Unfortunately, the experimentalists did not always give enough attention to monitoring the composition of the condensed phase. This accounts for the large discrepancies in the data on both the pressure of the dissociation products of lithium hydride and the individual components of the mixture. The principal studies dealing with the measurement of the vapor pressure at equilibrium in the Li-LiH (LiD, LiT) systems are given in Table 3.5.

The first measurements of the equilibrium pressure of the dissociation products of a lithium hydride sample in an iron cell were carried out by Guntz.[134,135] At the melting point of the sample, which was 680 °C during the measurements, the pressure was 26.7 torr.

Hurd and Moore[150] studied the dissociation of LiH at phase equilibrium and measured the hydrogen pressure at temperatures between 780 and 930

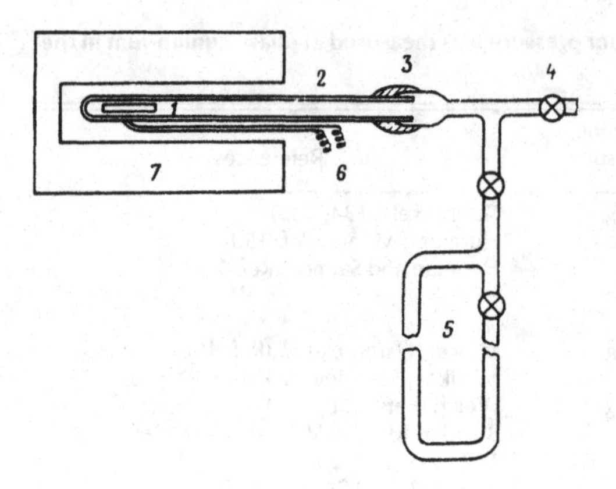

Figure 3.5. The experimental arrangement of Hurd and Moore.[150] 1) Cell; 2) quartz tube; 3) coupling to the external system; 4) stopcock for the gas or vacuum line; 5) mercury pressure gauge; 6) thermocouple; 7) furnace.

K. The measurement method can be described as follows. Lithium in a hydrogen atmosphere was placed into a nickel cell (Fig. 3.5) whose walls were coated with a thin layer of cobalt by electrodeposition. The cell was then sealed. The cell had the shape of a 1020-mm-long cylinder whose outside diameter was 9.6 mm and whose wall thickness was 1.27 mm. The cell wall was permeable to hydrogen and impermeable to other lithium gases and vapors. The cell was inserted into a quartz tube 1020 mm long and 16 mm in diameter which was placed into a furnace. The piezoelectric trace, i.e., the pressure in the system as a function of the gas consumption, was recorded in hydrogen at a constant temperature that was monitored by a Chromel–Alumel thermocouple. The equilibrium pressure was measured by a mercury gauge in combination with a cathetometer whose sensitivity was 0.1 mm. The procedure was carried out both in the hydrogen saturation direction and in the opposite direction. At high temperatures, lithium passed through the cell walls. This experiment did not have the capability of monitoring the composition of the condensed phase. Although it is not absolutely clear, Hurd and Moore seem to have defined the equilibrium pressure as the area under the piezoelectric curve. We assume, therefore, that this pressure refers to the two-dimensional (flat) region. A similar assertion was made by Gibb and Messer.[127] Hurd and Moore present the following results:

T, K...	782	870	871.5	922	926	953
P_{H_2}, mmHg...	0.1	1.85	1.75	7.3	7.8	26.7

The hydrogen pressure at 953 K was taken from Refs. 134 and 135.

Heuman and Salmon[144] carried out pioneering studies of the isotopic effects in Li-LiH (LiD, LiT) systems when the liquid and vapor are at equilibrium. These studies were carried out according to the Sieverts method. A certain amount of lithium was hermetically sealed in a metallic cell in a vacuum, and the cell was inserted into a quartz container that was connected to the external system. This system was used to create a vacuum in the quartz cell and to supply hydrogen (deuterium or tritium). The hydrogen diffused through the wall into the cell in which lithium was hydrogenated. After equilibrium was established (requiring several hours in each regime), the pressure, hydrogen consumption, and temperature were measured.

The purity of the original lithium was 99.26% and 99.75%. Before filling the apparatus, the hydrogen and its isotopes were purified by a method in which uranium hydride (deuteride, tritide) was produced and then thermally decomposed. The cell was fabricated from low-carbon steel, since this material is, in the authors' view, sufficiently corrosion-free with respect to lithium. Two cells were used—one was 12.7 mm in diameter and 38 mm high and the other was 25.4 mm in diameter and 41 mm high. The wall thickness varied from 0.25 to 0.50mm.

Heuman and Salmon[144] carried out a special study of the diffusion of hydrogen and deuterium through a steel wall and obtained the necessary dependences. The cell temperature was measured with a Chromel–Alumel thermocouple. The hydrogen pressure of the system was measured with a mercury gauge (at pressures > 10 mmHg) in conjunction with a cathetometer whose accuracy was within 0.05 mm. A McLeod gauge was used for measurements in the range of 0 to 10 mmHg.

Heuman and Salmon[144] studied the concentration dependence of the partial pressure of hydrogen (deuterium, tritium) at constant temperature when the liquid and vapor were in equilibrium. They also studied the temperature dependence of the partial pressure of hydrogen isotopes in the flat region.

The experimental data obtained by Heuman and Salmon[144] unfortunately have a large scatter, which indicates the presence of strong random-error sources. These data are presented in Tables 3.6 and 3.7. These experiments also seem to have a systematic error, since a single-wall quartz container cannot, as the experiments of Ref. 237 have shown, prevent the loss of hydrogen as a result of diffusion through the wall. The study of Heuman and Salmon, however, played an important role in the further evolution of the studies in this field.

Messer *et al.*[40] studied the equilibrium between the solid and liquid phases in the Li-LiH system. They measured the equilibrium vapor pressure of the dissociation products at 700 °C and with 37.0% and 64.8% mole concentrations of lithium hydride in the condensed phase, which amounted to 28.8 and 27.8 mmHg, respectively. They did not disclose the method they used for measuring these values. They noted, however, that their results are in good agreement with the data of Ref. 144, which were obtained at 700 °C. The flat pressure was found to be 28 ± 1 mmHg.

The pressure of the dissociation products of lithium hydride at high temperatures was measured by Novikov, Solov'ev, Suchkov, and Petrukhin in 1959 (see Ref. 70). The substance to be studied was placed into a thick-walled stainless-steel piezoelectric meter 600 cm^3 in volume. During the experiment, half of this volume was occupied by liquid lithium hydride. The channel of the meter was held at a constant temperature, slightly higher than the melting point of lithium hydride. The pressure of the decomposition products was cancelled in a bellows which was held at the initial position with the use of argon. The initial position of the bellows was monitored by an optical system, and the argon pressure was monitored by standard spring-loaded pressure gauges. The temperature of the piezoelectric meter was measured with four thermocouples placed along its height. The temperature deviation was within 10–15 deg.

Table 3.6. Dependence of the equilibrium vapor pressure on the composition of the liquid phase in the Li-LiH (LiD, LiT) systems [x is the atomic concentration of hydrogen or of its isotope in the liquid phase (%), and P is the pressure of hydrogen or of its isotope (mmHg)] from the experimental data of Heuman and Salmon (Ref. 144).

x	P	x	P	x	P	x	P	x	P
Li-LiH		88.02	47.0	16.02	58.0	70.00	246.0	Set II	
700 °C		94.36	47.0	19.38	72.0	86.40	254.0	2.30	3.35
2.08	2.23	96.74	78.0	24.63	84.0	92.60	276.0	3.00	4.50
3.24	1.96	97.04	175.0	26.77	84.0	Li-LiT		3.77	8.09
5.44	4.00	Set II		Set II		700 °C		5.57	12.14
8.41	7.55	2.67	3.0	23.90	79.0	Set I		7.70	18.20
12.63	14.80	3.83	4.0	34.30	97.0	2.0	6.93	750 °C	
12.41	14.90	5.02	7.0	42.34	99.0	2.91	8.50	Set I	
17.94	22.65	7.18	11.5	63.67	97.0	4.65	12.88	4.86	10.62
18.80	23.50	10.47	20.0	85.00	99.0	6.8	16.65	8.71	24.60
21.14	25.60	13.85	24.5	89.25	99.0	9.68	24.92	13.64	49.70
24.90	26.50	17.45	30.0	32.27	99.5	11.84	24.43	19.40	75.42
27.48	27.30	19.04	34.0	94.99	114.0	15.20	34.40	27.30	104.65
29.53	26.50	55.75	43.0	96.06	183.0	22.04	39.51	32.87	120.00
29.57	28.70	75.55	45.0	800 °C		23.20	44.10	39.81	131.79
49.20	28.90	Set III		Set I		29.75	50.80	50.69	131.56
69.91	27.80	8.76	10.0	9.75	42.0	29.83	49.10	51.90	132.94
90.18	30.10	10.70	19.0	14.40	88.0	38.20	49.36	59.57	132.94
96.67	42.50	16.42	30.0	19.02	176.0	42.10	48.90	Set II	
97.55	91.40	18.96	36.0	38.63	230.0	44.42	53.24	88.85	230.00
99.11	144.20	26.18	44.0	Set II		46.58	51.18	90.30	349.00
Li-LiD		750 °C		11.10	76.0	55.83	48.80	800 °C	
700 °C		Set I		22.50	184.0	61.00	56.00	54.48	250.20
Set I		5.43	10.0	23.80	150.0	61.10	54.20	63.97	251.65
19.48	39.0	6.68	16.0	30.60	206.0	67.10	54.34	66.30	257.80
24.29	40.0	8.23	20.0	41.4	222.0	77.46	53.90	74.04	245.20
27.18	41.0	9.45	31.0	42.30	229.0	92.06	57.30	82.64	262.11
31.53	44.0	12.87	43.0	49.50	243.0	92.47	66.10	84.45	346.60
65.63	47.0	15.35	56.0	66.20	253.0	94.63	143.00	85.54	411.60

Table 3.7. Vapor pressure in the two-dimensional region in the Li-LiH (LiD, LiT) systems [x is the atomic concentration of hydrogen or of its isotope in the liquid phase (%), P is the pressure of hydrogen or of its isotope (mmHg), and T is the temperature of the system (K)] from the experimental data of Heuman and Salmon (Ref. 144).

T	x	P	T	x	P
	Li-LiH		1023	85.00	99
973	29.5	26.5	1073	66.20	253
1001	90.0	51.05		Li-LiT	
1001.5	88.7	52.7	973	62.4	50.55
1029	90.0	88.0	1001	74.13	88.72
1047	84.2	109.1	1029	77.78	137.24
1058	80.5	131.7	1039	78.67	148.80
1077	69.1	179.6	1044	45.50	175.30
	Li-LiD		1074	67.80	262.00
973	75.55	45			

Table 3.8. Pressure of the dissociation products in the Li-LiH system from the experimental data of Novikov and co-workers.

T,K	P,atm	T,K	P,atm	T,K	P,atm	T,K	P,atm
963	0.217	1239	9.8	1074	1.32	1133	3.38
1063	1.00	1364	28.4	1023	0.62	1179	5.06
1166	5.06	1038	0.526	1063	1.44	1218	7.3
1333	27.8						

The following procedure was used to fill the system: An appropriate amount of lithium hydride in the form of finely divided granules 2 mm in diameter was poured into the piezoelectric meter. The meter was then connected to the system which was evacuated. The hydride was then melted down and poured into the entire system, including the bellows. Unfortunately, the initial composition of the hydride and its variation during the evacuation and during the experiments were not monitored. In the initial stage of the experiments, the sample liberated sorbed foreign gases which were released from the system through a valve. The principal measurements were carried out after this procedure.

The experimental results obtained are presented in Table 3.8. The data obtained by these experimentalists are described by the interpolation equation

$$\log P = -(7100/T) + 6.73, \tag{3.11}$$

where P is the pressure (atm) and T is the temperature (K).

The maximum relative measurement error was estimated to be 8%.

The data on the pressure of the dissociation products for lithium hydride obtained by Welch[250,251] are

T, °C...	615	639.7	664.6	671.3	691.3	727.9	766.0	787.5
P, mmHg...	3.7	7.37	14.76	21.5	27.3	62.6	113	175

Without giving the composition of the vapor mixture, Welch reported only that the measurements were carried out by the transpiration method, i.e., by saturating the flowing gas with vapor and then freezing out the saturated vapor.

Several studies published some time ago (Refs. 126, 161, and 162), which dealt with the theoretical and experimental determination of the equilibrium pressure of the dissociation products of lithium hydride, are of only a limited value, principally because of the low accuracy of the results.

A marked increase in the study of the vapor pressure at phase equilibrium in the Li-LiH (LiD, LiT) systems during the past decade is attributable principally to the growth in controlled fusion research.

Many experimental studies of the vapor pressure at phase equilibrium in the Li-LiH (LiD, LiT) systems were carried out by Blander, Maroni, Veleckis, and others (see Refs. 102, 103, 181–183, 233–239, 241, and 242) at Argonne National Laboratory. Like Heuman and Salmon, these researchers used two approaches in the experiments. The first approach involved the measurement of the partial equilibrium pressure of the hydrogen isotopes above the Li-LiH (LiD) systems at a constant temperature as a function of the composition of the liquid phase, which varied from 2% to 99.8% hydride. The second approach, which was used for measurements in the flat region, involved the determination of the functional dependence $P_{H_2(D_2,T_2)} = f(T)$ when the composition of the solution was clearly in the immiscibility zone.

The accuracy of some of the results published by these authors has been improved. We shall analyze the results of these experimental studies on the basis of the experiments of Veleckis[235] and Veleckis *et al.*,[237] drawing the reader's attention, when necessary, to the other papers published by this group. The Sieverts method was used in these studies. A cell made from Armco iron contained ~2 g of lithium. The cell, 57 mm long, 0.38 mm thick, and 12.7 mm in diameter, was hermetically sealed in a vacuum by an electron beam and then inserted into a double-wall fused-quartz tube. Armco iron was chosen because of its good resistance to corrosion in the case of the interaction with lithium and also because of its high permeability and low solubility with respect to hydrogen. During fabrication, the cell was heated in hydrogen and outgassed at temperatures between 850 and 900 °C.

Each sealed cell containing lithium was tested for leaks and outgassed again at 900 °C. The double wall of the quartz tube prevented loss of hydrogen from the tube by establishing a hydrogen counterpressure in the annular gap equal to the pressure inside the tube. A test showed that at 815 °C the permeability of hydrogen through a single-wall fused-quartz tube can be as high as 0.5 std cm^3 /(h atm).

The amount of hydrogen supplied to the tube depended on the thermodynamic state of the system. In the homogeneous region, where the partial hydrogen pressure on the isotherm depends essentially on the concentration of the liquid phase, the hydrogen supply was low (~2 mmole), while in the immiscibility region, the hydrogen supply was increased (~20 mmole). The hydrogen supplied to the tube diffused through the cell wall and interacted with the lithium inside the tube. To achieve equilibrium, the system was held at a constant temperature for 12–28 h, depending on the amount of hydrogen. The measurements were then carried out. The readings were taken when the hydrogen pressure remained constant within ±0.01 mmHg for 6–8 h.

The hydrogen consumption was monitored with a calibrated 500-cm^3 container held at a constant temperature. The volume of the connecting lines was calibrated beforehand with helium. The hydrogen concentration in lithium was calculated on the basis of the following data: the lithium mass in the cell, the total amount of hydrogen added to the system, the amount of hydrogen in the connecting lines and in the vapor space above the sample inside the cell, and the amount of hydrogen dissolved in the cell walls. The temperature of the isothermal experiments was monitored by three Pt/Pt-Rh thermocouples which were placed in the gap between the quartz-tube walls around the sample in a helical configuration spaced 120 deg. apart. These thermocouples were calibrated against a standard thermocouple which was inserted into lead that replaced lithium in the cell. Chromel-Alumel thermocouples were used for the measurements in the flat region. The pressure in the calibrated container was determined with a spring-loaded pressure gauge over the pressure ranges 0–20 ($\pm$ 0.05), 0–200 ($\pm$ 0.1), and 0–800 ($\pm$ 0.2) mmHg. The pressure inside the tube was measured within $\pm$ 0.05 mmHg.

The mass fraction of the original lithium was 99.98% (99.9 for the Li-LiD system). The principal impurities were sodium and potassium. The ratio $^6Li/^7Li$ was 0.0815 $\pm$ 0.0007, which corresponded to the ratio for natural lithium. The hydrogen purity was not given. A 99.99% pure deuterium was used in the experiments with the Li-LiD system. According to the data of the mass-spectrographic analysis, the content of the HD molecules is 1.1 $\pm$ 0.2%.

The hydrogen for hydrogenation of lithium was obtained by decomposing titanium hydride TiH_2 which was produced beforehand from a titanium sponge. In the experiment, a gradual measuring out of hydrogen along the isotherm leads, in the case of this method, to a summing of the errors in the determination of the quantity of hydrogen that reacts with lithium, i.e., in the calculation of the concentration of the liquid phase. As a result, the calculated mole concentration of hydrogen in lithium exceeded 100% LiH in the limiting cases. Corrections were introduced with the help of a normalizing factor, which made it possible to reduce the calculated data to physically sound results.

In the region of low concentration of lithium hydride, lithium was seen to affect the results of the measurement of the residual hydrogen (deuterium) because of the incomplete outgassing or draining of the cell. This circumstance required the introduction of corrections. The corrections to the mole fraction of lithium hydride were appreciable. Upon removal of hydrogen from the sample, the control measurements of the pressure above the system showed no evidence of an apparent hysteresis. The measurement error for the pressure, which was found to be 0.5%, was due principally to the effect of the temperature oscillation on the external volume of the system.

The dependence of the partial pressures of hydrogen and deuterium on the composition of the liquid phase was measured over the temperature interval 700–905 °C (Tables 3.9 and 3.10). Experiments in the flat region were carried out over the temperature interval 600–860 °C. In this case, approximately 35 mg of lithium and approximately 40 std cm^3 H_2, D_2, or T_2 were admitted into the cell in order to obtain ~60 mole % concentration of hydride in the condensed phase.

The experimental data on the pressure in the two-dimensional region in

Table 3.9. Partial pressure of hydrogen P_{H_2} (mmHg) at phase equilibrium in the Li-LiH system from the data of Ref. 103 (x is the mole concentration of LiH, %).

x	P_{H_2}	x	P_{H_2}	x	P_{H_2}	x	P_{H_2}	x	P_{H_2}	x	P_{H_2}
710 °C		99.54	329.0	98.67	266.4	97.73	368.6	95.81	494.5	92.45	600.6
2.153	0.53	99.58	393.0	98.84	347.2	98.00	467.4	96.18	549.6	93.08	661.0
3.698	1.49	99.64	539.6	98.96	433.3	98.17	563.0	96.47	603.4	93.52	680.3
6.390	4.26	99.69	669.8	99.05	506.0	98.28	641.9	96.73	659.7	93.87	720.8
8.286	6.84	99.70	744.5	99.13	576.9	98.36	708.0	96.92	715.0	903 °C	
10.42	10.80	759 °C		99.19	640.9	847 °C		878 °C		2.595	6.37
12.41	14.58	4.632	4.17	99.25	702.2	1.596	1.38	1.163	1.04	4.528	18.17
14.20	18.08	6.749	8.64	803 °C		2.858	4.30	3.11	7.10	6.375	34.73
16.11	22.16	9.598	16.53	1.686	0.97	4.688	11.57	5.033	18.08	8.581	60.98
18.74	27.40	13.37	29.83	3.104	3.23	7.212	26.58	7.003	34.01	11.37	101.8
20.96	31.90	17.88	47.53	4.927	8.08	10.05	49.21	9.030	54.60	14.71	160.1
23.57	35.52	22.32	67.79	7.485	17.93	13.34	81.85	11.17	79.49	17.96	224.3
27.54	36.45	25.44	79.09	10.42	33.00	17.59	129.5	13.44	110.4	21.21	292.6
33.27	37.10	28.10	85.66	13.71	53.22	22.32	184.3	15.83	144.8	25.52	378.6
39.38	37.75	30.84	86.77	17.40	77.90	27.67	244.7	18.77	191.3	30.36	474.3
50.13	37.38	34.97	87.29	22.83	117.5	31.74	284.2	22.11	254.7	36.63	571.3
60.95	38.38	50.15	87.94	26.72	142.6	35.24	310.7	25.98	316.2	42.67	629.2
71.67	39.24	69.85	88.53	32.19	166.8	39.02	323.2	30.37	380.9	49.86	653.6
82.39	39.16	89.50	88.54	37.64	169.7	42.87	329.4	35.31	439.1	57.78	657.8
87.39	38.84	92.26	88.71	47.25	171.6	48.76	329.9	40.81	486.1	68.74	658.3
90.88	39.26	92.81	88.73	61.99	172.1	64.76	331.1	43.70	499.0	75.95	662.0*
92.68	38.87	95.46	89.03	76.67	173.1	76.26	332.3	46.92	500.1	81.28	662.3
94.45	39.05	96.59	89.41	91.31	173.3	92.61	332.0	66.21	502.2	85.46	670.8*
96.22	39.88	97.38	104.5	95.16	168.5*	92.76	329.4	85.96	504.1	87.14	685.6
97.55	38.93	97.66	115.2	96.04	190.4	93.21	343.8*	89.30	504.0	87.96	705.2
98.92	74.18	97.97	141.0	96.36	284.8	93.88	363.2	89.96	513.7*		
99.30	242.8	98.18	166.6	96.79	232.6	94.53	392.2	91.06	541.6		
99.47	256.3	98.47	214.3	97.12	263.8	95.33	447.3	91.73	563.8		

*The experimental points were obtained by pumping off the hydrogen from the cell containing the sample, after reversing the experimental procedure.

Table 3.10. Partial pressure of deuterium P_{D_2} (mmHg) at phase equilibrium in the Li-LiD system from the data of Ref. 234 (x is the mole concentration of LiD, %).

x	P_{D_2}	x	P_{D_2}	x	P_{D_2}	x	P_{D_2}	x	P_{D_2}
705 °C		99.62	745.7	99.26	644.8	97.12	372	95.15	496.5
1.1113	0.200	756 °C		99.29	732.4	97.63	464.8	95.56	542.3
2.534	1.056	1.154	0.367	805 °C		97.98	571.9	95.83	574.7
4.220	2.891	2.703	2.006	0.972	0.482	98.21	677.9	96.27	648.0
6.234	6.154	4.329	5.216	2.790	4.103	98.32	747.7	96.57	718.1
8.545	11.03	6.304	11.10	4.684	11.26	840 °C		871 °C	
11.00	17.22	8.283	18.45	6.468	20.72	1.057	0.873	1.544	2.06
13.54	24.18	11.10	31.56	8.177	31.77	2.492	4.591	3.778	12.48
16.14	32.90	14.49	40.39	10.42	49.09	4.248	12.69	6.195	32.70
18.69	39.87	18.16	71.20	13.00	71.87	6.305	26.69	8.697	61.79
21.43	46.48	21.92	92.16	15.95	100.9	8.292	44.10	11.34	100.0
24.18	48.58	25.34	105.2	19.07	130.5	10.27	64.79	13.82	141.0
36.34	50.64	32.88	114.8	22.39	163.8	12.55	91.94	16.86	197.2
47.90	49.20	48.11	115.6	26.10	197.8	15.10	125.3	20.75	271.6
96.44	47.78	61.86	115.6	30.17	228.6	18.16	167.7	25.15	354.9
97.88	49.30	86.97	115.7	37.66	239.4	21.78	219.9	29.58	454.0
98.45	66.76	91.51	115.8	53.97	243.7	26.20	291.6	34.99	536.6
98.81	98.72	96.19	116.2	64.52	245.4	29.53	333.3	41.54	596.9
99.05	141.2	97.35	122.3	88.99	244.6	33.41	373.3	54.75	603.2
99.22	201.0	98.14	173.4	90.53	246.3	61.31	400.8	75.13	606.2
99.35	289.3	98.60	251.9	92.40	245.7	89.22	403.3	90.74	607.8
99.46	391.6	98.84	329.3	99.42	252.1	92.42	404.4	92.06	624.2
99.50	497.4	99.03	436.2	95.97	276.8	94.01	425.5	92.92	658.8
99.56	606.1	99.14	540.6	96.58	314.6	94.62	462.8	94.03	731.1

Table 3.11. The pressure *P* (mmHg) of hydrogen or of its isotope in the two-dimensional region in the Li-LiH (LiD, LiT) systems from the data of Ref. 236 (*T* is the temperature, °C).

T	*P*	*T*	*P*	*T*	*P*
Li-LiH		647.4	7.97	Li-LiT	
858.3	380.9	630.6	5.16	844.4	461.4
841.2	300.1	Li-LiD		820.0	327.9
839.6	295.8	850.6	461.2	792.2	221.4
820.3	223.8	827.5	337.7	767.6	155.1
818.9	217.8	799.7	225.6	742.2	104.3
800.0	166.8	774.5	155.0	732.7	90.53
785.4	129.2	762.1	127.1	720.7	71.27
768.8	100.6	751.6	109.9	710.7	60.51
755.5	81.10	738.7	88.15	694.0	46.70
743.1	65.74	726.9	71.10	688*	41.3
731.5	54.24	712.3	56.50	680.5	34.66
718.5	43.84	702.3	47.40	667.6	24.39
716.7	42.18	692.0	38.78	659.9	20.17
704.6	34.09	690*	37.9	647.6	14.16
694*	27.8	677.5	27.37	636.4	10.17
693.5	27.65	666.4	20.64	621.1	6.79
682.9	20.80	653.1	13.94	616.6	5.92
669.5	14.96	641.9	10.35	611.0	4.94
657.3	10.76	627.7	7.00	595.8	3.28

*The monotectic temperatures were taken from Ref. 236.

the Li-LiH (LiD, LiT) systems are presented in Table 3.11.

The use of this method for measurements below 700 °C gives rise, according to Katsuta *et al.*,[168] to basic difficulties stemming from the fact that the diffusion rate of hydrogen through the wall of the cell containing hydride decreases sharply with decreasing temperature. For this reason, it is basically difficult, in the first place, to establish an equilibrium between the hydrogen pressure inside and outside the cell and, secondly, the impurities contained in lithium have a considerable effect at low temperatures on the equilibrium if the hydrogen content in the gas space is low.

From the analysis of their experimental data the authors obtained some useful additional information: They defined the boundaries of the immiscibility zone in the Li-LiH (LiD) systems, which turned out to be essentially the same for these two systems; they determined the parameters of the critical solubility points [in the Li-LiH system $T_{cr} = 994$ °C, $P_{cr} = 2.6$ kg/cm^2, and mole concentration $x_{cr} = 57.2\%$; in the Li-LiD system, $T_{cr} = 1000 \pm 10°$ C and $x_{cr} = 61 \pm 3\%$]; they found the temperature of the monotectic (694 °C for Li-LiH, 690 °C for Li-LiD, and 688 °C for Li-LiT); and they found the relations for calculating the activities of the components of the Li-LiH (LiD) systems, the Sieverts constants, standard heats of formation of the isotopic modifications of lithium hydride, and several other characteristics.

Veleckis[235] attempted to determine the melting point of pure lithium deuteride from the concentration of the right boundary of the immiscibility zone by using the heat of fusion of lithium hydride obtained by Shpil'rain *et al.*[71] The temperature he obtained was 692.4 °C.

A similar method was used by another group of researchers from Oak Ridge National Laboratory to measure the partial pressure of hydrogen isotopes in the Li-LiH (LiD, LiT) systems.[215–217,219] The measurements were carried out using two separate apparatuses with iron or nickel cells containing between 0.2 and 2.0 g of lithium. The range of temperatures over which the measurements were carried out was 700–1000 °C at pressures of 0.1–760 mmHg.

The authors found that the Sieverts equation $P_{H_2}^{1/2} = Cx$ (where C is the Sieverts constant) holds for a mole fraction x of liquid-phase hydride up to 0.1. The boundary of the immiscibility zone for the Li-LiH system was the same as that for the Li-LiD system and was compatible with the data of Veleckis *et al.*

An experiment was also carried out at a temperature of 800 °C on the measurement in the flat region of the pressure of a mixture of hydrogen isotopes H_2, D_2, and HD for different ratios of lithium hydride and lithium deuteride in the solution. The content of hydrogen components in the gas phase was analyzed with a mass spectrometer. Hydrogen or deuterium in the amount necessary to achieve a liquid-phase concentration in the flat region was first introduced into the system. A small portion of deuterium or hydrogen was then added to it. The equilibrium pressure did not depend on the order in which the procedure was carried out or on the relative quantities of the two existing liquid phases with concentrations corresponding to the left-hand and right-hand boundaries of the immiscibility zone. This pressure P (mmHg) depends solely on the ratio of lithium hydride and lithium deuteride in the solution:

$$P_{H_2}^{1/2} = 12.9\, x_{LiH}/(x_{LiH} + x_{LiD}); \tag{3.12}$$

$$P_{D_2}^{1/2} = 15.9\, x_{LiD}/(x_{LiH} + x_{LiD}); \tag{3.13}$$

$$P_{total}^{1/2} = 15.9 - 3x_{LiH}/(x_{LiH} + x_{LiD}). \tag{3.14}$$

The partial equilibrium pressure of HD vapor was determined from the data on the equilibrium constant of the reaction (these data were found in the literature)

$$H_2 + D_2 = 2HD. \tag{3.15}$$

At 800 °C this constant is

$$K_e = P_{HD}^2/P_{H_2}P_{D_2} = 3.8. \tag{3.16}$$

When the solution contained a mixture of LiH and LiD, the boundaries of the immiscibility zone were the same as those for the Li-LiH and Li-LiD systems. In particular, at 800 °C the coordinate (mole fraction of LiH, LiD) of the left-hand boundary is 0.33 and that of the right-hand boundary is 0.96.

An interesting measurement of the partial pressure of hydrogen and deuterium in the flat region was carried out by Smith and Webb.[218] A hydride or deuteride sample was inserted into a tube made from niobium with 1 wt.% zirconium. The authors noted that this material is very useful from the standpoint of hydrogen permeability. The cell had the following dimensions: length 75 mm, outside diameter 6.5 mm, and wall thickness 0.4 mm. The mass of the sample was 0.5 g. The conditions under which the experiment was performed were controlled carefully. To reduce the temperature oscillations, the sealed cell with the hydride was placed into a furnace containing two concentrically mounted heaters that provided a temperature stability within ± 0.5 °C. The temperature was measured with Chromel–Alumel thermocouples, which are

Table 3.12. Pressure of hydrogen (Pa) in the two-dimensional region in the Li-LiH system from the experimental data of Smith and Webb (Ref. 218).

T	P_{2D}	T	P_{2D}	T	P_{2D}	T	P_{2D}
650	1015.04	534	26.90	483.5	3.32	716.5	4 942
651	1050.77	549	45.17	486	3.57	725	5 870.7
651	1056.74	569	89.17	472	2.19	726	5 930.8
651.5	1060.49	583.5	143.13	676	2197	726	5 946.6
652	1131.85	596	213.00	666	1649.1	725.5	5 880.9
657	1280.93	609	313.17	658.5	1347.9	736	7 102
657	1282.05	618.5	420.87	668.5	1721.6	736	7 182
659	1362.31	619	423.70	681.5	2549	736	7 186
658	1325.28	633	635.73	681.5	2541	746	8 450
659	1355.92	644	888.63	679	2357	746	8 413
643.5	867.8	654	1178.18	673	1977	755	9 937
643.5	868.12	653	1148.79	674	2033	755	9 899
643.5	867.80	653	1145.96	691	3179	765.5	11 754
632	623.12	606.5	298.06	686.5	2836.3	766	11 828
616	385.60	607	296.99	683.5	2668	768.5	12 318
602	252.38	582	143.91	683.5	2662	755	9 683
589	168.92	584	143.12	701.5	3831.7	752.3	9 343
571	94.71	562	72.07	700	3688.2	743.5	7 939
553.5	52.70	565	78.34	698.5	3599.7	738	7 361.8
539.5	32.13	565	78.07	698	3562.6	727	5 969.4
529	20.81	561	48.42	697.5	3511.4	727	5 847.8
515.5	13.04	532	27.86	696.2	3439.7	727	5 898.4
506	9.25	536	28.46	695	3364.5	727	5 939.8
485	4.04	522	16.99	694.5	3342.1	721	5 380.9
472.5	2.20	521.5	16.72	694	3335.7	721	5 405
463	1.45	522	16.93	694	3323.5	710	4 480
488	4.00	513	11.84	694	3335.5	701	3 754.1
487	3.93	514	12.17	704	3986.3	700	3 702.4
497	6.17	504	8.57	704.5	4010.3	694	3 345.4
507.5	9.60	505.5	8.83	704	3966.6	692	3 269.1
523	17.16	499	6.68	716.2	4926	692	3 231.7
		500	6.67				

more stable in hydrogen than platinorhodium–platinum thermocouples. A thorough analysis of the impurities in the original sample produced the following results, 10^{-3}%:

LiH: 95 Na; 48 K; 10 Ca.

LiD: 49 Na; 25 K; 467 Ca.

In addition, the presence of oxygen (wt.%) was detected: 0.28 in LiH and 0.58 in LiD.

The equilibrium pressure of hydrogen or deuterium on the outside of the cell was measured with a differential pressure gauge with a sensitivity on the order of 25 Pa. Experiments were carried out during the decomposition of hydride and during hydrogenation. The partial pressure of hydrogen and deuterium in the flat region was measured above and below the monotectic temperature over the temperature range 460–770 °C. The measurement results, listed in Tables 3.12 and 3.13 in the order in which they were obtained in the experiment, are described[218] by a general formula of the type

Table 3.13. Pressure of deuterium (Pa) in the two-dimensional region in the Li-LiD system from the experimental data of Smith and Webb (Ref. 218).

T	P_{2D}	T	P_{2D}	T	P_{2D}	T	P_{2D}
499	9.89	641.5	1172.44	578.5	164.54	712.5	6 352
485.5	5.63	642	1196.30	590	237.63	719.5	7 333
486.5	5.84	563	108.68	589	232.22	719.5	7 330
460.5	1.73	563.5	108.16	601	338.75	723.5	7 822
461	1.85	578.5	174.05	661	1 976	723	7 815
472	3.00	579	175.48	655.2	1 703	728.8	8 377
472	3.16	593.5	277.36	655	1 701	726.5	8 287
487	5.83	594	288.80	647.5	1 408	735.5	9 702
486.5	5.84	594	286.14	646.4	1 356	735.5	9 733
499.5	10.21	605.5	403.81	658.8	1 857	735.5	9 722
499.5	10.35	606	408.89	667	2 312	743	10 984
512.5	17.44	623	671.68	667	2 335	743	10 972
512.5	17.56	622	660.62	675	2 806	749.2	12 290
522.5	25.92	637	1021.81	674	2 798	749.2	12 314
523	26.02	507	18.76	676	3 034	759.2	14 311
537	42.34	470	2.77	680	3 384	759	14 207
537	42.80	470	2.81	688	4 088	755.5	13 387
551.5	71.09	461	1.87	683	3 594	755	13 322
552	71.66	460	1.97	687.5	4 077	754	12 995
562	99.46	461.5	1.92	690	4 382	754	12 966
562.5	100.30	460.5	1.79	691	4 453	751	12 372
577	171.59	401	6.72	690.5	4 348	751	12 324
580.5	185.05	491.5	6.92	690	4 302	746	11 232
579.5	184.11	502	10.97	688	4 022	746	11 247
577	167.45	502	11.20	687	3 970	746	11 270
589.5	251.05	513.5	17.27	687	3 962	742.5	10 679
589.0	239.98	513.5	17.51	686.5	3 861	742.5	10 687
589.0	246.67	527	28.74	686	3 814	739	9 958
497.5	9.31	526.8	28.50	694	4 677	739	9 953
602	357.30	538	42.85	695	4 724	735.5	9 327
602	367.25	537	41.66	695	4 724	727.2	8 097
602	355.25	548.5	63.46	709	6 061	724.5	7 714
614	523.49	548.8	64.34	708	5 944	702	5 272
614	512.36	558.5	87.75	700	5 178	705.5	5 618
623	684.49	559.5	89.33	700	5 180	704.5	5 549
623.5	685.14	570.5	132.19	705	5 620	706.5	5 736
640	1121.91	571.0	133.50	705	5 620	708	5 902
640	1137.24	571.5	134.34	718.5	71 138	708.2	5 909
641	1164.70	579.5	176.60	718.5	71 138	699	5 020
		577	159.69	715.8	6 846		

$\log P_{2D} = A - (B/T)$, where P_{2D} is the partial pressure in the flat region (Pa), and T is the temperature (K). The coefficients of this equation are listed in Table 3.14. The values of the monotectic temperature, which turned out to be the same for the Li-LiH system (688.6 °C) and the Li-LiD system (687.7 °C), were found from the point of intersection of the corresponding lines $\ln P_{2D} = f(1/T)$.

McCracken *et al.*[128,185] studied the phase equilibrium in the Li-LiD system at low concentrations of lithium deuteride in the liquid solution. Basically, the Sieverts method was used in the experiment. The cell, which initially contained 1 g of pure lithium, was fabricated from niobium and sealed in a

Table 3.14. Coefficients *A* and *B* based on the measurements of Smith and Webb.

	Below the monotectic temperature		Above the monotectic temperature	
System	*A*	*B*	*A*	*B*
Li-LiH	14.243 ± 0.022	$10\,355 \pm 18$	11.456 ± 0.031	7675 ± 30
Li-LiD	14.362 ± 0.019	$10\,328 \pm 16$	11.409 ± 0.039	7491 ± 39

vacuum by an electron beam. The diameter of the cell was 15 mm, its length was 30 mm, and its wall thickness was 1 mm. The temperature of the cell was measured with Chromel–Alumel thermocouples and its pressure was measured using a mass spectrometer, an ionization pressure gauge, and a McLeod gauge. Equilibrium was maintained for about 1 h before the instrument readings were taken. The pressure measurements at temperatures between 500 and 800 °C were reproducible within 10%. Corrections to the absorption of deuterium by the cell walls, adsorption by the inside surfaces of the supply lines, and a possible pressure drop between the internal and external volumes of the cell were taken into account. Analysis of the data obtained by these investigators showed that at 700 °C the value of the Sieverts constant is 3×10^{-2} mmHg$^{-1/2}$ in the range $10^{-3}<x<10^{-1}$. The heat of solution was found to be 21 ± 2 kcal/mole from the measurements of the temperature dependence of the partial pressure of deuterium for three constant values of the atomic fraction of deuterium in the liquid phase in the range 3.4×10^{-4}–5.6×10^{-3} at temperatures between 500 and 800 °C.

Katsuta *et al.*[168] measured the pressure of the dissociation products in the Li-LiH system by a method that is fundamentally different from the various versions of the Sieverts method. Three sets of different experiments were carried out. The first set involved the following procedure. A sample in the form of a Li-LiH solution was placed into an open molybdenum container which was inserted into a quartz vessel. Samples of two original compositions with initial masses of 50 and 100 g and molar contents of lithium to lithium hydride ratios of 30:70% and 53:47%, respectively, were tested.

The apparatus was supplied with argon whose pressure was held constant at 800 mmHg. The partial pressure of hydrogen in argon was determined with the help of a chromatograph. The minimum limit of partial hydrogen pressure measured by the chromatograph was 3×10^{-3} mmHg. The measurements were carried out at temperatures between 540 and 740 °C. The temperature difference between the surface and the bottom of the solution under study was several tens of degrees. The original lithium was 99.8% pure. Hydride was produced separately in a commercial apparatus.

The second set of experiments involved the study of the partial pressure of hydrogen in the vapor as a function of the concentration of the dilute solution of hydrogen in liquid lithium. A 4.1-liter container held about 1.8 kg of lithium with dissolved hydrogen. The slope of the $P_{H_2}^{1/2}$ curve was determined as a function of the variation in the concentration of hydrogen at a constant temperature, and the Sieverts constant was determined from the tangent of the slope angle. Although the method used to determine the change in hydrogen concentration in the melt was not indicated, it seems to have been determined from the measurable amount of hydrogen that was added to the system.

A sensor tube with a niobium membrane (0.2 mm thick and 5.8 cm^2 in area) or a stainless-steel membrane (0.15 mm thick and 122 cm^2 in area) was immersed into the melt. The hydrogen in the melt diffused through the membrane and its pressure was measured with a pressure gauge containing a thermocouple. The apparatus was evacuated beforehand to 10^{-7} mmHg. In the experiment, the partial pressure of hydrogen was varied between 1×10^{-5} and 1.0 mmHg as a function of the temperature. The temperature of the melt was measured with thermocouples which were placed at the level of the membrane. In this experiment the temperatures were measured within the range 350–650 °C.

The third set of measurements was an extension of the second set. A cold trap was connected to the system and the experiment was carried out in the hot container at a constant temperature of 473 ± 3 °C with the temperature of the cold trap varying between 200 and 470 °C. The composition of the liquid phase in the container was determined from the partial hydrogen pressure and from the Sieverts constant which was found in the preceding set of measurements.

It should be pointed out that in this method it is important to maintain a uniform heating of the internal volume of the system, which is filled with hydrogen that enters it through a permeable membrane. The authors did not stipulate this requirement and, judging from the design of the experimental apparatus, it was difficult for them to maintain uniform heating. Although the experimental method used by Katsuta *et al.* is quite original, the results obtained by them are rather crude, differing dramatically from other data. It appears that the results were affected primarily by the significant temperature drop (to 50°) of the sample that was tested and by the errors in the measurement of the hydrogen pressure (due to the lack of equilibrium between the system and the pressure gauge).

In contrast with the recent experimental studies considered above, which were carried out by the Sieverts static method, Ihle and Wu[153–156,258] from the Institute of Nuclear Chemistry (German Democratic Republic), studied the thermodynamic properties of the Li-LiH (LiD) systems by the Knudsen diffusion method, with the help of a mass spectrometer. The Sieverts method can be used to measure the partial pressure of only the hydrogen isotopes in the vapor phase. Ihle and Wu used the dynamic method to determine the concentration of the entire spectrum of the Li, LiD, D_2, Li_2, Li_2D, and other components contained in the Li-LiD system in the vapor phase. Furthermore, Ihle and Wu measured the partial pressure of hydrogen above the Li-LiH solution in the flat region.

The molybdenum effusion cell containing the sample was placed into the working zone of the apparatus. The cell had a 0.2-mm-diam. outlet when the Li-LiH system was studied and a 0.8-mm-diam. outlet when the Li-LiD system was studied. The ratio of the area of the outlet to the effective area of the cell was 1.6×10^{-3}. The temperature of the cell was measured with a platinorhodium–platinum thermocouple (an optical pyrometer was also used in the case of the Li-LiD system). The temperature was measured within ± 2 K. The

Table 3.15. Partial pressures (atm) of the components in the Li-LiD system from the data of Ihle and Wu (Ref. 155) for an atomic fraction of deuterium in the liquid solution $x_D = 10^{-5}$.

T,K	P_{Li}	P_{Li_2D}	P_{LiD}	P_{D_2}
773	0.457×10^{-5}	0.123×10^{-12}	0.132×10^{-11}	0.308×10^{-11}
823	0.196×10^{-4}	0.126×10^{-11}	0.851×10^{-11}	0.841×10^{-11}
873	0.708×10^{-4}	0.987×10^{-11}	0.443×10^{-10}	0.204×10^{-10}
923	0.222×10^{-3}	0.619×10^{-10}	0.192×10^{-9}	0.451×10^{-10}
973	0.620×10^{-3}	0.321×10^{-9}	0.720×10^{-9}	0.918×10^{-10}

solution of hydrogen or deuterium in lithium was produced directly in the cell. A certain amount of gas was supplied to the apparatus, and the escape of hydrogen (deuterium) from the gas space due to absorption by lithium was monitored by a mass filter.

The original lithium (0.8 g) was 99.99% pure. The gas (hydrogen or deuterium) was purified by producing and subsequently decomposing uranium hydride (uranium deuteride). This gas was used to fill a standard vessel 40.7 cm^3 in volume. The gas flow was controlled within 0.1% with a diaphragm micro pressure gauge. The relative error in determining the atomic fraction, particularly that of deuterium in liquid lithium, was within 1%. The partial pressure of hydrogen above the Li-LiH solution in the flat region was measured at temperatures between 700 and 800 K. The authors used the equation

$$\ln P = 35.51 - (23\,945/T) \tag{3.17}$$

to describe the experimental data obtained by them:

T, K...	725	730	740	745	763	773	790	800
P, Pa...	9.25	11.919	19.29	23.26	51.18	69.01	127.12	137.33

When the Li- LiD system was in equilibrium, the partial pressures of the components and the vapor composition were determined when the solution of deuterium in liquid lithium was in an infinitely dilute state, with a deuterium concentration (in atomic fractions) of 10^{-6}–10^{-4} at temperatures between 773 and 973 K (Table 3.15). This concentration range is considered to be the most effective range for the use of a lithium blanket in a fusion reactor.

The measurements in the flat region were carried out under the assumption that in the mass-spectrometric experiments the partial pressure P of the gas sample is proportional to the product of the ion current i^+ and the absolute temperature T:

$$P_{H_2} \sim I^+_{H_2}\,T. \tag{3.18}$$

In the flat region below the monotectic temperature, the reaction

$$LiH_T \rightleftarrows Li_l + \tfrac{1}{2}\,H_{2,gas} \tag{3.19}$$

has the equilibrium constant

$$K = a_{Li} P^{1/2}_{H_2} / a_{LiH}, \tag{3.20}$$

where a is the activity.

In this system, the liquid lithium in the condensed phase is in the saturated state with the solid lithium hydride. Writing the expression for the thermal effect of reaction (3.19) in the form

$$\Delta H_T = -R[d \ln K/d(1/T)] \tag{3.21}$$

and assuming that $a_{Li} = a_{LiH} = 1$, the authors determined from the slope of the curve $\log(I^+, T) = f(1/T)$ that $\Delta H_{750} = 24.0 \pm 0.5$ kcal/mole.

Ihle and Wu have then determined experimentally the activity of lithium from the relation

$$a_{Li} = P_{Li}/P_{Li}^0 = I_{Li}^+/I_{Li}^{+0}, \tag{3.22}$$

where P_{Li} and P_{Li}^0 are the partial pressure of lithium in the mixture in the flat region and the pressure of pure lithium at the given temperature, and I_{Li}^+ and I_{Li}^{+0} are the ion currents occurring under the same conditions. From this expression, they obtained the relation

$$\log a_{Li} = (401.5/T) - 0.55. \tag{3.23}$$

From the calibration experiments and the data on the cross section of ionization of molecular hydrogen, obtained in the literature, Ihle and Wu found the proportionality coefficient in (3.18) and, accordingly, the partial pressure of hydrogen. As for the partial pressure of the components of the Li-LiH system at low hydrogen concentration in the liquid phase, Wu and Ihle[258] found that, at low hydrogen concentration in liquid lithium ($\sim 10^{-3}\%$) and a temperature of ~ 900 K, the partial pressure of LiH vapor is equal to the partial pressure of Li_2H.

Wu and Ihle[258] also studied in detail the reactions of the interaction of the components in the gas phase

$$Li_2H + Li \rightleftarrows LiH + Li_2 \tag{3.24}$$

and

$$Li_2H + Li_2 \rightarrow LiH + Li_3. \tag{3.25}$$

It should be pointed out that the accuracy of determining the equilibrium constants by the mass-spectrometric method is relatively low. In the case of reactions (3.24) and (3.25), for example, the relative error of the data was estimated to be 30%–40%.

Ihle and Wu obtained extensive information in the study of the Li-LiD system in the range of low concentrations of deuterium in the liquid phase. They found the partial pressures of the components LiD, Li_2D, LiD_2, and Li_2D_2 as functions of the composition of the liquid phase and the temperature. They also found the partial pressures of Li, Li_2, and Li_3 and the ratio of the atomic concentrations of deuterium in the vapor and liquid phases, which turned out to be virtually independent of the concentration of deuterium in the liquid phase.

The accuracy of these data, as indicated above, is relatively low. This is evident from a comparison of the values of the two-dimensional pressure region in the Li-LiH system, obtained by Ihle and Wu, with the data found in the literature. Although the derivative $d \ln P_{H_2}/d(1/T)$ is approximately equal to the results of other authors, the absolute values of the partial pressure of hydrogen differ markedly.

Analysis of the data found in the literature on the experimental range of pressures of the dissociation products in the Li-LiH (LiD, LiT) systems shows that, of the many available measurements, only a few studies can claim to have reliable results. Among these studies are principally those of Veleckis, H. M. Smith, F. J. Smith, Novikov, and Goodal *et al.* At the same time, with the exception of the measurements of Novikov *et al.*, these studies are restricted to a maximum pressure of 10^5 Pa and temperature of 1200 K. The relatively low dissociation pressures account for the fact that the maximum composition of the liquid phase, in equilibrium with the vapor, differs markedly from the stoichiometric composition of lithium hydride.

The measurements carried out by Shpil'rain *et al.*[76–78] have substantially improved the information on the equilibrium vapor pressure in the Li-LiH system both in the immiscibility zone and in the region slightly to the right of the immiscibility zone. This is particularly important at high concentrations of lithium hydride and also in the high-temperature region above the critical dissolution point.

Experiments carried out at high concentrations of lithium hydride at $V = \text{const}$

In designing and calibrating a specific apparatus, it is important to have on hand data on the equilibrium pressure when the apparatus, which initially contains lithium hydride of a particular purity, begins to operate in a given temperature regime. An experiment carried out in a closed volume is distinguished by the fact that the composition of the sample, and hence its properties depend essentially on the external conditions—the temperature, the free space above the sample, the material of the wall through which the hydrogen from the vessel containing the sample diffuses.

The experimental arrangement for measuring the pressure of the dissociation products of lithium hydride at constant volume of the system is shown schematically in Fig. 3.6. The principal component is a 1X18H9T stainless-steel cylindrical container (4) which holds the substance to be tested. The inside diameter of the container is 24 mm, the height is 95 mm, and the wall thickness is 3 mm. The top cover of the container becomes a 150-mm-long capillary which is connected to a vacuum-gas system and to one of the chambers of a membrane differential pressure gauge (2). The internal volume of the container, which is determined at room temperature (22 °C) by comparing the mass of the empty container and the mass of the container when it is filled with distilled water under a vacuum, is 43.60 ± 0.01 cm^3. The internal volume of the capillary, which is determined from the design dimensions, is 1.55 ± 0.25 cm^3, and the volume of the adjacent chamber of the differential pressure gauge is 0.076 cm^3.

The temperature was measured by three Chromel–Alumel thermocouples (3), which were placed at 35-mm intervals along the height of the container and shielded from the furnace by a stainless-steel foil. The thermocouples were first calibrated in helium relative to a second-class platinorhodium-platinum thermocouple. The temperature was measured to within ± 2 K over the working range. The temperature profile of the furnace (5) was regulated by auxiliary end heaters. The difference in the thermocouple readings was no greater than 0.5 K during the measurements.

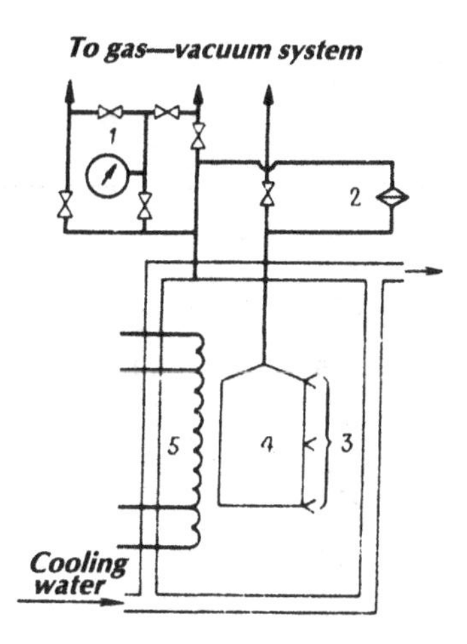

Figure 3.6. Schematic of an experimental setup for measuring the partial pressure of hydrogen in the Li-LiH system at high concentrations of liquid lithium hydride.

The pressure in the container was measured with a diaphragmed differential pressure gauge (2) (a null detector) and a standard spring-loaded pressure gauge (1). The position of the diaphragm was controlled by a microscope with a sensitivity of 1 μm, using an indicator which was connected to the diaphragm with a sensitivity of 0.6 mmHg.

The experiment can be summarized as follows. At room temperature an initial hydrogen pressure was established in the container and furnace. The container was separated from the furnace by a valve and the temperature of the furnace was raised in steps. The hydrogen pressure in the furnace, which was maintained at the same level as that in the container, was measured after the temperature regime was established. The same procedure was used to take the readings when the process was reversed, i.e., when the temperature was lowered.

Three sets of measurements were carried out at different initial hydrogen pressures and hence different concentrations of lithium hydride (i.e., the measurements were carried out on three isochores). The experiment was begun with lithium hydride in the solid state. A pressure jump caused by the expansion of the condensed phase upon melting and by a change in its composition occurred at the melting point. The melting point of hydride was approximately 692 °C, which suggested that the composition of the tested sample was nearly 100% lithium hydride. The measurement results are given (in the order in which the experiment was carried out) in Table 3.16 and in Fig. 3.7.

These results were analyzed by the method of least squares on the basis of the equation

$$P = a + bT + cT^2 + dT^3, \tag{3.26}$$

where P is the pressure (kgf/cm^2) and T is the temperature (K). The coefficients of Eq. (3.26) are given in Table 3.17.

Table 3.16. Experimental data of Ref. 76 on the vapor pressure in the Li-LiH system at V = const.

T,K	P,kg/cm²	T,K	P,kg/cm²	T,K	P,kg/cm²
Isochore I		928.2	9.92	955.9	6.93
Before melting		949.0	10.09	During melting	
891.3	15.36	During melting		964.9	7.17
907.5	15.56	962.1	10.29	963.6	7.66
940.1	16.01	967.0	10.93	After melting	
950.5	16.14	After melting		994.0	8.09
958.2	16.27	979.4	11.39	1006.8	8.24
958.7	16.28	993.3	11.57	1020.3	8.24
During melting		1007.5	11.77	1032.0	8.63
964.5	16.81	1022.6	11.99	1045.7	8.89
963.2	17.04	1037.0	12.26	1059.4	9.17
964.6	17.68	1048.6	12.46	1068.1	9.38
946.9	17.64	1058.6	12.67	1061.5	9.20
After melting		1070.9	12.95	1047.2	8.90
966.1	17.91	1066.2	12.87	1040.4	8.76
977.9	18.09	1049.1	12.43	1029.5	8.55
986.6	18.21	1040.8	12.29	1020.9	8.42
997.0	18.41	1030.0	12.11	1010.8	8.28
1004.1	18.51	1009.7	11.79	1000.4	8.13
1012.3	18.66	1002.7	11.70	993.1	8.05
1022.7	18.84	988.0	11.51	990.2	8.03
1029.5	19.0	974.1	11.35	987.4	8.00
1047.2	19.46	Isochore III		984.8	7.97
Isochore II		Before melting		982.0	7.94
Before melting		915.1	6.69	974.2	7.86
891.1	9.60	931.0	6.77	970.3	7.82
909.6	9.76	949.6	6.88	966.7	7.78

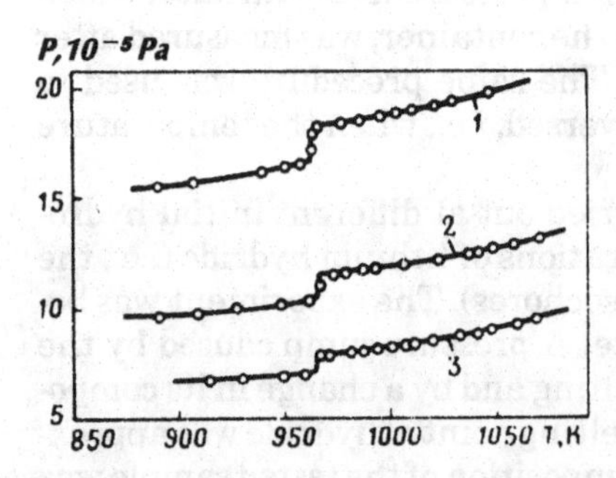

Figure 3.7. Partial pressure of hydrogen in the Li-LiH system with high concentrations of liquid lithium hydride at V = const (curves 1, 2, and 3 are isochores; see Table 3.16).

Experiments carried out to the left of the immiscibility zone and in the two-dimensional region

When this experiment was being planned, the data on the partial pressure of hydrogen as a function of temperature and concentration of the liquid phase in the region to the left of the immiscibility zone of the Li-LiH system and also the data on the temperature dependence of the partial pressure in the flat region were highly contradictory. This situation provided the incentive in the beginning of the 1970s to carry out special experiments. This study[77] was

Table 3.17. Coefficients in Eq. (3.26).

Isochore	State	a,kg/cm²	b,kg/(cm²×K)	c,kg/(cm²×K²)	d,kg/(cm²×K³)
I	Before melting	3.367	13.562×10^{-3}	0	0
	After melting	74.76	-129.699×10^{-3}	73.681×10^{-6}	0
II	Before melting	2.081	8.443×10^{-3}	0	0
	After melting	68.171	-126.318×10^{-3}	69.798×10^{-6}	0
III	Before melting	1.359	5.819×10^{-3}	0	0
	After melting	309.388	987.959×10^{-3}	1035.697×10^{-6}	365.268×10^{-9}

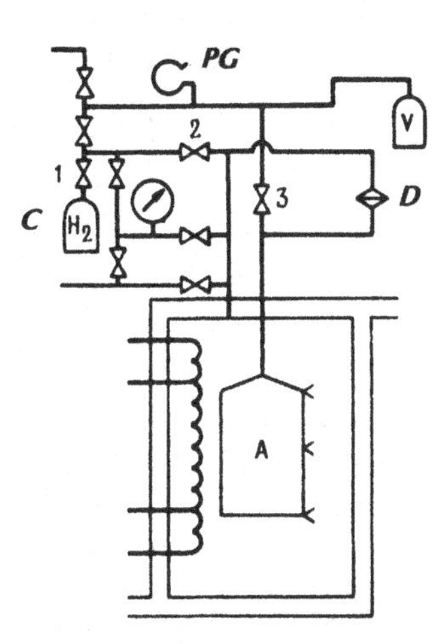

Figure 3.8. Schematic of an experimental setup for measuring the partial pressure of hydrogen in the Li-LiH system to the left of the immiscibility zone and in the flat region.

completed and published in 1977. Measurements carried out in this region at the same time by other investigators confirmed the results. This has made it possible to formulate a theory on the behavior of the partial pressure of hydrogen in the Li-LiH system at different temperatures and concentrations of the condensed phase.

To carry out an actual experiment, an apparatus which could determine the concentration of lithium hydride in the liquid phase was built (Fig. 3.8). The pressure of the dissociation products of the lithium–lithium hydride system was measured at temperatures higher than the monotectic temperature. The composition of the condensed phase of the system was determined by introducing in measured doses the hydrogen into the container with hot lithium. The molar concentration x of lithium hydride in the condensed phase was calculated from the H_2 dose:

$$x = 2n_{H_2}/n_{Li}, \tag{3.27}$$

where n_{H_2} is the hydrogen (in gram-molecules) absorbed by the condensed phase, and n_{Li} is the lithium (in gram-molecules) loaded into the container. To determine the hydrogen dose, they used a special steel vessel (V) (see Fig. 3.8), whose volume at room temperature was determined from the difference

Table 3.18. Experimental data of Ref. 77 on the vapor pressure in the Li-LiH system to the left of the immiscibility zone and in the flat region.

Point	T,K	p_{H_2}, mmHg	x,%	Point	T,K	p_{H_2}, mmHg	x,%
1	1043.3	64.5	18.6	20	1088.6	234.7	41.7
2	1040.4	61.8	18.6	21	1088.4	234.8	45.9
3	1041.7	67.4	20.0	22	1087.8	230.3	48.9
4	1043.7	84.2	23.9	23	1160.0	623.0	48.9
5	1041.2	82.2	23.9	24	1159.8	622.8	52.7
6	1043.2	102.7	28.0	25	1159.8	623.2	57.7
7	1037.2	94.0	28.0	26	1122.8	382.0	57.7
8	1042.1	112.7	32.1	27	1053.0	137.2	57.7
9	1039.6	108.8	32.1	28	984.6	42.5	57.7
10	1040.8	111.2	34.8	29	968.9	31.2	57.7
11	1042.1	116.1	34.8	30	1001.0	57.8	57.7
12	1042.1	112.4	37.6	31	1035.2	104.2	57.7
13	1038.6	106.1	37.6	32	1070.6	181.5	57.7
14	981.7	40.3	37.6	33	1106.6	304.7	57.7
15	963.3	28.6	37.7	34	1142.0	493.1	57.7
16	1006.0	62.0	37.6	35	1178.2	783.1	57.7
17	1015.0	70.2	37.6	36	1196.0	976.5	57.7
18	1015.8	71.2	37.6	37	1196.2	970.5	70.9
19	1088.4	230.2	37.6	38	1028.8	93.2	70.9

in the masses between the empty vessel and the vessel filled with water under a vacuum.

The experiment can be described as follows. Lithium was heated under a vacuum to the specified temperature. The hydrogen was then introduced in measured doses, while maintaining the isothermal regime (points 1–13 in Table 3.18), from volume V through valve 3 into the container with lithium. To eliminate losses through the heated walls of the container, hydrogen was simultaneously supplied from a cylinder (C) into the furnace through valves 1 and 2, so that the diaphragmed differential pressure gauge (D) was held at null state. The dose was monitored by a pressure gauge (PG). After valve 1 was shut upon establishing equilibrium pressure in the container and a corresponding hydrogen pressure in the furnace, the hydrogen pressure in the furnace was measured with a cathetometer and a U-shaped mercury pressure gauge, one elbow of which was connected to the furnace and the other was exposed to air. After recording the 1042-K isotherm, several operations were performed involving a change in the temperature regime and the dosage (the sequence of these operations corresponds to the order in which the experimental data are listed in Table 3.18), in which the temperature was raised and lowered. No increase in pressure during isothermal monitoring indicated that the liquid-phase composition had reached the level corresponding to the immiscibility region. The readings were taken after the temperature and pressure were stabilized. The equilibrium isothermal state was reached in 1 h on the average. All points were obtained in three steps: points 1–14 in the first set of measurements, points 15 and 16 in the second set, and points 17–38 in the third. Equation (3.27) was used to calculate the molar concentration x of liquid-phase hydride for each point. The relative error of the calculation of x

resulting from the inaccurate measurement of pressure during the dosing was within 0.1%.

Points 1–7 in Table 3.18 are situated in the homogeneous region of the Li-LiH solution slightly to the left of the immiscibility zone, and the remaining points correspond to the immiscibility zone. According to these data, the partial pressure of hydrogen P_{H_2} (mmHg) in the immiscibility zone is described by the equation (T is the temperature, K)

$$\log P_{H_2} = 21.538 - (17\,515/T). \tag{3.28}$$

Measurements at high temperatures

Experimental study of the vapor pressure above the liquid solution of lithium–lithium hydride at high temperatures is greatly complicated because of the evaporation of lithium which condenses on the cold parts of the apparatus and clogs the connecting lines. In the literature, the experimental data on the partial pressure of hydrogen above the Li-LiH system were limited by the temperature of $\sim$1200 K because of this circumstance. The only measurements carried out at higher temperatures (up to 1330 K) were those of Novikov and co-workers (see Ref. 70).

The method used by Shpil'rain *et al.*[78] to measure vapor pressures at high temperatures can be described as follows. The cell used to hold the melt was divided by a thin membrane into two parts (Fig. 3.9). The membrane was permeable to hydrogen and impermeable to lithium and lithium hydride. The lower, hermetically sealed part of the cell initially held lithium, and the upper part was connected to the gas container that was filled with hydrogen. The volume of the gas container, including the part that belongs to the cell, was measured before the experiment. The concentration of lithium hydride in the condensed phase and the corresponding equilibrium partial pressure of hydrogen were determined from the known quantity of lithium, from the known quantity of hydrogen which was transmitted through the membrane and which reacted with lithium in forming the hydride, and from the measured pressure in the gas phase in the equilibrium state at the working temperature (Table 3.19).

The temperature was measured with three tungsten–rhenium thermocouples that were placed along the height of the cell.

Comparison of the results with the data on the vapor pressure in the Li-LiH system

As mentioned above, most of the data in the literature on the measurement of the partial pressure of hydrogen in the Li-LiH system include the region up to the temperature 1200 K and pressure 10^5 Pa. The exceptions are Gibb's[126] data on measurements carried out up to the temperature of 1300 K and pressure up to 1.7×10^6 Pa (see Ref. 70) and the measurements of Novikov and co-workers carried out roughly in the same temperature range and at pressures up to 2.8×10^6 Pa (see also Ref. 70).

Analysis has shown that at large parameter values Gibb's[126] data behave irregularly [at $T=$ const, the curve $P_{H_2} = f(x)$ crosses the ordinate $x=1$] and seem to be erroneous. Novikov's data raised the level of knowledge consider-

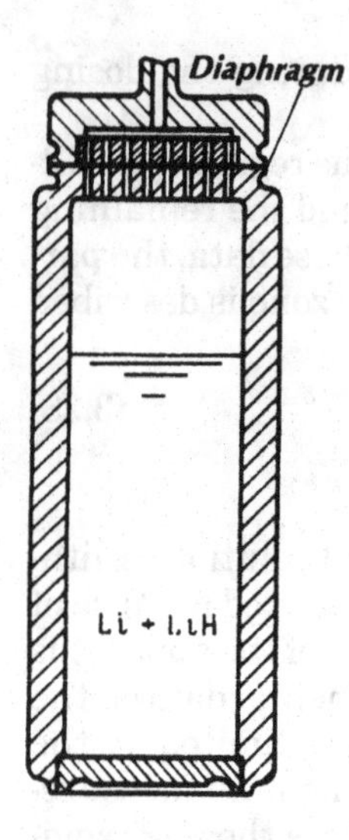

Figure 3.9. A cell for measuring the partial pressure of hydrogen in the Li-LiH system at high temperatures.

Table 3.19. Experimental data of Ref. 78 on the vapor pressure in the Li-LiH system at high temperatures.

Point	T,K	P,kg/cm²	x mole fraction of LiH
1	1609.3 ± 16	7.28 ± 0.04	0.299 ± 0.065
2	1654.5 ± 16	8.38 ± 0.04	0.298 ± 0.065
3	1609.9 ± 16	14.86 ± 0.07	0.576 ± 0.150

ably in this field at the time, but the measurement method used by him did not provide for the determination of the composition of the liquid phase which coexists with the equilibrium vapor. It is clear from the analysis of the study, nonetheless, that the measurements were carried out with use of a lithium hydride sample whose composition was very close to the stoichiometric composition. A thermodynamic method, which makes it possible to find the concentration of the liquid phase in these systems at equilibrium at the specified parameters (temperature and pressure), is proposed in Secs. 2.2 and 2.3.

Figure 3.10 shows the data on the partial pressure of hydrogen measured to the right of the immiscibility zone. In the temperature range 1000–1100 K the liquid phase is closer, in terms of lithium impurity, to the stoichiometric composition of lithium hydride in Ref. 76 than in other studies, which accounts for the higher equilibrium pressure of hydrogen in the system. The results on the vapor pressure obtained in Ref. 77 in the immiscibility zone of the Li-LiH system are in good agreement with the data of Veleckis.[236]

Of particular interest are the measurements at high temperatures carried out in Ref. 78. Although little data were obtained in this experiment (only three data points) because of extraordinary experimental difficulties, these three data points nonetheless nearly doubled, in terms of the temperature, the previously studied range of the equilibrium state of the Li-LiH system and were used to predict the behavior of the partial pressure of hydrogen as a

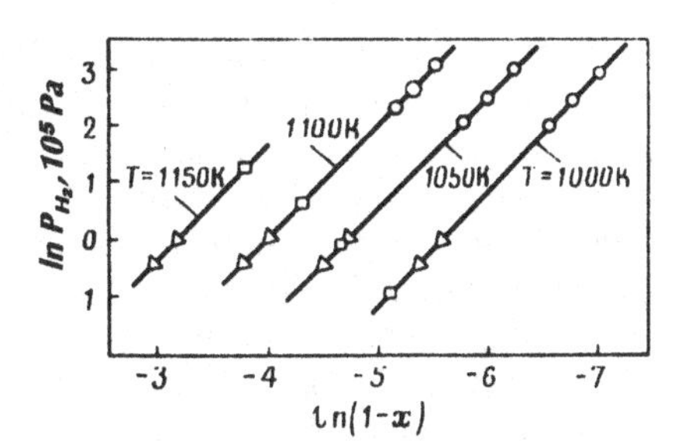

Figure 3.10. Comparison of the experimental data on the partial pressure of hydrogen in the Li-LiH system in the liquid phase at high concentrations of lithium hydride. △—Data of Blander *et al.*[103]; □—data of Novikov *et al.* (see Ref. 70); ○—data of Shpil'rain *et al.*[76]

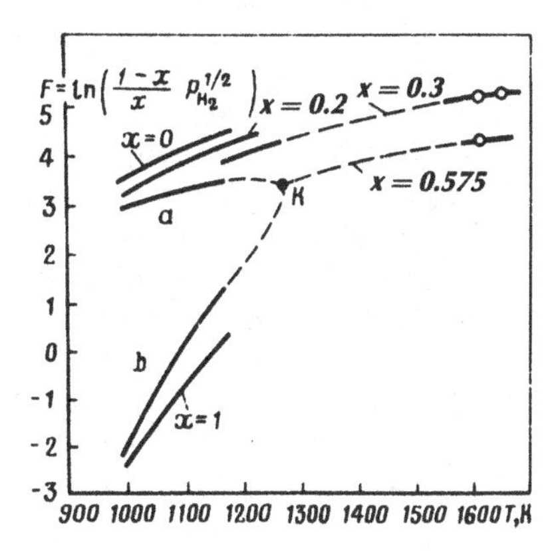

Figure 3.11. Comparison of the experimental data of Shpil'rain *et al.*[78] on the partial pressure of hydrogen in the Li-LiH system at high temperatures with the low-temperature data found in the literature (curves *a* and *b* are respectively the left and right boundaries of the immiscibility zone; K is the critical melting point; $P_c = 2.6$ kg/cm^2; $T_c = 1267$ K; and $x_c = 0.572$).

function of the temperature and composition of the liquid phase at these large parameter values.

Figure 3.11 is a phase diagram of the Li-LiH system plotted in the coordinates $\ln\{[(1-x)/x]\,P_{H_2}^{1/2}\} = f(T,x)$ on the basis of a method described in Secs. 2.2 and 2.3. We see from this figure that the results obtained at high temperatures in Ref. 78 are in satisfactory agreement with the data obtained at lower temperatures reported in the literature.

2.2. The procedure for calculating the activities of the components and other thermodynamic characteristics of Li-LiH solutions from plots of the partial pressure of hydrogen versus composition of the liquid phase

Li-LiH solutions are identified principally by the presence of the dissociation reaction in the vapor phase [see Eq. (3.8)] and also by the presence of the immiscibility zone in the liquid phase. In the concentration region in which the fraction of hydride in liquid lithium is very small, several other reactions, which account for the formation of Li_2 and Li_2H molecules, can also be seen in the vapor phase. The extensive collection of recently accumulated experimental data on the partial pressure of hydrogen isotopes above Li-LiH solutions, as functions of temperature and concentration, has made it possible to calculate quite accurately the thermodynamic characteristics of these systems. Such attempts have already been made, but in their efforts, the authors have made several assumptions that impose certain constraints on the common features of the suggested calculation methods. In calculating the activity of the components, Ihle and Wu,[153] for example, used the Hildebrand–Scott[146] equation, which can be used for regular solutions, i.e., for a class of solutions for which the condition $h_{exc} \gg Ts_{exc}$ holds (here, h_{exc} and s_{exc} are, respectively, the excess heat of displacement and the excess entropy of displacement, and T is the absolute temperature). This theory works best for binary systems in which the molecules of the components are nonpolar and spherical in shape (see Ref. 66). Since Ihle and Wu did not have at their disposal any experimental or theoretical data on the caloric functions for the Li-LiH solutions, there was no sufficient justification at the outset to apply the theory of regular solutions to these systems.* Additionally, the hydride molecule of an alkali metal is known to be polar and clearly nonspherical.

Blander *et al.*[102,103] have measured the partial pressures of hydrogen and deuterium above Li-LiH and Li-LiD solutions, respectively, over broad temperature and concentration intervals and analyzed the results using the Margules equation, which expresses the concentration dependence of the logarithm of the activity coefficient in terms of an integral power series. Analysis has shown that this evaluation does not meet the requirements of certain limiting conditions for the concentration dependence of the activity coefficient of the components of the solution. Cherkasov and Pankrat'eva[65] carried out essentially a simple extrapolation of the equations, suggested by Blander *et al.*,[102,103] to lower temperatures, without introducing any additional physical considerations.

Accordingly, a method for calculating the activity coefficients of the components of such solutions was developed (see Ref. 82). This method, which eliminated the need of specifying the form of the equation for the concentration dependence, will be described for the particular case of the Li-LiH system.

The problem can be studied in the following way. We have a two-phase equilibrium liquid-vapor system: In the liquid phase, the actual solution is Li-LiH and the vapor phase is an ideal mixture of ideal Li, LiH, and H_2 gases that interact through the reaction (3.8).† The dependence of the partial pressure of hydrogen on the composition of the liquid phase at a constant temperature has been determined experimentally. It is clear that this thermodynamic system has only two independent components, and since the problem is considered under phase equilibrium conditions, the system has two degrees of freedom. As independent variables, we will consider the equilibrium temperature of the system and the concentration of lithium hydride in the liquid phase.

For the reaction (3.8), the equation for the balance of the chemical potentials of the interacting components can be written in the form

$$\mu_2'' = \mu_1'' + \tfrac{1}{2}\mu_3, \tag{3.29}$$

*At temperatures of 1000 K and higher h_{exc} is, as indicated in Chap. 3, Sec. 2.5, similar in magnitude to Ts_{exc} for the systems under consideration.

†The other reactions in the gas phase that were mentioned above have no effect on the following discussion.

where μ_1'', μ_2'', and μ_3 are the chemical potentials of lithium, lithium hydride, and molecular hydrogen, respectively, in the gas phase. The chemical potentials of the components of the liquid solution can be written in the form

$$\mu_1' = \mu_1^{0'} + RT \ln \gamma_1(1-x) \tag{3.30}$$

for lithium and

$$\mu_2' = \mu_2^{0'} + RT \ln \gamma_2 x \tag{3.31}$$

for lithium hydride. Here $\mu_1^{0'}(T)$ and $\mu_2^{0'}(T)$ are the chemical potentials of pure liquid lithium and pure liquid lithium hydride, respectively, when the liquid and vapor are at phase equilibrium; $\gamma_1(T,x)$ and $\gamma_2(T,x)$ are the activity coefficients of these components in the solution; x is the mole fraction of lithium hydride; T is the equilibrium temperature; and R is the gas constant.

An essential feature of systems of this type is that at $T = \text{const}$ and $x \to 1$, the partial pressure of hydrogen $P_3 \to \infty$. Accordingly, the stoichiometric composition of lithium hydride is essentially unattainable in the condensed phase. Equation (3.31), written in the form

$$\mu_2^{0'} = \lim_{x \to 1}(\mu_2' - RT \ln \gamma_2 x), \tag{3.32}$$

in this case should be considered as an equation which formally determines the chemical potential of pure lithium hydride. Here we mean that

$$\lim_{x \to 1} \gamma_2 \to 1.$$

Since the relations $\mu_1'' = \mu_1'$ and $\mu_2'' = \mu_2'$ should hold at phase equilibrium, Eq. (3.29) can be transformed to the form

$$\mu_2^{0'} + RT \ln \gamma_2 x = \mu_1^{0'} + RT \ln \gamma_1(1-x) + \tfrac{1}{2}\mu_3^0 + \tfrac{1}{2} RT \ln P_3, \tag{3.33}$$

where $\mu_3^0(T)$ is the chemical potential of pure hydrogen at standard pressure $P_3^0 = 1$ atm (1 kgf/cm^2).

This equation can also be written in the form

$$K = \frac{\gamma_1(1-x)}{\gamma_2 x}\left(\frac{P_3}{P_3^0}\right)^{1/2} = \frac{\gamma_1(1-x)}{\gamma_2 x} P_3^{1/3}, \tag{3.34}$$

where

$$K = \exp[(\mu_2^{0'} - \mu_1^{0'} - \tfrac{1}{2}\mu_3^0)/RT]. \tag{3.35}$$

Here K is the equilibrium constant of the chemical reaction (3.8), and P_3 is the pressure, kgf/cm^2.

In our discussion below, we will divide, for convenience, the entire concentration range $0 \leqslant x \leqslant 1$ into three parts. The initial part, from $x = 0$ to $x \simeq 0.1$, spans the region in which pure lithium vapor dominates the vapor phase, but other molecules like LiH and H_2 are present in appreciable quantities. Furthermore, an important feature of this region is that the partial pressure of hydrogen satisfies, as has been found experimentally (see Ref. 217, for example), the Sieverts equation:

$$P_3^{1/2} = Cx, \tag{3.36}$$

where C is the temperature-dependent Sieverts constant.

This relation can also be obtained thermodynamically. In this connection, we need only consider the system in the state of infinitely dilute solution when lithium is a solvent and lithium hydride is a dissolved component. Maintaining equilibrium isothermal conditions in the system as $x \to 0$, we can write the following expression on the basis of Eq. (3.33):

$$\mu_2^* + RT\ln x = \mu_1^{0'} + RT\ln(1-x) + \tfrac{1}{2}\mu_3^0 + \tfrac{1}{2}RT\ln P_3. \tag{3.37}$$

Here $\mu_2^*(T)$ is the standard chemical potential of the dissolved component,

$$\mu_2^*(T) = \mu_2^{0'} + RT\ln\, \dot{\gamma}_2, \tag{3.38}$$

where

$$\dot{\gamma}_2 = \lim_{x\to 0} \gamma_2.$$

The activity coefficient of the solvent in this case is $\gamma_1 = 1$. Equation (3.37) implies that

$$P_3^{1/2} = C[x/(1-x)]. \tag{3.39}$$

Since $x \ll 1$, we can write this equation in the form (3.36), where

$$C = \exp[(\mu_2^* - \mu_1^{0'} - \tfrac{1}{2}\mu_3^0)/RT] \tag{3.40}$$

is a function of only the temperature at phase equilibrium.*

It has been suggested that at low concentrations, hydrogen in a solution does not exist in the form of hydride but rather as atomic or even ionic hydrogen (see Ref. 144). However, this does not alter Eq. (3.36) in any way. Furthermore, Privalov[50] has shown experimentally with the use of a calorimeter that hydrogen in a Na-H solution exists in the form of sodium hydride. Accordingly, there are justifiable reasons to assume that hydrogen even in a Li-H solution exists in the form of lithium hydride.

The main concentration region is one in which the vapor-phase molecular hydrogen is dominant without an excessively high pressure. By assuming that $P_3 \lesssim 10$ atm, we find that the main concentration region lies between $x_1 \simeq 0.1$ and x_2 0.96–0.995, depending on the temperature. This region also includes the immiscibility zone.

Using the general equation (3.34), we obtain the following expression for the interval $0 \leqslant x \leqslant x_2$:

$$\left(\frac{\partial \ln \gamma_2/\gamma_1}{\partial x}\right)_T = \left(\frac{\partial F}{\partial x}\right)_T, \tag{3.41}$$

where

$$F(T,x) = \ln\left(\frac{1-x}{x} P_3^{1/2}\right). \tag{3.42}$$

From a comparison of Eqs. (3.39) and (3.42) we see that

$$\lim_{T=\text{const},\, x\to 0} F(T,x) = \ln C.$$

Using the general Gibbs–Duhem equation

*Although the constant C does not have a dimensionality, the dimensionality-maintaining rule in Eq. (3.39) is not broken, since the left-hand side of this equation, strictly speaking, must be written in the form $[P_3/(P_3^0 = 1\text{ atm})]^{1/2}$, as Eq. (3.33) implies. This remark also pertains to Eq. (3.36).

$$(1-x)\left(\frac{\partial \ln \gamma_1}{\partial x}\right)_{P,T}+x\left(\frac{\partial \ln \gamma_2}{\partial x}\right)_{P,T}=0, \tag{3.43}$$

from Eq. (3.41) we find (ignoring the pressure)

$$\left(\frac{\partial \ln \gamma_1}{\partial x}\right)_T=-x\left(\frac{\partial F}{\partial x}\right)_T \tag{3.44}$$

and

$$\left(\frac{\partial \ln \gamma_2}{\partial x}\right)_T=(1-x)\left(\frac{\partial F}{\partial x}\right)_T. \tag{3.45}$$

Integrating Eq. (3.44) by parts and taking into account that $\gamma_1 = 1$ at $x = 0$, we find

$$\ln \gamma_1 = -xF + \int_0^x F\,dx. \tag{3.46}$$

The activity coefficient of lithium hydride can be determined by taking this relation from Eq. (3.34) into account:

$$\ln \gamma_2 = (1-x)F + \int_0^x F\,dx - \ln K. \tag{3.47}$$

Since the experimental data on the partial pressure of hydrogen as a function of the composition of the solution in the Li-LiH system suggest that the Sieverts equation is valid over a broad concentration range, Eq. (3.37) must also be valid in this concentration range. At the same time, analysis of the experimental data on the partial pressure of hydrogen in the Li-LiH system[103] and in the Li-LiD system[234] shows (see Sec. 2.3) that the activity coefficient of lithium in the concentration region to the left of the immiscibility zone is described at $T=$ const by a power series of the type

$$\ln \gamma_1 = ax^n + bx^m, \tag{3.48}$$

where $m>n>2$.

From these arguments, we conclude that in the system under consideration

$$\lim_{x\to 0}\left(\frac{\partial \ln \gamma_1}{\partial x}\right)_T=0 \tag{3.49}$$

and [when combined with Eqs. (3.39), (3.42), and (3.45)]

$$\lim_{x\to 0}\left(\frac{\partial \ln \gamma_2}{\partial x}\right)_T=0. \tag{3.50}$$

These equations are clearly in agreement with Eq. (3.37).

In the Li-LiH system, the concentration region $x_2 < x < 1$ is characterized by a considerable increase of the equilibrium pressure as x is increased. If in this concentration region the liquid phase is represented as an infinitely dilute solution of lithium in lithium hydride, then by analogy with Eq. (3.37) we can write

$$\mu_2^{0'} + RT\ln x = \mu_1^* + RT\ln(1-x) + \tfrac{1}{2}\mu_3^0 + \tfrac{1}{2}RT\ln P_3. \tag{3.51}$$

Here $\mu_1^*(T)$ is the standard chemical potential of lithium

$$\mu_1^*(T) = \mu_1^{0'}(T) + RT\ln \dot{\gamma}_1, \tag{3.52}$$

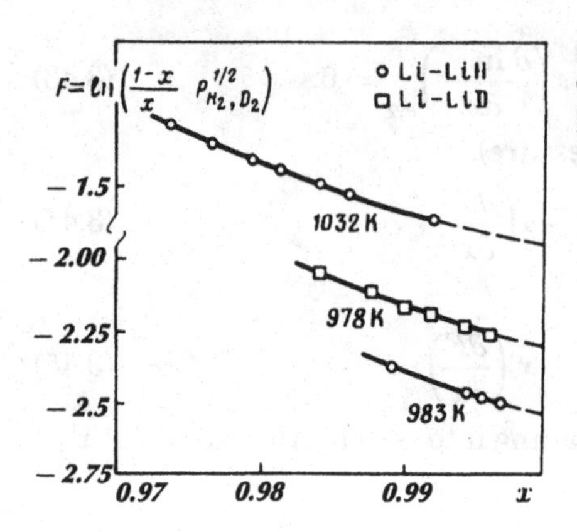

Figure 3.12. Partial pressure of hydrogen and deuterium versus the composition of the liquid phase as x→1, based on the data of Veleckis[235] and Veleckis *et al.*[237]

where

$$\dot{\gamma}_1 = \lim_{x \to 1} \gamma_1 .$$

As noted above, at phase equilibrium the partial pressure of hydrogen P_3 at $T=$ const tends toward infinity as $x \to 1$. The same situation occurs in the Li-LiH system (see Fig. 3.2). To find practical solutions under these conditions, we can use as the limiting state of an equilibrium Li-LiH system (instead of the hypothetical state of a stoichiometrically pure liquid lithium hydride, where $x=1$ and $P_3 = \infty$) the state with $x = 1 - \epsilon$, where $\epsilon \ll 1$ and is dependent upon equilibrium pressures of hydrogen that are actually attainable under real conditions. The equilibrium pressure depends on the temperature of the system. If 50 atm is arbitrarily assumed to be the limiting pressure, then a rough estimate based on the experimental data yields $\epsilon < 10^{-4}$ at a temperature of 700 °C and $\epsilon < 10^{-3}$ at a temperature of 900 °C.

Analysis of the experimental data on the partial pressure of hydrogen in the Li-LiH system at large concentrations of lithium hydride beyond the right boundary of the immiscibility zone shows that within the limits of the attainable state variables P, T, and x, the function $F(T,x)$ tends at $T=$ const to a particular limit (Fig. 3.12) as $x \to 1$. By assuming that this function behaves in a similar manner in the limit $x \to 1$ even outside the concentration range studied and using Eq. (3.51), we can write (at least in the limits up to $x = 1 - \epsilon$)

$$\lim_{x \to 1} F = \lim \ln\left[\frac{1-x}{x}\left(\frac{P_3}{P_3^0}\right)^{1/2}\right] = \ln H. \tag{3.53}$$

Equation (3.53) implies that in the limit $x \to 1$ ($P_3^0 = 1$ kg/cm^2)

$$(1-x)P_3^{1/2} = Hx, \tag{3.54}$$

where

$$H = \exp[(\mu_2^{0'} - \mu_1^* - \tfrac{1}{2}\mu_3^0)/RT]. \tag{3.55}$$

Here H is a function of temperature. The coefficient H is thus a unique analog of the Sieverts constant (concerning the dimensionality of H, see the footnote on page 172).

From Eqs. (3.35), (3.38), (3.40), (3.52), and (3.55) we see that

$$\text{for } x \to 0 \quad \dot{\gamma}_2 = C/K; \tag{3.56}$$

$$\text{for } x \to 1 \quad \dot{\gamma}_1 = K/H. \tag{3.57}$$

By extending Eq. (3.47) to the third concentration region* we will obtain the following interesting corollary, since

$$\lim_{x \to 1} \gamma_2 = 1$$

and

$$\lim_{x \to 1}(1-x)F = \lim_{x \to 1} \ln(\{[(1-x)/x]P_3^{1/2}\}^{1-x}) = 0;$$

$$\int_0^1 \ln\left(\frac{1-x}{x} P_3^{1/2}\right) dx = \ln K. \tag{3.58}$$

In sum, this result is first a direct proof of the existence of an integral and an indirect proof that Eq. (3.51) can legitimately be regarded as the result of taking the limit of Eq. (3.33) with $x \to 1$ at $T =$ const and, secondly, it gives rise to an interesting possibility of determining the thermodynamic properties of the equilibrium constant K. The relations obtained above and the conclusions drawn from them follow from the specific features of the types of systems considered and can be explained principally by the presence of an equilibrium vapor phase with a condensed binary solution of the heterogeneous chemical reaction between the components of the solution. This situation, in particular, has made it possible to use the equations for infinitely dilute solutions in the limiting cases ($x \to 0$ and $x \to 1$) and to obtain the characteristic features.

These circumstances were ignored by Veleckis[235] and Veleckis *et al.*,[237] causing certain errors in the analysis of the data on the partial pressure of hydrogen and deuterium in the Li-LiH and Li-LiD systems. In their studies,

$$\lim_{x \to 0}(\partial \ln \gamma_2 / \partial x)_T,$$

for example, depends, in contrast with Eq. (3.50), on the second component of the system (i.e., LiH or LiD) and on the temperature. Specifically, for the Li-LiH system it was found that

$$\lim_{x \to 0}(\partial \ln \gamma_2 / \partial x)_T = (-1.2) - (-1.5)$$

in the temperature interval 710–900 °C.

*The principal error in this equation for this concentration region is in disregarding the term of order $(v_2^0 - v_2)(\partial P/\partial x)_T/RT$ on going from Eq. (3.43) to Eqs. (3.44) and (3.45) (v_2^0 and v_2 are, respectively, the molar volume and the partial molar volume of pure lithium hydride). Experimental data have shown that, within the limits of the pressure and concentration regions indicated above, the value of $(\partial P/\partial x)_T/RT$ is 2×10^{-5} mole/cm^3 at a temperature of 1100 K.

Table 3.20. Coefficients in Eqs. (3.59) and (3.60).

T, °C	C	H	b	d	B
		Li-LiH			
710	33.44	0.0794	−2.360	−4.020	38.52
759	44.55	0.1840	−2.320	−2.493	31.52
803	58.40	0.3366	−2.376	−2.285	25.28
847	73.40	0.6353	−2.355	−1.762	19.14
878	86.27	1.1106	−2.522	−1.068	13.03
903	95.75	1.7611	−2.487	−1.105	9.20
		Li-LiD			
705	40.45	0.0996	−2.194	−5.388	37.73
756	52.52	0.1738	−1.726	−5.918	43.00
805	72.61	0.3789	−2.692	−1.630	28.66
840	86.84	0.6769	−3.182	0.754	19.57
871	94.10	0.9930	−2.484	−0.951	16.18

2.3. Determination of the activity coefficients, the functions *C(T)* and *H(T)*, and the equilibrium constant for the reaction $LiH_l \rightleftarrows Li_l + \frac{1}{2}H_{2,gas}$

The experimental data on the vapor pressure at phase equilibrium in the Li-LiH system[103] and on the vapor pressure in the Li-LiD system[234] were analyzed on the basis of the relations obtained in Sec. 2.2. This analysis has made it possible to obtain simpler equations that satisfy the thermodynamic conditions indicated in the preceding section. Since the Li-LiH systems have an immiscibility zone, for specific calculations, the concentration interval $0 \leqslant x \leqslant 1$ can be divided into three parts: left-hand part $0 \leqslant x \leqslant x'$, immiscibility zone $x' \leqslant x \leqslant x''$, and right-hand part $x'' \leqslant x < 1$. Here x' and x'' are, respectively, the left and right boundaries of the immiscibility zone. The position of these boundaries was determined from the data of Refs. 235 and 237.

A mathematical analysis (see Ref. 84) of the experimental data of Refs. 103 and 234 by the method of least squares on the basis of the function $P_3 = f(x)$ showed that in the case of the Li-LiH and Li-LiD systems, at $T=$ const, the data of the left part of the concentration region are best described [within $\pm$ (1–2)% scatter] by the equation

$$F = \ln C + bx^{1.2} + dx^{2.4}, \tag{3.59}$$

and the data of the right part are best described by the equation

$$F = \ln H + B(1-x)^{1.2}. \tag{3.60}$$

Here $F = \ln\{[(1-x)/x]P_3^{1/2}\}$ (P_3 is in mmHg).

The results of the analysis are presented in Table 3.20 and in Figs. 3.13 and 3.14. In Fig. 3.13, we see that the experimental data for the Li-LiD system is not quite on a par with the experimental data for the Li-LiH system, which is fully consistent with the higher degree of difficulty of the analysis of the Li-LiD system.

For the concentration region $0 \leqslant x \leqslant x'$, we find from Eqs. (3.46) and (3.47), with allowance for Eq. (3.59), the expressions

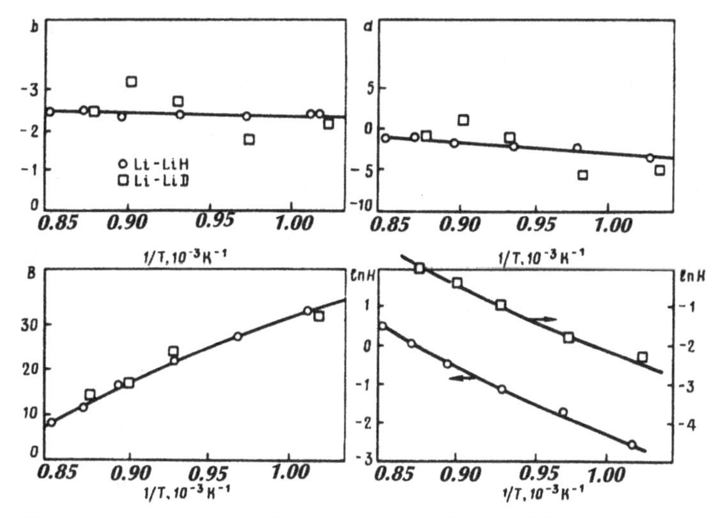

Figure 3.13. The temperature dependence of the coefficients *b, d, B,* and *H* for the Li-LiH system (○) and the Li-LiD system (□) [— –Eq. (3.71) for the coefficients *b* and *d* and Eq. (3.72) for *B* and ln *H*].

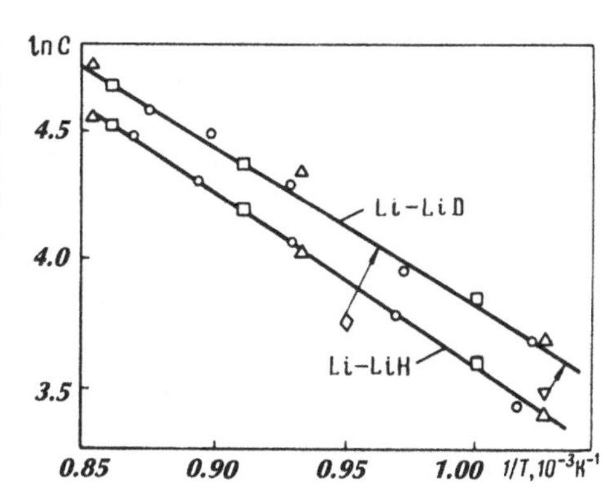

Figure 3.14. The temperature dependence of the Sieverts constant, based on the data of Blander, Veleckis *et al.* (□) (Ref. 234), the data of Smith et al. (△) (Ref. 215), the data of Goodal and McCracken (▽) (Ref. 128), and the data of Ihle and Wu (◇) (Ref. 155); Table 3.20 (○).

$$\ln \gamma_1 = -0.5455\, bx^{2.2} - 0.706\, dx^{3.4}; \tag{3.61}$$

$$\ln \gamma_2 = bx^{1/2} - 0.5455\, bx^{2.2} + dx^{2.4} - 0.706 dx^{3.4} + \ln \dot{\gamma}_2. \tag{3.62}$$

In the immiscibility zone ($x' \leqslant x \leqslant x''$), we find from the general thermodynamic considerations that

$$\ln \gamma_1 = \ln \gamma_1' \ln \frac{1-x'}{1-x} = \ln \gamma_1'' + \ln \frac{1-x''}{1-x} \tag{3.63}$$

and

$$\ln \gamma_2 = \ln \gamma_2' + \ln \frac{x'}{x} = \ln \gamma_2'' + \ln \frac{x''}{x}. \tag{3.64}$$

In the region $x'' \leqslant x < 1$, we find from Eqs. (3.44), (3.15), and (3.53) the expressions

Table 3.21. Results of calculations of the activity coefficients and the equilibrium constant for the Li-LiH and Li-LiD systems.

T,°C	el $\dot{\gamma}_1$	el $\dot{\gamma}_2$	K'	K''
		Li-LiH		
710	4.200	1.874	5.13	5.29
759	3.753	1.765	7.63	7.85
803	3.461	1.714	10.52	10.72
847	3.129	1.632	15.35	15.52
878	2.773	1.587	17.65	17.78
903	2.453	1.543	20.47	20.47
		Li-LiD		
705	4.065	1.918	5.94	5.80
756	3.939	1.979	8.71	8.93
805	3.523	1.749	12.63	12.84
840	3.183	1.809	14.23	16.33
871	2.960	1.581	19.34	19.16

$$\ln \gamma_1 = \ln\dot{\gamma}_1 - Bx(1-x)^{1.2} - 0.4545B(1-x)^{2.2}; \tag{3.65}$$

$$\ln \gamma_2 = 0.5455\, B(1-x)^{2.2}. \tag{3.66}$$

Equations (3.61), (3.63), and (3.65) can be used to determine $\dot{\gamma}_1$:

$$\ln \dot{\gamma}_1 = \ln \frac{1-x'}{1-x''} 0.5455b(x')^{2.2} - 0.706d(x')^{3.4}$$

$$+ Bx''(1-x'')^{1.2} + 0.4545B(1-x'')^{2.2} \tag{3.67}$$

and Eqs. (3.62), (3.64), and (3.66) can be used to determine $\dot{\gamma}_2$:

$$\ln \dot{\gamma}_2 = \ln \frac{x''}{x'} + 0.5455B(1-x'')^{2.2} - b(x')^{1.2}$$

$$+ 0.5455b(x')^{2.2} - d(x')^{2.4} + 0.706d(x')^{3.4}. \tag{3.68}$$

Using Eqs. (3.56) and (3.57), we can determine the equilibrium constant K by two methods from these values of $\dot{\gamma}_1$ and $\dot{\gamma}_2$ and from $C(T)$ and $H(T)$ found by an independent analysis of the experimental data. The calculation results presented in Table 3.21 show that the agreement between K' [calculated from Eq. (3.56)] and K'' [calculated from Eq. (3.57)] is satisfactory.

Figure 3.15 shows the temperature dependence of $\dot{\gamma}_1$ and $\dot{\gamma}_2$ for the Li-LiH and Li-LiD systems. We see that within the spread of the initial data there is no difference between the activity coefficients of $\dot{\gamma}_1$ and of $\dot{\gamma}_2$ in the Li-LiH and Li-LiD systems.

Analyses[217,235,237] of the data in the literature on the temperature dependence of the concentrations at the left and right boundaries of the immiscibility zone for the Li-LiH and Li-LiD systems at temperatures above the monotectic temperature show that the locations of the boundaries coincide within experimental error. A joint analysis of these data led to the following equations:

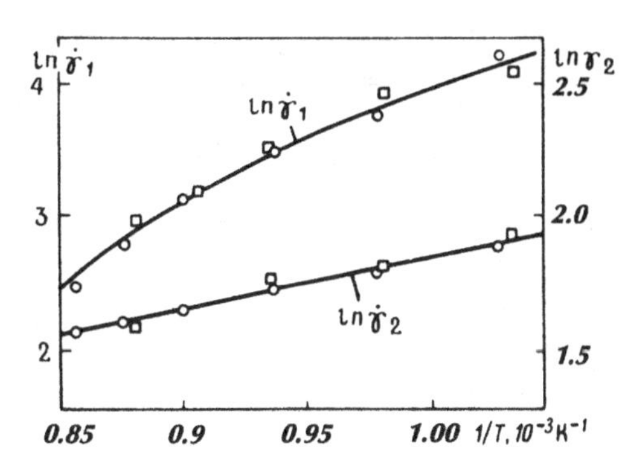

Figure 3.15. The temperature dependence of $\dot{\gamma}_1$ and $\dot{\gamma}_2$ for the Li-LiH system (O) and the Li-LiD system (□), based on the data of Table 3.21 and Eqs. (3.71) and (3.72). Solid line—with use of coefficients from Table 3.23.

Table 3.22. Comparison of experimental and calculated data at the boundaries of the immiscibility zone in the Li-LiH and Li-LiD systems.

T,°C	x'_{exp}	x'_{calc}	$\delta x'$,%	x''_{exp}	x''_{calc}	$\delta x''$,%
			Li-LiH			
710	0.252	0.289	+5.4	0.984	0.980	+0.4
759	0.285	0.287	−0.7	0.971	0.977	−0.6
803	0.332	0.334	−0.6	0.957	0.956	+0.1
847	0.386	0.383	+0.8	0.923	0.921	+0.2
878	0.418	0.419	−0.2	0.888	0.890	−0.2
903	0.454	0.450	+0.9	0.856	0.861	−0.8
800	0.330	0.330	0.0	0.960	0.958	+0.2
			Li-LiD			
705	0.228	0.235	−0.3	0.980	0.978	+0.1
756	0.280	0.284	−1.4	0.972	0.978	−0.6
805	0.332	0.336	−1.2	0.953	0.954	−0.2
840	0.377	0.375	+0.5	0.931	0.927	+0.4
871	0.411	0.411	0.0	0.905	0.897	+0.9

$$\ln x' = 2.416 - (3780/T) \tag{3.69}$$

for the left boundary and

$$\ln x'' = 85.235 - (10\,803/T) - 10.778 \ln T \tag{3.70}$$

for the right boundary.

Table 3.22 compares the calculated values at the boundaries of the immiscibility zone with the values obtained experimentally. We see that there is essentially no discrepancy between the positions of the boundaries for the Li-LiH and Li-LiD systems.

The above circumstance, the agreement between $\dot{\gamma}_1$ and $\dot{\gamma}_2$ for the Li-LiH (LiD) systems and the similarity between the concentration dependences of the activity coefficients of the components of these systems [see Eqs. (3.61), (3.62), (3.65), and (3.66)] all lead to the conclusion that the temperature and concentration dependences of the activity coefficients of the corresponding components of the Li-LiH and Li-LiD systems are essentially the same over the entire concentration interval $0 \leqslant x < 1$ within experimental error.

The data in Tables 3.20 and 3.21 were analyzed by the method of least

Table 3.23. Coefficients in Eqs. (3.71) and (3.72).

Y	System	α	β	δ	Scatter,%
b	Li-LiH (LiD)	– 3.281	952.4	—	3
d	Li-LiH (LiD)	14.396	– 17 875	—	15
B	Li-LiH (LiD)	9574	– 459 294	– 590	3
$\ln \dot{\gamma}_1$	Li-LiH (LiD)	416.45	– 47 325	– 52.845	2.5
$\ln \gamma_2$	Li-LiH (LiD)	0.131	1 979	—	2.5
$\ln C$	Li-LiH	9.989	– 6 376	—	0.5
	Li-LiD	9.785	– 5 958	—	1
$\ln H$	Li-LiH	– 342.453	30 087	44.89	5
	Li-LiD	– 470.399	48 282	60.81	5
$\ln K$	Li-LiH	10.007	– 8 217	—	1
	Li-LiD	10.028	– 8 075	—	1

squares. For the quantities b,d, ln C, and ln K, the temperature dependence is

$$Y = \alpha + \beta/T; \tag{3.71}$$

while the expression

$$Y = \alpha + \beta/T + \delta \ln T \tag{3.72}$$

was used for the functions B and ln H.

Since the activity coefficients of the corresponding components of the Li-LiH and Li-LiD systems are the same, and since the errors of the data of Ref. 234 for the Li-LiD system are, as indicated above, slightly higher, only the data of Ref. 103 for the Li-LiH system were used to approximate the coefficients b, d, and B.

The values of K' and K'' were analyzed jointly for each system (Li-LiH and Li-LiD). This approach has made it possible, in principle, to improve the accuracy of the equation for the equilibrium constant K by drawing upon a larger experimental data base, i.e., the data on the boundaries of the immiscibility zone plus the experimental measurements of the partial pressure of hydrogen and deuterium to the left and to the right of the immiscibility zone.

The coefficients of Eqs. (3.71) and (3.72) are given in Table 3.23. For the activity coefficients $\dot{\gamma}_1$ and $\dot{\gamma}_2$, the equations for the temperature dependence were chosen so that they would correspond to Eqs. (3.72) and (3.71), respectively. This process involved the use of Eqs. (3.56) and (3.57) and appropriate averaging of the coefficients. The agreement between the equations obtained in this manner and the initial data is illustrated in Fig. 3.15. A comparatively large scatter in the data obtained in the approximation of the coefficient d is not important, since the corresponding terms of the equations that contain this coefficient are an order of magnitude smaller than the other terms.

This comparative analysis of the thermodynamic properties of the Li-LiH and Li-LiD systems has not only made it possible to obtain simple standard working relations for determining the thermophysical characteristics of these systems but also led to finding similar properties of the Li-LiT system for which there are virtually no experimental or theoretical data available in the literature. Attention can be called to only the following studies: a study by Heuman and Salmon,[144] which has many errors; the experimental studies of

Table 3.24. Test of Eq. (3.73).

T,K	$(P_{H_2}/P_{D_2})^{1/2}$	K_{LiH}/K_{LiD}	C_{LiH}/C_{LiD}	$\delta(K/P)$,%	$\delta(C/P)$,%
1000	0.835	0.850	0.791	1.0	−5.3
1100	0.856	0.861	0.823	0.6	−3.9
1200	0.875	0.870	0.848	−0.6	−3.1

Smith *et al.*,[215–217,219] in which the results of measurements of the Sieverts constant in the Li-LiH (LiD, LiT) systems are presented; and the data of Veleckis[236] on the measurement of the partial pressure of tritium above the Li-LiT solution in the immiscibility zone.

Because of the equality of the activity coefficients of the corresponding components of the Li-LiH and Li-LiD systems, and because the boundaries of the immiscibility zone of these systems are the same, it can be assumed that similar conclusions also apply to the Li-LiT system, since general thermodynamic considerations show that the difference in the thermodynamic properties for the Li-LiD and Li-LiT systems is smaller than the difference in these properties for the Li-LiH and Li-LiD systems. From this we can obtain, on the basis of Eqs. (3.34) and (3.56), the relation (the numerator gives the values for the Li-LiH system and the denominator gives the values for the Li-LiD system)

$$\ln \frac{C_{\mathrm{LiH}}}{C_{\mathrm{LiD}}} = \ln \frac{K_{\mathrm{LiH}}}{K_{\mathrm{LiD}}} = \tfrac{1}{2} \ln \frac{P_{\mathrm{H_2}}}{P_{\mathrm{D_2}}}. \tag{3.73}$$

The use of this equation for the boundaries of the immiscibility zone gives rise to an interesting possibility involving independent measurements of the partial pressure of hydrogen, deuterium, and tritium, respectively, for the Li-LiH, Li-LiD, and Li-LiT systems in the flat region (see Sec. 2.1). Using these data, we can find from Eq. (3.73) the Sieverts constant C and the equilibrium constant K for the Li-LiT system.

The temperature dependence of the partial pressure P (mmHg) of the hydrogen isotopes of the systems under consideration in the flat region at temperatures above the monotectic temperature, according to the data of Ref. 236, is

$$\ln P_{\mathrm{H_2}} = 21.34 - (17\,420/T); \tag{3.74}$$

$$\ln P_{\mathrm{D_2}} = 21.15 - (16\,870/T); \tag{3.75}$$

$$\ln P_{\mathrm{T_2}} = 20.92 - (16\,530/T). \tag{3.76}$$

The values of $(P_{H_2}/P_{D_2})^{1/2}$ found from these equations are compared in Table 3.24 with the ratios C_{LiH}/C_{LiD} and K_{LiH}/K_{LiD} obtained from Table 3.22.

For the equilibrium constants, the agreement is good. The agreement is slightly worse for the Sieverts constants, but allowing for the absence of reliable data for the Li-LiT system, the agreement seems quite satisfactory for calculating the properties of this system.

Using the available data of Ref. 236 on P_{H_2}, P_{D_2}, and P_{T_2}, the equilibrium constant K and the Sieverts constant C for the Li-LiH (LiD) systems (see Table

3.23), we can obtain the following relations for the Li-LiT system:

$$\ln K = 9.797 - (7772/T); \quad (3.77)$$

$$\ln C = 9.670 - (5780/T). \quad (3.78)$$

These equations are of vital importance, because in practical applications involving the use of tritium and lithium tritide, knowing $K(T)$ and $C(T)$ makes it possible to determine the partial pressure of tritium outside the immiscibility zone, the caloric properties of liquid LiT (see Sec. 1.4), and several other thermodynamic characteristics which cannot otherwise be obtained.

2.4. Component partial pressures, total pressures, and vapor compositions of the Li-LiH (LiD, LiT) systems when the liquid phase is in equilibrium with the vapor phase

The results of theoretical calculations have made it possible to determine the partial pressures of the components when the vapor is at equilibrium with the solution in the Li-LiH (LiD, LiT) systems under the assumption that the vapor phase is an ideal phase. As Ihle and Wu[153–155] have shown experimentally, at low hydride concentrations (in the particular case of lithium hydride), the important reactions in the vapor phase, in addition to the dissociation reaction

$$LiH \rightleftarrows Li + \tfrac{1}{2} H_2, \quad (3.79)$$

are the reactions

$$Li \rightleftarrows \tfrac{1}{2} Li_2 \quad (3.80)$$

and

$$LiH + Li_2 \rightleftarrows Li_2H + Li. \quad (3.81)$$

The gas phase therefore has five components—Li, LiH, H_2, Li_2, and Li_2H. The concentrations and partial pressures of these components can be determined from the following relations.

We will first write the expressions for the equilibrium constants of the chemical reactions in the gas phase (3.79), (3.80), and (3.81) in the form

$$K_{P_1} = \frac{P_1 P_3^{1/2}}{P_2}; \quad (3.82)$$

$$K_{P_2} = P_4^{1/2}/P_1; \quad (3.83)$$

$$K_{P_3} = \frac{P_5 P_1}{P_2 P_4}. \quad (3.84)$$

The subscripts of the pressures correspond to the vapor-phase components with the order indicated above. The pressure P_3 is assumed specified. [For the Li-LiH (LiD) systems, it was found from the equations derived above, (3.59) and (3.60), with allowance for Table 3.23; and for the Li-LiT system, it was found from Eqs. (3.77) and (3.78), with allowance for the dependences $\gamma_1(x, T)$ and $\gamma_2(x, T)$ and (Table 3.23)]. The partial pressure of lithium above the solution can be determined from the equation

$$P_1 = P_1^0 \gamma_1 (1 - x), \tag{3.85}$$

where P_1^0 is the saturation vapor pressure of pure lithium at a given temperature.

The partial pressure of lithium hydride can be determined from Eq. (3.82):

$$P_2 = P_1 P_3^{1/2} / K_{P_1}. \tag{3.86}$$

From Eqs. (3.83) and (3.85) we find

$$P_4 = (K_{P_2} P_1^0)^2 \gamma_1^2 (1 - x)^2. \tag{3.87}$$

Analogously, from Eqs. (3.84)–(3.87) we find

$$P_5 = (K_{P_3} K_{P_2}^2 (P_1^0)^2 / K_{P_1}) P_3^{1/2} \gamma_1^2 (1 - x)^2. \tag{3.88}$$

The results of calculations of the partial pressures of the components and of the composition ($x_i = P_i / P$, where x_i is the mole fraction of the component in the vapor phase, and $P = \Sigma P_i$) of the equilibrium vapor in the Li-LiH (LiD, LiT) systems at various temperatures are summarized in Tables 5.7–5.12. The initial data on K_{P_1} and K_{P_2} (taken from Ref. 14) and on P_1^0 (taken from Refs. 68 and 136) were used in the calculations. The equilibrium constant K_{P_3} was determined by Aslanyan and Tsirlina, with allowance for the data on the thermodynamic functions of molecules of the type Li_2H, found in Ref. 3.

It should be noted that some isotherms exhibit a minimum on the $P = f(x)$ curve at low concentrations of x. This point is discussed in more detail in Sec. 3.2. A certain discrepancy in the relations $\Sigma P_i = P$ and $\Sigma x_i = 1$, found in Tables 5.7–5.12, stems from an unavoidable scatter resulting from rounding off the numbers.

Table 5.13 gives values for the pressure in the two-dimensional region for the Li-LiH (LiD, LiT)systems on the basis of the data of Ref. 236. At temperatures above the monotectic temperature, the partial pressures P (mmHg) were calculated from Eqs. (3.74)–(3.76); and at temperatures below the monotectic temperature, the partial pressures were calculated on the basis of an analogous equation $\ln P_{H_2(D_2,T_2)} = A - B/T$, with coefficients having the following values:

	Li-LiH	Li-LiD	Li-LiT
A	27.50	28.13	27.87
B	23 380	23 590	23 210

The monotectic temperatures in Table 5.13 are based on the data found in Sec. 1. Figure 3.16 shows the 1100-K isotherm of $P_3 = f(x)$ in the region to the left of the immiscibility zone for the Li-LiH (LiD, LiT) systems. An isotopic effect is clearly evident.

The relative errors of the data (in Tables 5.7–5.12) on the partial pressures of the components are rather difficult to determine, since a large variety of temperature and concentration dependences was used in the calculations. A comparison of the calculated partial pressures of hydrogen in the Li-LiH (LiD) systems with the experimental data obtained in Refs. 76, 77, 235, and 236 shows, however, that the calculated values are in good agreement with the experimental values [within $\pm$ (1–2)%]. The errors of the calculated

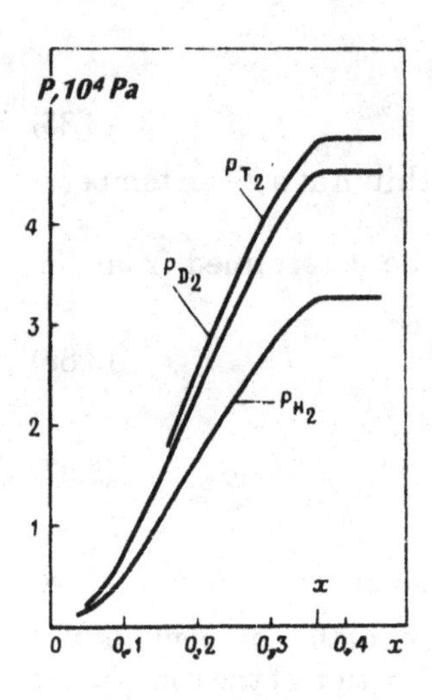

Figure 3.16. Comparison of the partial pressures of hydrogen isotopes in the Li-LiH (LiD, LiT) systems in the region to the left of the immiscibility zone at a temperature of 1100 K.

data for the Li-LiH (LiD) systems can therefore be estimated on the basis of the analysis of the experimental error for the Li-LiH system carried out in Refs. 76 and 77.

As was pointed out in Sec. 2.1, the maximum error in measuring the pressure in these experiments was ± 1 mmHg, or ± 1.0–1.5%, on the 1040-K isotherm. The maximum relative error in determining the partial pressure of hydrogen and deuterium in the Li-LiH (LiD) systems at concentrations to the left of the immiscibility zone is estimated to be 2%, and at concentrations to the right of the immiscibility zone is estimated to be 5%.

The error in determining the partial pressure of tritium in the Li-LiT system is larger, since the partial pressure could not be measured directly (except in the immiscibility region) and since the relations obtained from the similarity condition had to be used.

Equation (3.77), which was used to determine the partial pressure of tritium, was obtained on the basis of Eq. (3.73), which is valid within 2% in the temperature interval under study, as indicated in Table 3.19.

Taking into account the errors of the data on the partial pressure of hydrogen and the equilibrium constant of the reaction (3.71) for the Li-LiH system, which were used to find Eq. (3.77), we estimate the maximum relative errors of the calculated data on the partial pressure of tritium, given in Table 5.9, to be 5%.

2.5. Excess caloric functions in the Li-LiH system

The enthalpy and entropy of the binary Li-LiH system, when the liquid phase is in equilibrium with the vapor phase, were calculated by Shpil'rain and Yakimovich.[74] They constructed enthalpy-composition and entropy-composition diagrams indicating the position of the vaporization and condensation lines. This analysis, however, was carried out under the assumption of an ideal vapor phase and an ideal coexisting liquid solution. The temperature and concentration dependences of the activity coefficients of lithium and lithium hydride obtained in Sec. 2.3 and the caloric functions of liquid lithium hydride, lithium deuteride, and lithium tritide determined above make it possible to determine the caloric functions of the Li-LiH (LiD, LiT) solutions, with corrections for deviation from ideal-solution behavior and with allowance for the data on pure lithium found in the literature.

On the basis of the standard relations for the thermodynamics of solutions (see Refs. 21 and 51, for example) we can write the following expression for the molar enthalpy of the solution

$$i = \bar{i}_1(1-x) + \bar{i}_2 x, \tag{3.89}$$

where x is the molar composition of the solution, and $\bar{i}_1$ and $\bar{i}_2$ are the partial molar enthalpies of lithium and lithium hydride (deuteride, tritide), respectively.

The partial molar enthalpy of lithium (ignoring the pressure) is

$$\bar{i}_2 = i_1^0 - RT^2\left(\frac{\partial \ln \gamma_1}{\partial T}\right)_x \tag{3.90}$$

and the partial molar enthalpy of lithium hydride is

$$\bar{i}_2 = i_2^0 - RT^2\left(\frac{\partial \ln \gamma_2}{\partial T}\right)_x . \tag{3.91}$$

Here i_1^0 and i_2^0 are, respectively, the enthalpies of pure first component and pure second component (of lithium and hydride).

The excess enthalpy can be determined from the relation

$$\Delta i_{\text{exc}} = -RT^2\left[\left(\frac{\partial \ln \gamma_1}{\partial T}\right)_x (1-x) + \left(\frac{\partial \ln \gamma_2}{\partial T}\right)_x x\right]. \tag{3.92}$$

Such expressions were also used for calculating the entropy of the solutions and the excess entropy (the notation is similar to that used for the enthalpy):

$$s = \bar{s}_1(1-x) + \bar{s}_2 x; \tag{3.93}$$

$$\bar{s}_1 = s_1^0 - RT\left(\frac{\partial \ln \gamma_1}{\partial T}\right)_x - R\ln[(1-x)\gamma_1]; \tag{3.94}$$

$$\bar{s}_2 = s_2^0 - RT\left(\frac{\partial \ln \gamma_2}{\partial T}\right)_x - R\ln(x\gamma_2); \tag{3.95}$$

$$\Delta s_{\text{exc}} = -RT\left[\left(\frac{\partial \ln \gamma_1}{\partial T}\right)_x (1-x) + \left(\frac{\partial \ln \gamma_2}{\partial T}\right)_x x\right] - R[(1-x)\ln \gamma_1 + x\ln \gamma_2]. \tag{3.96}$$

Table 3.25 gives the calculated results of the excess caloric functions of a Li-LiH solution, and Fig. 3.17 is a plot of Δi_{exc} and $\Delta \varphi_{\text{exc}}$ as functions of the composition of the solution in the left region of the concentrations. Notice that the temperature has virtually no effect on the excess functions in the temperature interval studied. Since the temperature and concentration dependences of the activity coefficients of the components in an Li-LiH solution do not change, as was pointed out in Sec. 2.3, upon isotopic replacement of the

Table 3.25. Excess caloric functions of liquid binary solutions of the Li-LiH type.

T,K	x	Δi_{exc}, J/mole	ΔS_{exc}, J/(mole deg)	$\Delta\varphi_{exc}$, J/mole
1000	0.10	1 650	0.173	1477
	0.15	2 454	0.353	2101
	0.20	3 211	0.429	2782
	0.98	947	0.288	659
	0.99	448	0.111	337
1100	0.10	1 652	0.173	1460
	0.20	3 212	0.429	2740
	0.30	4 463	0.617	3784
	0.93	11 730	8.96	1872
	0.96	6 129	4.52	1152
	0.99	1 455	1.05	299
1200	0.10	1 651	0.173	1443
	0.30	4 461	0.617	3720
	0.40	5 142	0.542	4492
	0.80	53 990	41.9	3669
	0.90	18 708	13.8	2136

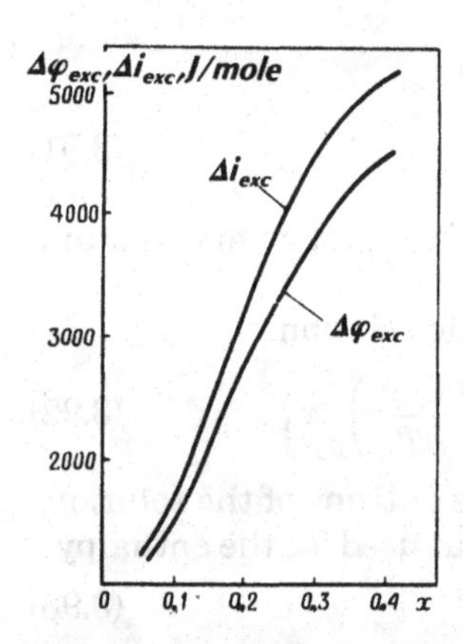

Figure 3.17. Excess enthalpy Δi_{exc} and excess thermodynamic potential $\Delta\varphi_{exc}$ in a solution of the Li-LiH type.

second component for LiD or LiT, this situation should also hold for the excess thermodynamic functions.

If the values of x are large beyond the right boundary of the immiscibility zone, the calculated values of the excess functions are less dependable because of the relatively low accuracy of the derivatives of the activity coefficients which are calculated in this region from the temperature.

3. Thermodynamic features of Li-LiH systems when the liquid phase is in equilibrium with the vapor phase

The characteristic features of the systems under consideration—the dissociation of lithium hydride in the vapor and the immiscibility zone in the con-

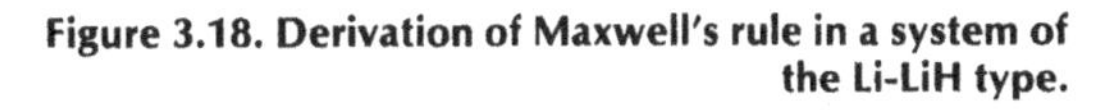

Figure 3.18. Derivation of Maxwell's rule in a system of the Li-LiH type.

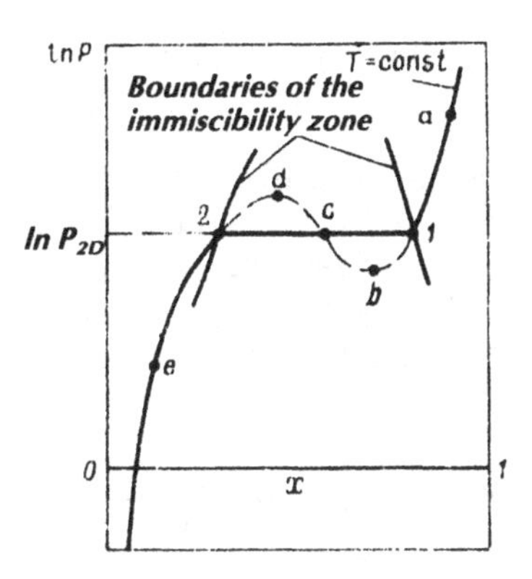

densed phase—account for several curious thermodynamic peculiarities. An in-depth study of these peculiarities is not only of purely scientific interest but also of definite practical importance.

3.1. Behavior of the isotherm on a *P–x* diagram in the immiscibility zone. Maxwell's rule

If the condensed phase of a binary system has an immiscibility zone, it can be shown experimentally, as was pointed out above, that the isotherm on a P–x diagram is horizontal in this zone. By using the continuous liquid phase composition function x (the dashed curve in Fig. 3.18) to represent the pressure P, we can use a rule which is similar to Maxwell's rule in thermodynamics. However, instead of the conventional P–V coordinates for a pure substance in the two-phase liquid-vapor region, we must use $\ln P$–x coordinates in the three-phase liquid-liquid-vapor region (see Ref. 75).

The problem is considered under equilibrium conditions where the vapor phase consists of a single ideal gas component and the liquid phase consists of two coexisting phases with different contents of the given component.

Let us examine the lithium–lithium hydride system, which has an immiscibility zone in the condensed phase, and let us also examine a vapor phase consisting of virtually pure hydrogen, which is assumed an ideal gas.

The isotherm consisting of the logarithm of the vapor pressure as a function of the composition of the equilibrium condensed phase is represented by a solid line in Fig. 3.18. The horizontal area indicates that the pressure of the vapor phase remains constant over a certain concentration range of the components in the condensed phase. As we move from right to left along the isotherm, the condensed phase becomes hydrogen depleted.

In a closed lithium–lithium hydride system of volume V, the liquid phase of volume V' coexists with the vapor phase of volume V'':

$$V = V' + V''. \tag{3.97}$$

The liquid phase is a solution of lithium and lithium hydride and the vapor phase is essentially pure hydrogen, H_2. The mole concentration of hydride in the liquid phase can be expressed as

$$x = 2n'_{H_2}/n_{Li}, \tag{3.98}$$

where n_{Li} and n'_{H_2} are, respectively, the total amounts of lithium (in the form of pure lithium and LiH) and hydrogen (in the form of lithium hydride) in the liquid phase, expressed in moles. Since the gas phase, by definition, consists of only hydrogen, n_{Li} = const. On the basis of the assumption that the gas phase is ideal, we can write

$$PV'' = n''_{H_2}\, \mu RT, \tag{3.99}$$

where n''_{H_2} is the number of hydrogen moles in the gas phase and μR is the universal gas constant.

In the case of an isothermal transition of the system from point *1* to point *2* (see Fig. 3.18) and back again, the work performed must be equal to zero; if an isothermal cycle *12dcb1* can be arranged, then

$$\int_1^2 (P - P_{2D})dV = \int_1^2 (P - P_{2D})dV''. \tag{3.100}$$

Here P is the melting occurring in the curved part *1bcd2,* and P_{2D} is the melting in the horizontal ("2D ") part *1c2.* A change in the volume of the liquid phase in this process can be ignored.

At T = const, from Eq. (3.99) we find

$$P\,dV'' = \mu RT\,dn''_{H_2} - V''dP = \mu RT\,dn''_{H_2} - n''_{H_2}\mu RT \ln P. \tag{3.101}$$

Substituting Eq. (3.101) into Eq. (3.100), we find

$$(P_2V''_2 - P_1V''_1) - \mu RT \int_1^2 n''_{H_2} d\ln P - P_{2D}(V''_2 - V''_1) = 0. \tag{3.102}$$

The first and third terms on the left-hand side of Eq. (3.102) are the same in absolute value. The integral in Eq. (3.102) can be transformed and integrated by parts:

$$\int_1^2 n''_{H_2}\, d\ln P = \int_1^2 n''_{H_2}\, d\ln \frac{P}{P_{2D}}$$
$$= n''_{H_2} \ln \frac{P}{P_{2D}}\bigg|_1^2 - \int_1^2 \ln \frac{P}{P_{2D}}\, dn''_{H_2}. \tag{3.103}$$

The ratios of the pressures under the logarithm at points *1* and *2* are equal to unity. Since the system is closed, for relation (3.98) we have

$$dn''_{H_2} = -\,dn'_{H_2} = -\tfrac{1}{2}\, n_{Li}\, dx. \tag{3.104}$$

After substituting Eq. (3.104) into Eq. (3.103), we find

$$\int_1^2 (\ln P - \ln P_{2D})dx = 0. \tag{3.105}$$

This equation is an analog of Maxwell's rule in the $\ln P - x$ coordinates.

A geometric interpretation of this expression is shown in Fig. 3.18. Points *1* and *2* lie at the boundary of the immiscibility zone. The real isotherm of the vapor pressure as a function of the concentration of the condensed phase on the liquid–vapor equilibrium diagram goes through the points *a-1-c-2-e,* with the horizontal part being in the immiscibility zone. It follows from Eq. (3.105) that the area of the region *1-b-c-1* is equal to the area of the region *2-d-c-2*.

3.2. Thermodynamic features in the homogeneous region. Azeotropy

Shpil'rain and Yakimovich[70] showed that the equilibrium liquid–vapor phase diagram of the Li-LiH system (even under the assumption that each coexisting phase is an ideal phase) has a peculiarity in the nature of azeotropy. This peculiarity, which stems from the presence in the gas phase of a dissociation reaction

$$LiH \rightleftarrows Li + \tfrac{1}{2} H_2 \tag{3.106}$$

can be physically accounted for by the fact that at a constant temperature of the system, the partial pressure of lithium in the vapor decreases, while the partial pressure of hydrogen increases sharply as the concentration of lithium hydride or, equivalently, the concentration of hydrogen is increased in the solution.

Let us identify this peculiarity, taking into account the real properties of the liquid solution. We call attention to the fact that, in the concentration region in which the fraction of hydride in the liquid phase is small, the vapor phase has two other important reactions. First, the reaction leading to the dimerization of lithium

$$Li \rightleftarrows \tfrac{1}{2} Li_2 \tag{3.107}$$

and, second, the reaction leading to the formation of Li_2H, which was studied experimentally by Ihle and Wu[153]:

$$LiH + Li_2 \rightleftarrows Li_2H + Li. \tag{3.108}$$

In the gas phase, there are accordingly five components: Li, LiH, H_2, Li_2, and Li_2H, while in the liquid phase there are two components: Li and LiH. As an example, we point out that at a temperature of 700 °C the ratio of the partial pressures of LiD, Li_2D, and D_2 in the Li-LiD system is, according to the data of Ihle and Wu,[156] 80:30:1, respectively, when the liquid phase is at equilibrium with the vapor phase (the atomic fraction of deuterium in the liquid phase is 10^{-6}).

The problem can be summarized as follows (see Ref. 79): The liquid phase is a nonideal solution of Li-LiH, while the coexisting vapor phase is an ideal mixture of ideal gases. Chemical reactions (3.106)–(3.108) occur between the components. In the thermodynamic system under consideration, the displacement along the liquid-vapor phase-equilibrium line can be described by several equations, which include two Gibbs-Duhem equations (for each of the coexisting phases that are at equilibrium separately and jointly), and by three stoichiometric equations for chemical reactions (3. 106)–(3. 108):

$$d\Phi'' = -S''dT + V''dP + \mu_1'' dn_1'' + \mu_2'' dn_2'' + \mu_3'' dn_3'' + \mu_4'' dn_4'' + \mu_5'' dn_5'' = 0; \tag{3.109}$$

$$d\Phi' = -S'dT + V'dP + \mu_1' dn_1' + \mu_2' dn_2' = 0; \tag{3.110}$$

$$\mu_1'' + \tfrac{1}{2}\mu_3'' - \mu_2'' = 0; \tag{3.111}$$

$$\tfrac{1}{2}\mu_4'' - \mu_1'' = 0; \tag{3.112}$$

$$\mu_5'' + \mu_1'' - \mu_2'' - \mu_4'' = 0. \tag{3.113}$$

Here and elsewhere in the text, the primes and double primes denote the liquid phase and the vapor phase, respectively; the subscripts refer to the components in the order in which they are listed above; Φ, V, and S are the thermodynamic potential, volume, and entropy of each of the phases; n_i and μ_i are the number of moles and the chemical potentials of the corresponding components; and P and T are the pressure and temperature of the system.

Equations (3.111)–(3.113) (allowing for the equality of the chemical potentials $\mu_1' = \mu_2''$ and $\mu_2' = \mu_2''$) imply that

$$d\mu_3'' = 2\,d\mu_2' - 2\,d\mu_1'; \tag{3.114}$$

$$d\mu_4'' = 2\,d\mu_1'; \tag{3.115}$$

$$d\mu_5'' = d\mu_2' + d\mu_1'. \tag{3.116}$$

Using the relation $\Phi = \Sigma n_i\,\mu_i$, with $d\Phi = \Sigma \mu_i dn_i + \Sigma n_i d\mu_i$ and Eqs. (3.114)–(3.116), we can transform Eqs. (3.109) and (3.110) as follows:

$$-S''dT + V''dP - (n_1'' - 2n_3'' + 2n_4'' + n_5)d\mu_1'$$
$$- (n_2'' + 2n_3'' + n_5'')d\mu_2' = 0; \tag{3.117}$$

$$-S'dT + V'dP - n_1'\,d\mu_1' - n_2'\,d\mu_2' = 0 \tag{3.118}$$

or

$$-s''dT + v''dP - (x_1'' - 2x_3'' + 2x_4'' + x_5'')d\mu_1' - (x_2'' + 2x_3'' + x_5'')d\mu_2') = 0; \tag{3.119}$$

$$-s'dT + v'dP - (1 - x')d\mu_1' - x'd\mu_2' = 0. \tag{3.120}$$

Here s and v are, respectively, the molar entropy and volume of each phase; x' is the mole fraction of lithium hydride in the liquid solution; and x_i'' is the mole fraction of the corresponding component of the vapor phase. Eliminating from Eqs. (3.119) and (3.120) the chemical potential of one of the components, say, μ_2', we find the fundamental relation

$$[s'(x_2'' + 2x_3'' + x_5'') - s''x']dT - [v'(x_2'' + 2x_3'' + x_5'') - v''x']dP$$
$$+ [(x_2'' + 2x_3'' + x_5'')(1 - x') - (x_1'' - 2x_3'' + 2x_4'' + x_5'')x']d\mu_1' = 0. \tag{3.121}$$

Equation (3.121) implies that an isothermal displacement ($dT = 0$) in the system may give rise to conditions under which $dP = 0$.

These conditions are*

(1) $(x_2'' + 2x_3'' + x_5''(1 - x') - (x_1'' - 2x_3'' + 2x_4'' + x_5'')x' = 0$;

(2) $d\mu_1' = 0$.

* It follows from Tables 5.11–5.13 that the sum of the concentrations of the components, $x_2'' + 2x_3''x_5''$, is not greater than x'. At the same time, the molar volumes of the liquid and vapor phases differ by 3 to 4 orders of magnitude at temperatures up to 1300 K and pressures between 10^{-8} and 50 atm. Under these conditions the coefficient of dP in Eq. (3.121) cannot be zero.

The first condition leads to a particular ratio of concentrations of the components in the coexisting phases:

$$\frac{x'}{1-x'}=\frac{x_2''+2x_3''+x_5''}{x_1''-2x_3''+2x_4''+x_5''}. \tag{3.122}$$

Shpil'rain and Yakimovich[70] showed that in the analysis of systems of this sort, the concentration of the components (in mole fractions) can be replaced by the atomic concentration of hydrogen, which is defined as the ratio of the number of gram atoms of hydrogen to the total number of gram atoms in a given phase:

$$a'=\frac{n_{\mathrm{H}}'}{n_a'}=\frac{n_2'}{n_1'+2n_2'} \tag{3.123}$$

for the liquid phase and

$$a''=\frac{a_{\mathrm{H}}''}{n_a''}=\frac{n_2''+2n_3''+n_5''}{n_1''+2n_2''+2n_3''+2n_4''+3n_5''} \tag{3.124}$$

for the vapor phase.

The atomic concentration of hydrogen is related to the mole fraction of the components by

$$a'=x'/(1-x') \tag{3.125}$$

and by

$$a''=(x_2''+2x_3''+x_5'')/(2-x_1''+x_5''). \tag{3.126}$$

Equation (3.122) can be transformed to the following form on the basis of the Dalton law:

$$x'/(1-x')=(x_2''+2x_3''+x_5'')/(2-x_1''+x_5''). \tag{3.127}$$

It follows [see Eqs. (3.125) and (3.126)] that the first condition under which $dP=0$ gives rise to the relation

$$a'=a''.\ . \tag{3.128}$$

Analysis has shown that at $T=\mathrm{const}$, this point on the P–a diagram is the minimum point of the total pressure; i.e., an azeotropy-type anomaly arises and the atomic compositions of the two coexisting phases are the same.

The second condition, for which $dP=0\ (d\mu_1'=0)$, is the result of the presence of the immiscibility zone in systems of this sort. Since Eq. (3.121) can be used only in the regions in which the function $f(P,T,\mu)$ is smooth, this restriction can be reconciled with the condition $dP=0$ only at the critical solution point, for which at $dT=0$ we have $d\mu_1'=0$ as well.

By introducing the concepts of atomic volume v_a and atomic entropy s_a, respectively, as the volume and entropy per gram atom of a given phase we can transform Eqs. (3.117) and (3.118) to the form

$$-s_a''\,dT+v_a''\,dP-(1-2a'')d\mu_1'-a''d\mu_2'=0; \tag{3.129}$$

$$-s_a'\,dT+v_a'\,dP-(1-2a')d\mu_1'-a'd\mu_2'=0. \tag{3.130}$$

These equations imply that

$$-(s_a''a'-s_a'a'')dT+v_a''a'-v_a'a'')dP-(a'-a'')d\mu_1'=0. \tag{3.131}$$

From relation (3.131) we can derive two individual equations:

$$\left(\frac{\partial P}{\partial T}\right)_{a'=a''} = \frac{s''_a - s'_a}{v''_a - v'_a} \tag{3.132}$$

for the curve of the displacement of the azeotropic point and

$$\left(\frac{\partial P}{\partial T}\right)_{\mu'_1=\text{const}} = \frac{s''_a/a'' - s'_a/a'}{v''_a/a'' - v'_a/a'} \tag{3.133}$$

for the curve of the displacement of the critical solution point. For the cases under consideration, Eq. (3.131) is an analog of the van der Waals equation for solutions in the phase-equilibrium state (see Ref. 11, for example).

The use of Eq. (3.131) for the case in which the coexisting phases or at least one of them is a nonideal solution (the liquid phase in our case) compounds the problem, because the activity coefficients of the components must be known before the atomic volume or the entropy can be determined. A preliminary analysis has shown, however, that in a Li-LiH system, the azeotropic concentrations lie in the region $x < 10^{-3}$. For such conditions, advantage can be taken of the theory of dilute solutions to obtain several specific results.

Let us first consider the temperature dependence of the concentration of the azeotropic point. We express the mole fractions of the vapor-phase components in terms of the partial pressures $x''_i = P_1/P$ (where $P = \Sigma P_i$) and then in terms of the mole fraction of lithium hydride in the liquid phase, x'.

In a Li-LiH system, the partial pressure of hydrogen at low content of lithium hydride in the liquid phase is described by relation (3.39). The partial pressure of lithium (of the solvent) can be determined on the basis of the Raoult equation

$$P_1 = P_1^0(1 - x'), \tag{3.134}$$

where P_1^0 is the saturation vapor pressure of pure lithium. The partial pressures of the rest of the components are expressed in terms of the corresponding equilibrium constants of chemical reactions (3.106)–(3.108):

$$P_2 = \frac{P_1 P_3^{1/2}}{K_{P_1}} = \frac{P_1^0 C}{K_{P_1}} x'; \tag{3.135}$$

$$P_4 = K_{P_2}^2 P_1^2 = K_{P_2}^2 (P_1^0)^2 (1 - x')^2; \tag{3.136}$$

$$P_5 = \frac{K_{P_3} P_2 P_4}{P_1} = \frac{K_{P_2}^2 K_{P_3}}{K_{P_1}} C (P_1^0)^2 x'(1 - x'). \tag{3.137}$$

The total pressure in the system is

$$P = P_1^0(1 - x') + \frac{P_1^0 C}{K_{P_1}} x' + C^2 \left(\frac{x'}{1 - x'}\right)^2 + (P_1^0)^2 K_{P_2}^2 (1 - x')^2 + \frac{(P_2^0)^2 K_{P_2}^2 K_{P_3} C}{K_{P_1}} x'(1 - x'). \tag{3.138}$$

At the azeotropic point x^* or a^* (since at low concentrations of lithium hydride in the solution the concentration of x' is essentially the same as that of a'), the composition of the solution can be determined on the basis of Eqs.

Table 3.26. Temperature dependence of the concentrations of the azeotropic points of $a^*(10^{-5})$.

T,K	Li-LiH	Li-LiD	Li-LiT	T,K	Li-LiH	Li-LiD	Li-LiT
800	6.29	3.32	2.14	1100	21.1	15.5	15.1
900	10.11	6.75	5.87	1200	17.9	15.3	15.9
1000	17.4	11.5	10.5	1300	3.17	6.85	9.50

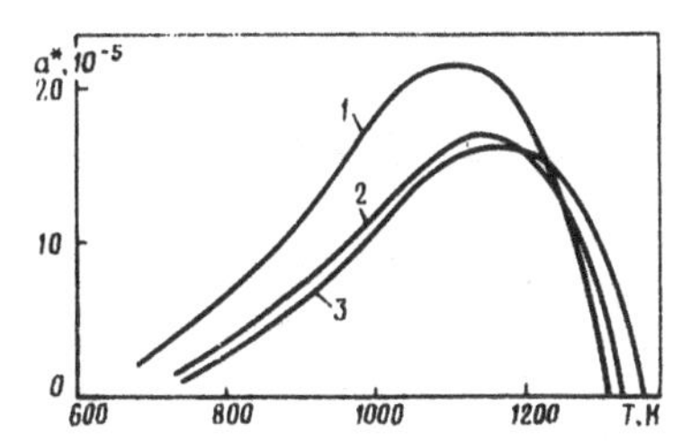

Figure 3.19. Concentration at the azeotropic point a^* versus temperature for the following systems: Li-LiH (1), Li-LiD (2), and Li-LiT (3).

(3.39) and (3.134)–(3.137) either from Eq. (3.138) by equating the derivative $(\partial P/\partial x')_T$ to zero or from Eq. (3.127) by inserting the mole fractions of the components of the vapor phase. As a result (ignoring insignificant terms), we find

$$x^* = a^* = \frac{K_{P_1} - C + 2P_1^0 K_{P_1} K_{P_2}^2 - P_1^0 K_{P_2}^2 K_{P_3} C}{2K_{P_1} C^2 P_1^0}. \tag{3.139}$$

The right-hand side of Eq. (3.139) has quantities that are functions only of temperature. Analysis of this equation gives rise to two important corollaries. First, a^* shifts to zero at a certain temperature; i.e., there is a limiting temperature at which the system exhibits azeotropic phenomena. This limiting temperature is given by the equation

$$K_{P_1} - C + 2P_1^0 K_{P_1} K_{P_2}^2 - P_1^0 K_{P_2}^2 K_{P_3} C = 0. \tag{3.140}$$

Secondly, there is a temperature at which $da^*/dT = 0$ and at which the azeotropic concentration in the system reaches a maximum.

The results of specific calculations of $a^* = f(T)$ carried out by us for the Li-LiH (LiD, LiT) systems are presented in Table 3.26 ($a^* \times 10^{-5}$) and in Fig. 3.19. In the calculations we used the initial data on P_1^0, K_{P_1}, K_{P_2}, and K_{P_3} from the same sources (using an extrapolation whenever necessary) as those we used in Sec. 2.4 to calculate the partial pressures of the components from $C_{Li\text{-}LiH(LiD)}$ (Table 3.23) and from $C_{Li\text{-}LiT}$ on the basis of Eq. (3.78).

The left branch of the curve approaches the abscissa asymptotically. At a temperature of 600 K, the value of a^*, in comparison with a^*_{max}, decreases by two orders of magnitude.

Figure 3.20 is a schematic representation of the left part of the P–a diagram of the system under consideration when the liquid phase is at equilibrium with the vapor phase. The diagram shows several typical isotherms. The solid curves represent the total saturated vapor pressure on the isotherms as

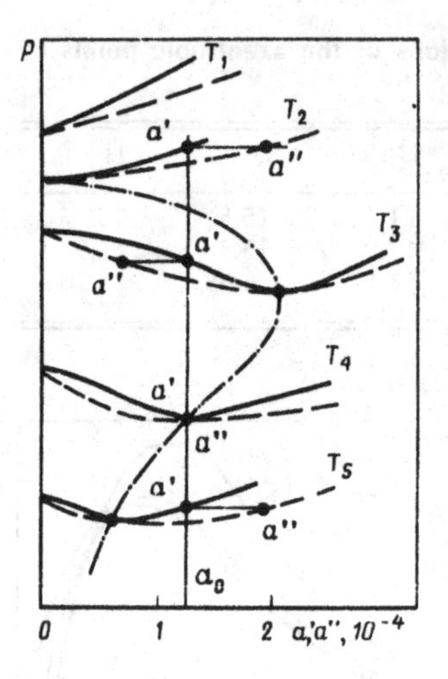

Figure 3.20. The total vapor pressure at phase equilibrium in a system of the Li-LiH type versus the atomic concentration of hydrogen in the liquid a' (solid lines) and in the vapor a'' (dashed lines).

a function of a', and the dashed curves denote the same pressure as a function of a''.

The isotherm T_2 describes the behavior of the pressure at the limiting temperature of the azeotropy. At a higher temperature (isotherm T_1), the azeotropic phenomena disappear in the system. The isotherm T_3 is characteristic of the maximum values of a^*, and the isotherms T_4 and T_5 characterize the behavior of $P(a')$ and $P(a'')$ with further decrease in the temperature. By using the atomic concentrations in the analysis of such systems, we can clearly show on the diagram the region of the liquid solution, the region of superheated vapor, and the two-phase liquid-vapor region. The line connecting the azeotropic points is described by Eq. (3.139). The curve defines the region, outside of which the equilibrium vapor is more highly enriched in hydrogen compared with the liquid solution.

Equation (3.138) can be used to find in the P–a diagram the slope of the isotherm of the total vapor pressure in the limit $a' \to 0$:

$$\left(\frac{dP}{da'}\right)_{T,a'=0} = -P_1^0\left(1 - \frac{C}{K_{P_1}} + 2P_1^0 K_{P_2}^2 - \frac{P_1^0 C K_{P_2}^2 K_{P_3}}{K_{P_1}}\right). \quad (3.141)$$

This relation gives rise to Eq. (3.140) under the condition that $(dP/da')_{T,a'=0} = 0$.

Let us analyze the vaporization process in such a system in the context of the general theory of solutions. We assume that the initial composition of the liquid solution is $a_0 = a'$. If the phase equilibrium sets in at a temperature T_5, the vapor turns out to be enriched in hydrogen compared with the equilibrium liquid solution. At a temperature T_3, the situation is exactly the reverse—the concentration of hydrogen in the liquid in this case is higher than that in the equilibrium vapor. At a temperature T_4, the atomic concentrations of hydrogen in the liquid and vapor are the same at equilibrium (the azeotropic point), and at a temperature T_2, the hydrogen concentration in the equilibrium vapor is again higher than that in the liquid. Note that the existence domain of the function $P = f(a')$ is the interval $0 \leqslant a' < 0.5$ and that of the function $P = f(a'')$ is the interval $0 \leqslant a' < 1$.

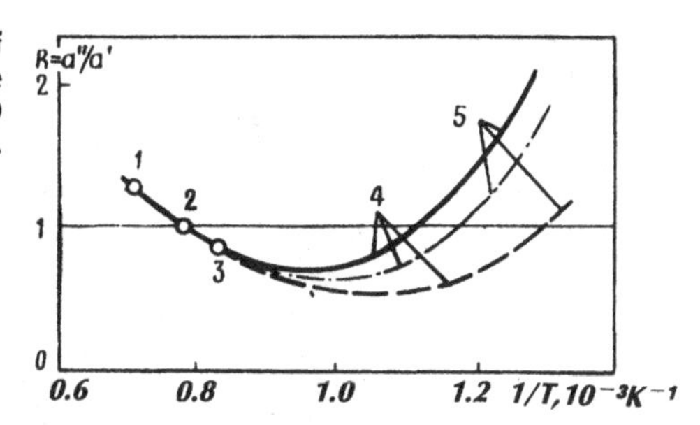

Figure 3.21. The temperature dependence of $R = a''/a'$ for a constant composition of the solution in the Li-LiH system (- - -), the Li-LiD system (-·-·-), and the Li-LiT system (—).

Figure 3.21 is a plot of $R = a''/a'$ as a function of temperature at $a' = 5\times10^{-5}$. Analytically, this plot can easily be found from Eqs. (3.125) and (3.126), with a subsequent conversion from mole fractions to partial and total pressures. Omitting the insignificant terms, we can write this functional dependence as

$$R = \frac{C + 2K_{P_1}C^2a'/P_1^0 + P_1^0K_{P_2}^2K_{P_3}^2C}{K_{P_1} + 2P_1^0K_{P_1}K_{P_2}^2}. \tag{3.142}$$

The calculation results for the Li-LiH (LiD, LiT) systems based on this equation are presented in Table 3.27. The initial data necessary for the calculation were taken from the same sources as those used to compile Table 3.26. Equation (3.142) shows the relative change in the concentration of hydrogen in the vapor and liquid phases with changing temperature and concentration of hydrogen in the solution.

Upon heating a Li-LiH solution with an initial composition a_0 to a two-phase state, the phase distribution of the components in gram-atoms (the total number of gram-atoms, $n_a = n_a' + n_a''$, in the system remains constant) can be determined on the basis of the lever rule (see Ref. 70):

$$n_a'/n_a'' = (a'' - a_0)/(a_0 - a'). \tag{3.143}$$

Substituting into Eq. (3.142) the values of a^* from Eq. (3.139), we accordingly find $R = 1$ at the azeotropic point.

Let us consider the curves in Fig. 3.21 in greater detail. At point *1*, the hydrogen concentration in the vapor is higher than that in the liquid ($R > 1$). With decrease in equilibrium temperature, the concentration ratio decreases, and at point *2*, the concentration in the liquid is equal to that in the vapor ($R = 1$). A further decrease in the temperature changes the ratio of hydrogen concentration in the vapor and liquid ($R < 1$).

Section 1–3 was studied experimentally by Ihle and Wu[157] for the Li-LiD system. They found that for an initial composition of the liquid $a' = 10^{-5}$, the temperature T_2 is $\sim$1240 K.

At high temperatures, the curves $R = f(1/T)$ are essentially the same for all three systems, Li-LiH (LiD, LiT). With decrease in the temperature, the curves begin to diverge. For each system, the curve $R = f(1/T)$ reaches a minimum and then begins to rise, crossing the point $R = 1$ on its way to the zone where $R > 1$ (region 5); i.e., it returns to the zone in which the hydrogen concentration in the vapor is higher than in the equilibrium liquid solution. The position of the minimum $R = f(1/T)$ is given by Eq. (3.142) on the basis of

Table 3.27. Dependence of $R = a''/a'$ in the Li-LiH (LiD, LiT) systems on the composition of the liquid phase.

T,K	a'	$R_{\text{Li-LiH}}$	$R_{\text{Li-LiD}}$	$R_{\text{Li-LiT}}$
800	0.000 01	0.268	0.393	0.446
	0.000 05	0.822	1.44	1.72
	0.000 10	1.51	2.75	3.31
	0.000 50	7.03	13.13	16.03
900	0.000 01	0.311	0.342	0.354
	0.000 05	0.583	0.800	0.885
	0.000 10	0.923	1.37	1.55
	0.000 50	3.63	5.93	6.86
1000	0.000 01	0.400	0.417	0.415
	0.000 05	0.553	0.638	0.662
	0.000 10	0.733	0.915	0.971
	0.000 50	2.17	3.12	3.44
1100	0.000 01	0.568	0.556	0.534
	0.000 05	0.654	0.678	0.666
	0.000 10	0.762	0.831	0.831
	0.000 50	1.62	2.055	2.15
1200	0.000 01	0.763	0.733	0.706
	0.000 05	0.891	0.808	0.785
	0.000 10	0.889	0.901	0.883
	0.000 50	1.45	1.64	1.670
1300	0.000 01	0.979	0.928	0.893
	0.000 05	1.02	0.997	0.944
	0.000 10	1.07	1.04	1.01
	0.000 50	1.45	1.53	1.52

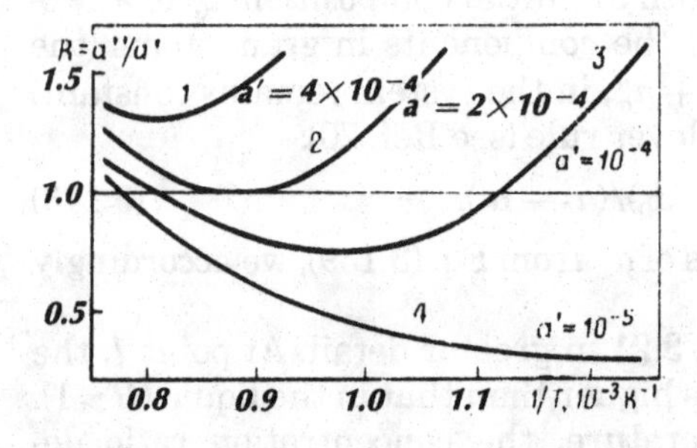

Figure 3.22. Dependence of $R = a''/a'$ on the temperature and on the initial composition of the solution in the Li-LiH system.

the condition

$$\left(\frac{\partial R}{\partial T}\right)_{a'} = 0. \tag{3.144}$$

Figure 3.22 shows the behavior of $R = f(1/T)$ in the particular case of the Li-LiH system for various initial atomic concentrations of the liquid phase. Curve *1* corresponds to an isoconcentrate $a_0' > a^*_{\max}$, curve *2* corresponds to $a_0' = a^*_{\max}$, and curves 3 and 4 correspond to isoconcentrates $a_0' < a^*_{\max}$.

These results have specific practical importance. We point out two cases of practical importance. In nuclear reactor technology, where lithium is used as a coolant, special attention is given to the removal of dissolved gases, in particular, hydrogen from lithium. Different methods, including distillation of lithium, are used. Our studies have shown that, in the first place, a lithium-hydrogen system has a temperature and concentration region in which the hydrogen content in lithium cannot be reduced by distillation (the region enclosed by a curve connecting the azeotropic points; see Figs. 3.19 and 3.20). Second, the zone in which the hydrogen content in the vapor exceeds the hydrogen content in the equilibrium liquid (in which distillation thus produces a positive effect) can be reached not only by raising the temperature but also by lowering it.

A similar problem arises in dealing with the confinement and recovery of tritium in fusion reactors, for which distillation is considered as one of several competing methods (see Ref. 185). The results obtained in this section are also applicable in this case.

Chapter 4
Transport properties

1. Thermal conductivity of lithium hydride in the condensed phase

The thermal conductivity of lithium hydride is an intrinsic parameter of its composition and is highly sensitive to it. In the solid phase, the thermal conductivity is also sensitive to the structure of the sample. Since the particular factors cannot always be controlled, the considerable discrepancies found in the literature are attributed primarily to this problem. The principal experimental studies dealing with this problem are summarized in Table 4.1. The transport properties of lithium deuteride and lithium tritide in the condensed phase have not been studied.

1.1. Thermal conductivity of solid lithium hydride

Fieldhouse[122] studied the thermal conductivity of lithium hydride by the stationary radial heat-flux method. The measurements were carried out over the temperature interval 60–510 °C. The test sample was a cylinder constructed from 1.2- to 2.5-cm-thick, hollow disks with 1.6-cm inside diameter and 7.6-cm outside diameter. The heat flux from the internal heater was measured by means of potential leads and the temperature drop in the range of 15° to 25° was measured with thermocouples placed along the inner and outer surfaces of the disk to be studied. The sample, whose structure was not specified, was held in a helium atmosphere. The experimental error was estimated to be 5% by the author. The thermal conductivity of lithium hydride measured by Fieldhouse is given in Table 4.2.

Welch[250] measured the thermal conductivity of lithium hydride using crystal samples placed in various media. The measurement procedure can be briefly described as follows. One set of measurements

T,°C ...	50	100	150	200	250	300	350	400
λ, $\frac{\text{kcal}}{\text{m h deg}}$	6.70	6.01	5.44	4.94	4.50	4.10	3.78	3.56

was said to have been carried out in an inert gas by a relative method involving axial heat flux. The rest of the data (Table 4.3) were obtained by a method in which the heat flux was determined by measuring the rise in the temperature of gas that passed through a tube inside the lithium hydride test sample (the calorimetric method). The crystal sample in the shape of a cylinder was

Table 4.1. List of papers in which the thermal conductivity of lithium hydride was studied.

References	Temperature range, °C	Sample characteristics	Medium	Method
Fieldhouse (Ref. 122)	60– 510	Solid	Helium	Radial heat flux
Welch (Ref. 250)	50– 400	Solid	Inert	Axial heat flux
	120– 480	Solid	Hydrogen	Calorimetric, radial heat flux
	90– 450	Solid	Helium	The same
	150– 450	Solid	Argon	The same
	90– 480	Solid	Vacuum	The same
Welch (Ref. 251)	64– 407	Pressed	—	Axial heat flux
	654	Solid	—	—
	710	Liquid	—	—
Vetrano (See Refs. 190, 250, and 251)	50– 600	Pressed	Hydrogen	Axial heat flux
	50– 500	Crystalline	Hydrogen	The same
Novikov, Kraev, *et al.* (see Ref. 70)	700–1000	Liquid	Argon	Continuous heating

Table 4.2. Thermal conductivity λ [kcal/(m h deg)] of solid lithium hydride from the experimental data of Fieldhouse (Ref. 122).

T,°C	λ	T,°C	λ	T,°C	λ
62	6.26	266	4.61	450	3.75
101	5.58	3.4	4.36	513	3.51
172	5.09	351	4.06	—	—
236	4.78	407	3.87	—	—

Table 4.3. Thermal conductivity of lithium hydride obtained by the calorimetric method [kcal/(m h deg)] from the data of Welch (Ref. 250).

T,°C	In hydrogen	In argon	In a vacuum		In helium			
93	7.75	—	5.65	—	8.16	—	—	
121	7.75	—	—	8.64	—	7.53	8.50	8.20
149	7.05	4.90	5.07	7.95	6.74	6.98	7.53	7.45
204	6.26	4.39	4.39	6.62	5.80	5.94	6.01	6.01
260	5.50	4.04	3.89	5.51	5.00	5.07	4.90	4.90
315	5.00	3.70	3.32	4.57	4.32	4.32	4.18	4.10
371	4.61	3.42	2.92	4.14	3.78	3.78	3.78	3.70
426	4.32	3.20	2.45	0.37	3.42	3.42	3.60	3.53
454	—	3.06	2.16	—	3.28	3.28	3.56	—
482	4.17	—	—	3.49	—	—	—	3.49

Table 4.4. Thermal conductivity of solid lithium hydride [kcal/(m h deg)] from the data of Welch (Ref. 251).

T,°C	λ	T,°C	λ
64.4	6.48	300.9	3.96
96.8	6.12	349.1	3.60
155.4	5.04	354.2	3.96
203.7	5.04	406.8	3.60
250.5	4.68		

Table 4.5. Thermal conductivity of lithium hydride [kcal/(m h deg)] from the data of Vetrano.

T,°C	Pressed sample	Crystalline sample	T,°C	Pressed sample	Crystalline sample
50	5.94	10.75	350	4.21	4.04
100	5.61	9.15	400	4.04	3.60
150	5.25	7.55	450	3.85	3.28
200	4.93	6.36	500	3.71	3.10
250	4.64	5.40	550	3.60	—
300	4.43	4.71	600	3.54	—

inserted into an electric furnace. Such samples were used in several experiments: one in hydrogen, one in argon, two in a vacuum at a pressure of 0.025 mmHg, and four in helium. The measurements were carried out in the temperature range 93–482 °C.

In another study, Welch[251] obtained two more sets of measurement data on the thermal conductivity of lithium hydride. The first set of measurements was carried out at temperatures between 64 and 407 °C by the relative method of axial heat flux using a pressed sample (Table 4.4). Two points were given for the second set of data: one point, $\lambda = 0.0087$ cal/cm s deg, for a solid sample at a temperature of 654.4 °C and the other point, $\lambda = 0.0188$ cal/cm s deg, for a liquid phase at a temperature of 710 °C.

Data on the thermal conductivity of crystalline lithium hydride and pressed lithium hydride obtained by Vetrano in the temperature interval 50–600 °C were published in a review by Messer[190] and also in the two studies by Welch indicated above. These results (Table 4.5) were reported in a private communication by the author. The measurements were carried out in hydrogen by the relative method of axial heat flux. The pressed sample was fabricated under a pressure of 5000 atm. The density of the 95% pure pressed sample was 0.775 g/cm^3. The crystalline sample was 98% pure.

Analysis of the published studies has shown (Fig. 4.1) that the medium in which the solid sample was held during the experiment, whether it was hydrogen, inert gas (helium, argon), or a vacuum, did not substantially affect the measurement results. The effect of the medium seems to be within the limits of the measurement error, which is quite large for these data.

A larger discrepancy is seen in the structure dependence of the sample. The data for crystalline samples generally lie above those for the samples

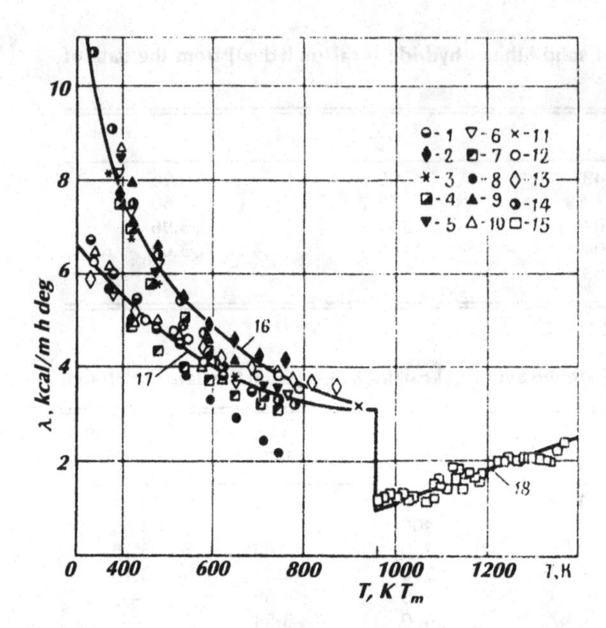

Figure 4.1. Thermal conductivity of lithium hydride based on the data of different authors. 1–9) Welch[250] [in inert gas (1), in hydrogen (2), in helium (3–6), in argon (7), in a vacuum (8,9)]; 10,11) Welch[251]; 12) Fieldhouse[122]; 13,14) Vetrano (see Refs. 190, 250, and 251); 13) for a pressed sample; 14) for a crystalline sample; 15) Novikov *et al.* (see Ref. 70); 16–18) curves were calculated on the basis of Eqs. (4.1), (4.2), and (4.4), respectively.

fabricated from finely divided lithium hydride pressed under a high pressure. Although the density of a pressed sample in some cases approaches the density of a crystalline sample (0.775 g/cm^3) (see the data of Vetrano), the thermal conductivity is considerably lower, especially at low temperatures. As the temperature is raised, the difference in the thermal conductivities of the pressed and crystalline samples diminishes and virtually disappears at the melting point.*

Of the ten sets of measurements of the thermal conductivity of crystalline lithium hydride found in the literature, two sets (1 and 10 in Fig. 4.1) do not conform to the general rule stated above. Since there is no adequate information about the systematic features of the experiments, the reason for these discrepancies cannot be explained.

The bulk of experimental data on the thermal conductivity, λ_{cr} (W/m deg), of crystalline lithium hydride can be satisfactorily described by the equation

$$\lambda_{cr} = [1.98\times10^3/(T-150)] + 1.09. \tag{4.1}$$

For the thermal conductivity, λ_{pr} (W/m deg), of pressed lithium hydride, the following equation, which, like Eq. (4.1), was obtained from a diagrammatic analysis of the experimental data, is valid:

$$\lambda_{pr} = (1.91\times10^3/T) + 1.51. \tag{4.2}$$

*An explanation of the data of Fieldhouse is in order. In his study he gives no clue as to the type of sample used. However, the procedure that he used [he stipulated the necessity of fabricating large-diameter (up to 80 mm) flat disks] and parallel experimental studies of other properties of lithium hydride carried out by him with the use of pressed samples, justify the assumption that here too he studied a pressed sample.

Table 5.14 gives the values of the thermal conductivity of solid lithium hydride calculated from Eqs. (4.1) and (4.2). The maximum discrepancy between the experimental data on the thermal conductivity of crystalline lithium hydride (with the exception of sets 1 and 10 in Fig. 4.1) and pressed lithium hydride (with the exception of set 8) and the data calculated on the basis of Eqs. (4.1) and (4.2), respectively, is within $\pm 15\%$.

1.2. Thermal conductivity of liquid lithium hydride

The thermal conductivity of liquid lithium hydride was measured at temperatures between 700 and 1100 °C by Novikov, Akhmatova, Kraevoi, Chirov, and Ukolov at the Engineering–Physics Institute in Moscow in 1959 (see Ref. 70). The measurements were carried out by the nonstationary continuous-heating method which was developed by Kraevoi.[17,26] The substance to be studied was placed in a narrow gap between two coaxial cylinders fabricated from Armco iron. The radial heat flux was directed inward toward the axis and the temperature increased very slowly. Under certain conditions the external surface of the cylinder was isothermal and the temperature of the external surface increased monotonically.

The working section had the following structural dimensions: the radius of the coaxial inner cylinder $R_1 = 8.5$ mm, the inner radius of the outer cylinder $R_2 = 9.5$ mm, and the height of the inner cylinder $h = 80$ mm. The thickness of the layer of the sample was accordingly 1 mm. The temperature was measured along the axis of the inner cylinder and at the wall of the outer cylinder.

During the experiment, the layer of lithium hydride to be studied was brought in contact with the loading chamber, which held argon at a pressure of 25–30 atm.

The thermal conductivity was calculated on the basis of the equation

$$\lambda = \frac{c\gamma[R_2^2 - R_1^2 - 2R_1 \ln(R_2/R_1)] + c_1\gamma_1 2R_1^2 \ln(R_2/R_1)}{4\Delta t(1-\psi)}(1 + x + \nu + \epsilon), \tag{4.3}$$

where c and γ are, respectively, the specific heat and the density of the test sample; c_1 and γ_1 are the specific heat and the density of the inner cylinder; Δt is the lag time of the temperature at the inner surface of the layer under study, in comparison with the temperature at the outer surface; ψ is a correction for the temperature drop in the outer and inner cylinders; x is a correction for the end effects; ϵ is a correction which takes into account the change in the heating rate (ϵ is 1–2% in the experiment); and ν is a correction which takes into account the heat loss through the inner cylinder supports. The maximum error of the data obtained by the authors was estimated to be 10%–15%. Table 4.6 gives the measured values of the thermal conductivity λ(kcal/m h deg) of liquid lithium hydride.

Within the error limits of the initial experiment, these data can be described by a linear dependence for λ(W/m deg):

$$\lambda = 3.72 \times 10^{-3} T - 2.37. \tag{4.4}$$

Table 4.6. Thermal conductivity of liquid lithium hydride from the data of Novikov *et al.*

T,°C	ψ	κ	ν	λ	T,°C	ψ	κ	ν	λ
		Set I			825	0.349	0.033	0.031	1.69
696	0.294	0.064	0.047	1.17	848	0.418	0.028	0.030	1.78
714	0.306	0.058	0.044	1.28	861	0.420	0.028	0.029	1.74
733	0.324	0.053	0.041	1.22	879	0.432	0.026	0.028	1.67
752	0.338	0.047	0.039	1.19	892	0.440	0.025	0.028	1.59
788	0.336	0.039	0.034	1.18	941	0.466	0.022	0.026	2.05
807	0.380	0.036	0.033	1.24	959	0.474	0.0215	0.024	2.08
825	0.392	0.032	0.031	1.32	980	0.480	0.020	0.024	2.12
843	0.407	0.030	0.030	1.34	993	0.484	0.020	0.024	2.11
870	0.426	0.028	0.030	1.33	1012	0.490	0.019	0.024	2.10
897	0.443	0.025	0.028	1.50	1028	0.494	0.019	0.024	2.07
950	0.470	0.022	0.025	2.00			Set III		
968	0.478	0.022	0.024	2.04	705	1.31	0.06	0.045	1.25
985	0.484	0.020	0.024	2.07	733	0.324	0.052	0.041	1.23
1002	0.488	0.019	0.024	2.10	761	0.344	0.045	0.037	1.21
1020	0.490	0.019	0.024	2.08	788	0.366	0.039	0.034	1.18
1037	0.495	0.019	0.024	2.08	825	0.394	0.033	0.031	1.45
1053	0.498	0.018	0.023	2.04	852	0.412	0.029	0.030	1.53
1071	0.501	0.018	0.022	2.05	888	0.438	0.026	0.028	1.58
1088	0.503	0.018	0.022	2.25	968	0.478	0.024	0.021	2.10
1105	0.505	0.018	0.022	2.40	1002	0.488	0.019	0.024	2.14
		Set II			1037	0.495	0.019	0.024	2.10
742	0.33	0.050	0.040	1.28					
817	0.386	0.034	0.032	1.57					

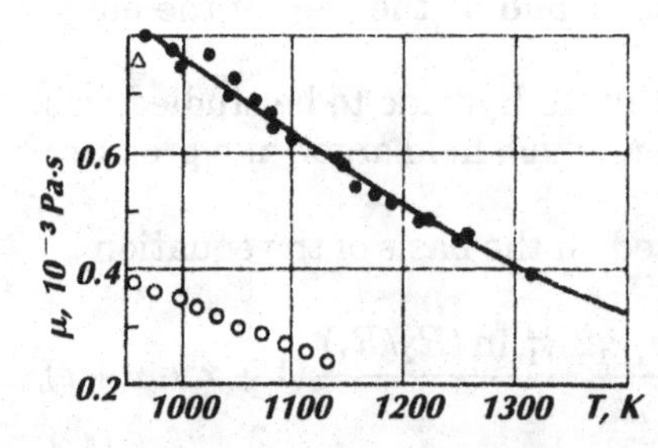

Figure 4.2. Dynamic viscosity of lithium hydride based on the data of Novikov, Solov'ev *et al.* (●) and the data of Welch (○): Δ—calculated from Eq. (4.5); — — calculated from Eq. (4.7).

Table 5.15 gives the values of the thermal conductivity of liquid lithium hydride calculated on the basis of Eq. (4.4).

2. Viscosity of liquid lithium hydride

The viscosity of lithium hydride is one of the least studied properties of this compound. The results of the two known studies of the viscosity of this compound differ from each other by a factor of more than 2.

The experimental studies of Welch *et al.*[249] and Welch[250] were carried out over the temperature interval 680–860 °C with the use of a so-called Brookfield viscometer. The results of the measurements, which were carried out in hydrogen at a pressure of 1.5 atm, are presented in Table 4.7 and in Fig. 4.2. Neither the experimental procedure nor the design of the apparatus was

Table 4.7. Dynamic viscosity of liquid lithium hydride from the data of Welch (Refs. 249, 250).

T,°C	μ,cP	T,°C	μ,cP
680	0.38	780	0.303
700	0.364	800	0.288
720	0.349	820	0.272
740	0.334	840	0.257
760	0.318	860	0.242

Table 4.8. Kinematic viscosity of liquid hydride from the data of Novikov *et al.*

T,°C	ν,cS	T,°C	ν,cS	T,°C	ν,cS	T,°C	ν,cS
Set I		Set II		805	1.146	943	0.849
715	1.290	693	1.316	836	1.073	950	0.848
764	1.173	722	1.237	867	1.012	975	0.803
809	1.092	744	1.287	890	0.954	983	0.810
871	0.991	767	1.230	916	0.902	1041	0.714
912	0.931	793	1.169				

described in these reports It is evident from the study of Tiede,[227] however, that the method used in conjunction with the Brookfield viscometer is a modified coaxial-cylinder method which is used to determine the viscosity from the angle of twist of an elastic filament. One end of the filament is vertically attached to the electric-motor shaft rotating at a constant velocity. A cylinder, which is lowered into a cylindrical vessel containing the substance to be studied at a given temperature, is suspended from the other end.

The viscosity of lithium hydride was measured over a broader temperature interval (690–1050 °C) by Novikov, Solov'ev, and Popov in the Engineering-Physics Institute in Moscow in 1959 (see Ref. 70). They used the method of torsional vibration of a beaker filled with the substance to be studied (this method was developed by Shvidkovskii[67]). The beaker was made of 1X18H9T steel and the suspension filament was fabricated from a Nichrome wire 0.2 mm in diameter. The moment of inertia of the system was determined by a preliminary calibration of the system with the sample disk and without it. The free space above the surface of lithium hydride in the beaker was filled with argon. The relative error of the data obtained by the authors was estimated to be 3%. The measurement results are given in Table 4.8 and in Fig. 4.2.

The results of the two studies are rather difficult to compare, since data on the dynamic viscosity are presented in the first study, while data on the kinematic viscosity are given in the second. A recalculation using values of the density of the liquid introduces an additional uncertainty. At the same time, the discrepancy between these data (see Fig. 4.2) is so great that it cannot be accounted for solely by the errors contained in the values of the density used in the recalculations.

Analysis of the experimental method used by Novikov and co-workers shows that, although the errors of the data obtained in this study may in fact be slightly higher than those found by the authors because of the inaccurate determination of the composition of the sample under study and because of some other factors, it cannot be of a magnitude that would explain the existing discrepancy. Unfortunately, the information about the work of Welch is quite limited, so that no reasonable conclusion about the quality of the data can be drawn. To find some additional justification for various recommendations, it was decided to use an indirect method—calculate the viscosity of lithium hydride at the melting point on the basis of the Andrade formula,[97] whose validity was confirmed by measuring many unassociated liquids—and then to compare the results with the data of each study. Andrade's formula is

$$\mu = 5.1\times10^{-4}(AT)^{1/2}/V_A^{2/3}, \tag{4.5}$$

where μ is the dynamic viscosity (P), A is the atomic or molecular mass, T is the melting point (K), and V_A is the atomic or molar volume (cm^3/mole).

A comparison of the dynamic viscosity of lithium hydride near the melting point clearly shows that the choice favors the data of Novikov and co-authors: $\mu = 0.38$ cP according to the data of Welch, $\mu = 0.77$ cP according to the data of Novikov, and $\mu = 0.78$ cP according to Andrade's formula.

On the basis of the data of Novikov, an interpolation equation was obtained for calculating the kinematic viscosity ν(m^2/s) of lithium hydride over the temperature interval from the melting point to 1300 K:

$$\ln \nu = 19.259 + (10.002\times10^3/T) - (4.320\times10^6/T^2). \tag{4.6}$$

This equation describes the experimental data of Novikov within $\pm 3\%$. The values of the kinematic viscosity, whose error is apparently about 5%, were calculated using Eq. (4.6). These values are presented in Table 5.16.

For engineering calculations, an interpolation equation can be formulated for calculating the dynamic viscosity μ(Pa s) of lithium hydride on the basis of Eq. (4.6) for the kinematic viscosity and on the basis of Eq. (2.111) (see Chap. 2, Sec. 2.4) for the density of liquid lithium hydride. The result takes the form

$$\ln \mu = -27.882 + (11.976\times10^3/T) - (5.121\times10^6/T^2). \tag{4.7}$$

The values of the dynamic viscosity of lithium hydride, calculated from Eq. (4.7), are presented in Table 5.16.

3. Electrical conductivity of lithium hydride and lithium deuteride

The electrical conductivity of lithium hydride and lithium deuteride in the condensed phase has been studied extensively (Refs. 4, 6, 166, 194, 201, 204, 230, and 231) using samples of various purity (principally in terms of lithium admixture). A table of experimental results, however, has been published only in a study by Moers.[194]

In ionic crystals the electrical conductivity is caused in most cases by the presence of initial defects in the crystal lattice (vacancies, divergence from stoichiometry, etc.)—the so-called extrinsic conductivity. At high temperatures, an additional source of defects appears as a result of thermal activation, anharmonicity, etc., giving rise to extrinsic conductivity which is associated with the thermodynamic features of the crystal. This temperature region is said to be part of the zone immediately preceding the melting point (in the case of lithium hydride, the zone near the melting point begins at temperatures on the order of 800–850 K).

It was shown in Refs. 6, 204, 230, and 231 that the electrical conductivity of solid-state LiH and LiD satisfies the Arrhenius law:

$$\bar{\sigma} = (n_1 + n_2)\omega \exp(-\Delta h_M/RT), \tag{4.8}$$

where n_1 and n_2 are, respectively, the molar concentrations of the initial (extrinsic) defects and the additional (intrinsic) defects; ω is the scale transport frequency of the defects (it is generally assumed that this frequency is on the order of the Debye frequency); Δh_M is the activation energy for defect migration; and R is the universal gas constant.

In these studies, it was pointed out that at $T \leqslant 800$ K, the activation energy Δh_M of both LiH and LiD is reasonably constant in the extrinsic conductivity region, lying in the range 0.52–0.53 eV (per mole). At the same time, it was shown experimentally that the coefficient of the exponential function in Eq. (4.8) depends essentially on the presence of impurities and other imperfections in the original crystal.

In the region of pronounced intrinsic conductivity of the crystal, the concentration of the defects that transport the electric current increases appreciably with temperature in accordance with the relation (see Ref. 22, for example)

$$n_2 = N_A \exp(-\Delta h_f/RT),$$

where N_A is Avogadro's number and Δh_f is the enthalpy of formation of intrinsic defects.

Varotsos[231] found that Δh_f is 2.50 eV for LiH and 2.54 eV for LiD. The entropies of formation of these defects are 15.9 and 14.68 e.u., respectively. The exponential part in Eq. (4.8), including its contribution to n_1, is thus essentially independent of the isotopic composition of lithium hydride. Although most authors consider the conductivity of lithium hydride to be ionic in nature, there is also a viewpoint which holds (see Ref. 6) that the electrical conductivity of lithium hydride is electronic in nature.

Let us analyze the work of Moers[194] in greater detail, since this work contains the only table of experimental results on the electrical conductivity of lithium hydride. Moers studied the electrical conductivity of lithium hydride in the temperature interval 440–750 °C. The measurements were carried out in a 400-mm-long steel tube with an inside diameter of 20 mm and a welded bottom. Two short butting tubes for hydrogen supply and for pumping were soldered into this tube slightly below the open end. The tube, doubling as one of the electrodes, was used for containing lithium hydride. The other electrode was a steel rod 450 mm long and 5 mm in diameter. This rod was inserted into a casing made of a thin glass tube. The upper plug was sealed in a thick layer of cement.

To completely eliminate short circuits, the electrode was centered relative to the tube by means of several concentric glass tubes which were inserted one inside the other and sealed in liquid glass. The electrode could be moved with a slight resistance in the central channel, and the glass tubes could be moved in the steel tube. To hold the glass tubes in place in the upper part of the steel tube, indentations were made in it with a vice in two places 110 mm below the point where the tube was attached to the apparatus.

The apparatus was heated in an electric furnace, into which it was inserted to a depth at which only the part that contained lithium hydride was heated before the establishment of the working temperature. The platinorhodium–platinum thermocouple used for measuring the temperature was soldered to the bottom of the tube. Since the thermocouple broke off frequently, it was subsequently moved to the upper end of the tube, instead of resoldering it each time. The thermocouple was carefully calibrated. The section of the reactor tube above the furnace was cooled intensively.

The apparatus was flushed with hydrogen two or three times before each experiment. It was then loaded with 1–2 g of lithium, limiting contact with air as much as possible, and lithium hydride was produced by reaching in hydrogen. The experiments were carried out at a hydrogen pressure slightly greater than atmospheric pressure. Lithium was melted and hydride was produced with the electrode in the raised position. The electrode was then lowered into the molten lithium hydride to the required depth. The initial measurements were carried out using direct current. At room temperature, the electrical conductivity initially was essentially zero, but then rose rapidly.

At intermediate temperatures, the instrument readings fluctuated sharply. The contact resistance was eliminated. To reduce the fluctuations in this temperature regime, the apparatus was held at a constant temperature for 2 h. However, the resistance did not remain constant even after taking this measure: It decreased as the current was increased and then varied as a function of time even at a constant current. These phenomena, in the author's view, were attributable to polarization. To eliminate these phenomena, the subsequent experiments were carried out with an alternating current. In these measurements, the furnace held the temperature constant within 1°. The resistance was controlled by means of Kohlrausch's alternating-current bridge with a null detector. The minimum signal was accurately determined at intermediate temperatures (roughly before the melting point). In the author's view, the readings are quite accurate in this temperature interval. The error is attributable solely to heat-flux oscillation.

Above the melting point, the conductivity was too high to be measured accurately with this apparatus. Furthermore, even at low temperatures the substance behaved as a dielectric in a vessel because of the large resistance, so that the accuracy of the data is low even in this case. The measured values of the electrical conductivity are given in Table 4.9. The slope of the conductivity versus the reciprocal of the temperature is nearly constant. Even the melting point, in author's view, has virtually no effect on this slope. The resistance increased only at a temperature of 761.5 °C after a long holding time "probably because of strong evaporation and loss of a large quantity of the conducting substance," according to the author. Moers described his experimental data on the temperature dependence of the conductivity by

Table 4.9. Experimental and calculated data of Moers (Ref. 194) on the electrical conductivity of lithium hydride (S/cm^{-1}).

T,°C	$\bar{\sigma}_{exp.}$	$\bar{\sigma}_{calc}$	T,°C	$\bar{\sigma}_{exp.}$	$\bar{\sigma}_{calc}$
443	2.124×10^{-5}	2.124×10^{-5}	638	0.011 39	—
507	2.113×10^{-4}	2.154×10^{-4}	656.5	0.018 02	1.811×10^{-2}
512.5	2.766×10^{-4}	—	661.5	0.020 18	—
514	2.968×10^{-4}	—	685	0.032 06	3.667×10^{-2}
556	8.447×10^{-4}	9.633×10^{-4}	692	0.040 49	—
570	1.491×10^{-3}	—	725	0.075 96	9.213×10^{-2}
588.5	2.381×10^{-3}	—	734	0.112 5	—
597	3.225×10^{-3}	3.665×10^{-3}	754	1.01	$17.96\ \times10^{-2}$

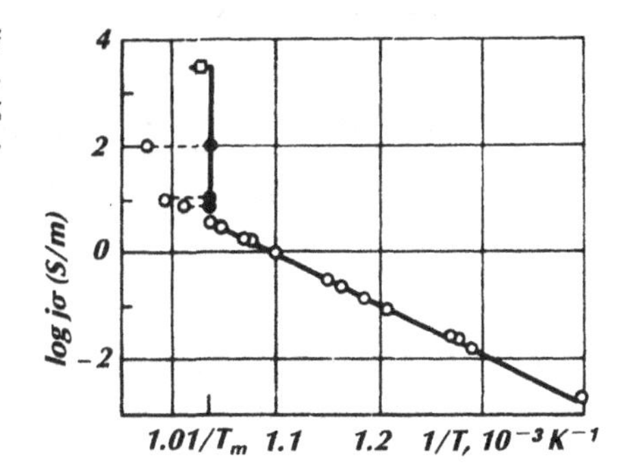

Figure 4.3. Electrical conductivity $\bar{\sigma}$ (Ω^{-1} m^{-1}) of lithium hydride based on the data of Moers (○), Johnson *et al.* (□), the recommendations of Bredig (●), and calculated from Eq. (4.9) (—).

means of a complex parabolic equation of eighth degree with a $\pm$ 10% scatter (except for the last point) (see Table 4.9).

Bredig[109] and Johnson *et al.*[166] reviewed the work of Moers. Bredig[109] pointed out that Moers's temperature measurements contain errors which are particularly large at high temperatures. In particular, the maximum values of the conductivity based on the data of Moers are identified, according to Bredig, with the melting point, rather than with a temperature of 725 or 754 °C (we mean here the calculated values listed in Table 4.9). Johnson *et al.*[166] regarded Bredig's criticism valid and further asserted that the conductivity changes abruptly at the melting point. As proof of this assertion, they cited a value of 30 mho/cm which was obtained by the authors for lithium hydride at a temperature of 700 °C. The temperature dependence of the conductivity in lithium hydride varies strongly in comparison with the other transport properties considered above.

Figure 4.3 is a comparison of the experimental data on the electrical conductivity of lithium hydride, corrected at the melting point, with allowance for Bredig's comment. These data were used by us to find the temperature dependence of the conductivity σ (mho/m) of solid lithium hydride in the form of Eq. (4.8) at temperatures between 710 K and the melting point:

$$\bar{\sigma} = 1.1494\times10^{10}\exp(-21\,069/T). \tag{4.9}$$

The scatter of the experimental points obtained by Moers relative to this equation is within $\pm$ 10% (with the exception of the points at temperatures of 556 and 558.5 °C).

Analysis of Eq. (4.9) shows that the use of a single Arrhenius-type equation with a constant coefficient of the exponential function to describe the conductivity of lithium hydride at intermediate and high temperatures gives the activation energy $\Delta h = 1.82$ eV. If it can be concluded on the basis of the data presented above that the argument of the exponential function in the equation for the conductivity is essentially independent of the isotopic modification of solid lithium hydride, then these data are enough to conclude that the coefficient of the exponential function is not affected by isotopy.

A table of smoothed values of the electrical conductivity of LiH has been compiled on the basis of Eq. (4.9), with allowance for the conductivity of liquid lithium hydride measured by Johnson *et al.*[166] at a temperature of 700 °C (see Table 5.17). These data fall in the category of pure (grade-A) lithium hydride, in which the mole content of metallic lithium impurity can be as high as 1%.

4. The kinetic coefficients of Li-LiH (LiD, LiT) systems in the vapor phase

4.1. Review of the experimental and theoretical situations

Extensive experimental studies of lithium–hydrogen (deuterium, tritium) gas mixtures have dealt solely with the thermodynamic properties and have not involved the transport properties of the vapor. Furthermore, even the transport properties of pure lithium vapor have not been studied experimentally until now, although reliable data are now available for other alkali-metal vapors (see Ref. 89, for example).

The theoretical methods of investigation should accordingly be considered first. Several studies involving calculations of the transport properties of a hydrogen and lithium gas mixture have been published in the first half of the 1960s. Krupenie *et al.*[173] studied a binary mixture of lithium and hydrogen atoms, which is a completely unacceptable approximation, since calculations of the vapor composition over the temperature and pressure ranges of interest show that the vapor contains hydrogen molecules, not hydrogen atoms.[84] Belov and Klyuchnikov[7] studied the viscosity, but ignored certain important vapor components such as Li_2 dimers. The presence of molecules in the vapor primarily effects the thermal conductivity but also has an appreciable effect on the viscosity. Krupenie *et al.*[173] and Belov and Klyuchnikov[7] used somewhat crude estimates for the collision integrals. As for the theoretical results for transport coefficients of pure lithium vapor that have been published (Refs. 68, 110, and 177), these results are even qualitatively at variance with the general experimental results for alkali metals.[89] It was found experimentally that the viscosity of vapor decreases with increasing pressure, while theory predicts the reverse. Furthermore, the calculations mentioned above considerably overestimate, in comparison with the experimental results, the effect of energy transfer from the dissociation reaction on the total heat flux. These drawbacks were attributed to the very low accuracy of calculations of the monomer–dimer collision integrals for lithium. It should be pointed out that atom–molecule collisions for lithium have an appreciable effect on the thermal conductivity and viscosity of lithium–hydrogen gas mixtures. In view of the increasing interest in the transport properties of alkali metal–hydrogen gas mixtures, these calculations cannot be considered satisfactory.

Shpil'rain and Polishchuk[80] considered the feasibility of calculating the potential of interaction between the hydrogen atoms and alkali metals. A calculation of this sort has made it possible to refine the variational calculations of the Li-H potential[120] at large spatial separations, which are of central importance for calculating the collision integrals in the temperature interval considered in this monograph. These results were subsequently used by Shpil'rain and Polishchuk[83] to calculate the potentials of interaction of a Li atom with H_2, LiH, and Li_2 molecules.

The interaction potentials for the components of lithium–hydrogen gas mixtures are strongly anisotropic. As calculations of the collision integrals in a very simple case of an anisotropic interaction (alkali dimer–helium atom) have shown, omission of the anisotropic nature of the potential may lead to a 20%–30% error.[47] It is therefore essential to be able to determine the accuracy of the calculation results. The potentials of interaction of lithium atoms with molecules in lithium–hydrogen gas mixtures are strongly attractive. This has been confirmed by experimental observations of stable Li_3, Li_2H, and LiH_2 molecules. The monomer–dimer interaction potentials for alkali-metal vapor are of exactly the same nature, as has been indicated by an experimental detection of trimers and more complex clusters in alkali-metal vapors.[141] The common nature of the potential surfaces of the atom–molecule systems mentioned above has made it possible to use a single method of calculating the collision integrals. This method, used by Polishchuk *et al.*,[48] is described below in greater detail.

The fact that general calculation methods can be used is of crucial importance, since the accuracy of approximate calculations is very difficult to determine *a priori* and it may be useful in this case to compare the calculation results of the collision integrals with the existing experimental data on alkali-metal vapor. In this respect, the presence of reliable experimental data on sodium, potassium, rubidium, and cesium vapor has made it possible to use alkali-metal vapor as a kind of reference point for choosing a reliable method of calculating the collision integrals. A good agreement between the calculated and experimental results for the atom–molecule collision integrals in the case of alkali metals[48] and for the thermal conductivity and viscosity of alkali-metal vapor[49] has made it possible to treat with confidence analogous results for a lithium–hydrogen gas mixture.[83]

4.2. Interaction potentials of the atomic particles in lithium–hydrogen gas mixtures

Calculation of the interaction potentials of atomic particles is one of the more complex problems in the theoretical determination of the transport properties of vapor. The need for various approximate methods for calculating the potentials accounts for the difficulty of the problem. In particular, the popularity of the asymptotic method of calculating the potentials has recently been rising.[60] According to this method, the interaction potential can be represented as a sum of the exchange and long-range interactions when the distance between atomic particles is considerably greater than their size. The long-range interaction is usually well known. Calculation of the exchange potential, on the other hand, generally requires a considerable effort. The asymptotic interaction potentials can be used most effectively at spatial separations considerably larger than the size of the interacting particles, where the variational calculations are generally not reliable.[142] The simple analytic form of the asymptotic potentials greatly simplifies the calculation procedure. If the intermediate and small spatial separations are the dominant components of the scattering processes, the use of the method of diatomic complexes in molecules, which utilizes the semiempirical atom–atom interaction potentials, would be more desirable.[42] In this section, we use these methods to analyze the atom–atom and atom–molecule interaction potentials for the components of lithium–hydrogen gas mixtures.

The procedure for calculating the interaction potentials between atoms

The method for calculating the atom–atom asymptotic interaction potentials was first introduced by Smirnov[56] for alkali-metal atoms and its use was extended by Shpil'rain and Polishchuk[80] to the case of the interaction of hydrogen and alkali-metal atoms. Since the results of studies by Smirnov have been discussed in detail in the literature, we shall discuss below mainly the interaction of hydrogen and lithium atoms.

The potential of interaction between the atoms indicated above can be found by solving the Schrödinger equation for the wave function of two valence electrons. The complete Hamiltonian H of a two-electron system can be represented in the form (atomic units are used in all cases in Sec. 4.2)

$$H = H_0 + V; \quad H_0 = -\tfrac{1}{2}\Delta_1 - \tfrac{1}{2}\Delta_2 + U_{a1} + U_{b2};$$
$$V = U_{a2} + U_{b1} + 1/R + 1/|\mathbf{r}_1 - \mathbf{r}_2|, \tag{4.10}$$

where U_{a1}, U_{a2}, U_{b1}, and U_{b2} are the potentials of interaction of electrons 1 and 2 with the atomic cores a and b, and $\mathbf{r}_1$ and $\mathbf{r}_2$ are the electron coordinates.

We introduce the wave functions $\Psi^s(\mathbf{r}_1, \mathbf{r}_2)$, and $\Psi(\mathbf{r}_1, \mathbf{r}_2)$, which satisfy the Schrödinger equations:

$$(H_0 + V)\Psi^s = E^s\Psi^s;$$
$$H_0\Psi = E\Psi = (E_a + E_b)\Psi, \tag{4.11}$$

where s is the total electron spin, and Ψ^s is the wave function of a two-electron system which takes into account the distortions caused by the interaction of atoms. Multiplying the first equation in Eqs.. (4.11) by Ψ and the second by Ψ^s and then subtracting one equation from the other, we find the following expression after integration over the volume Ω in $(\mathbf{r}_1, \mathbf{r}_2)$ space:

Figure 4.4. Coordinate system used in analysis of the interaction potentials between hydrogen atoms and alkali metals.

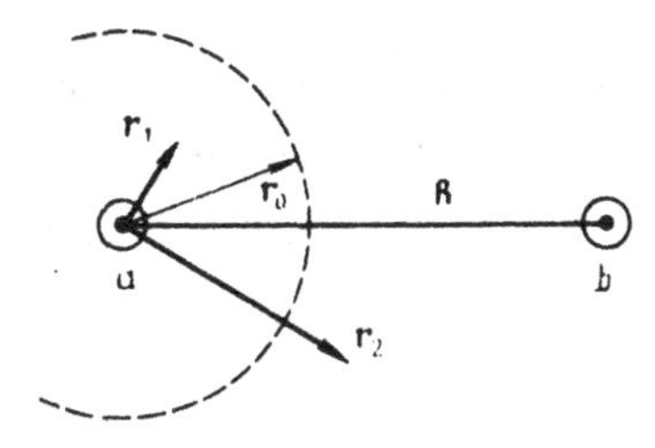

$$(E^s - E)\int_\Omega d^3r_1\, d^3r_2\, \Psi^s\Psi = \int d^3r_1\, d^3r_2 \Psi H_0\Psi^s - \Psi^s H_0\Psi) + \int_\Omega d^3r_1\, d^3r_2\, \Psi V\Psi^s. \tag{4.12}$$

Let a be the hydrogen atom and b the lithium atom with ionization energies that satisfy the inequality $\alpha^2/2 \gg \beta^2/2$. As Ω we use the following volume (Fig. 4.4): $\mathbf{r}_1$ is arbitrary and $|\mathbf{r}_2| \geqslant r_0$, where $1/\alpha \ll r_0 \ll R$. According to Ovchinnikova,[43] in the region $r_1 \sim 1/\alpha, r_2 \geqslant r_0 \gg 1/\alpha$, the wave function Ψ^s can be written in the form

$$\Psi^s = (1/\sqrt{2})\, \Psi^s_a(\mathbf{r}_1)\Psi^s_b(\mathbf{r}_2), \tag{4.13}$$

where Ψ^s_a and Ψ^s_b are single-electron wave functions which, in the absence of interaction, transform to the wave functions Ψ_a and Ψ_b of the unperturbed atoms. Substituting expression (4.13) into Eq. (4.12), we find

$$E^s - E = \tfrac{1}{2}\int_S dS(\Psi^s_b \nabla\Psi_b - \Psi_b \nabla\Psi^s_b) + \int_\Omega (\Psi V\Psi^s) d^3r_1\, d^3r_2. \tag{4.14}$$

In the calculation of Eq. (4.14) we have replaced the integral on the left-hand side of it by unity with an accuracy on the order of $\exp(-2\beta R)$ and transformed to an integral on the inner surface S of a sphere of radius r_0. The difference $E^s - E$ is the interaction potential which we are seeking.

The second integral in Eq. (4.14), denoted by V_g, is defined by the regions $r_1 \sim 1/\alpha \ll R$, $r_2 \sim 1/\beta \ll R$, so that for V we can use in this region the expression

$$V = -\mathbf{d}_a\mathbf{d}_b + 3(\mathbf{d}_a\mathbf{n})(\mathbf{d}_b\mathbf{n})/R^3, \quad \mathbf{n} = \mathbf{R}/R, \tag{4.15}$$

where $\mathbf{d}_a$ and $\mathbf{d}_b$ are the dipole-moment operators of the atoms.

Since the wave functions near their nuclei are distorted only slightly, we can use the perturbation theory in V. We then obtain

$$V_g = -(C_6/R^6) - (C_8/R^8) - (C_{10}/R^{10}), \tag{4.16}$$

where C_6, C_8, and C_{10} are constants of the long-range interaction for Li-H: $C_6 = 66$, $C_8 = 3127$, and $C_{10} = 2.1\times 10^5$ (Ref. 206). Since Eq. (4.16) is an asymptotic series, each successive term of the series is retained if it is not larger than the preceding term.[57] Analysis of expression (4.16) shows that for Li-H, the term C_8 should be taken into account, beginning with $R \approx 7$–$8\ a_0$, and the term C_{10} should be taken into account, beginning with $R \approx 8$–$9\ a_0$.

According to the results of a study by Shpil'rain and Polishchuk,[80] the wave function Ψ^s_b can be represented as a product $\Psi_{\rm Li} X^s_{\rm Li}$, where $X^s_{\rm Li}$ are

Table 4.10. Exchange part of the triplet potential of Li-H (a.u.) [comparison of the results obtained in Ref. 120 and those calculated on the basis of Eq. (4.17)].

R, a_0	Eq. (4.17)	Ref. 120	R, a_0	Eq. (4.17)	Ref. 120
6	4.52×10^{-3}	3.28×10^{-3}	8	5.22×10^{-4}	5.22×10^{-4}
6.5	2.68×10^{-3}	2.17×10^{-3}	8.5	2.99×10^{-4}	2.99×10^{-4}
7	1.65×10^{-4}	1.37×10^{-3}	9	$1.7\ \times10^{-4}$	$1.6\ \times10^{-4}$
7.5	9.04×10^{-4}	8.34×10^{-4}			

corrections due to the effect produced by a hydrogen atom. For the first integral in Eq. (4.14), we then find

$$V_{\text{exch}} = (B^s/2)|\Psi_{\text{Li}}(R)|^2, \tag{4.17}$$

where B^s is a factor set by the correction X^s_{Li}. Using the asymptotic expression for the function $\Psi_{\text{Li}}(R)$, which holds at large distances,[57]

$$\Psi_{\text{Li}}(R) = (A/\sqrt{4\pi})R^{(1/\beta)-1}\exp(-\beta R), \tag{4.18}$$

we find a definitive relation for the interaction potential of Li-H in the form of a sum of the exchange and long-range parts:

$$V^s(R) = (J^s/2)R^{2/\beta-2}\exp(-2\beta R) + V_g(R). \tag{4.19}$$

In considering an arbitrary alkali atom, Shpil'rain and Polishchuk[80] calculated the correction X^s and analyzed the X^s dependence of J^s. The results obtained by them need not necessarily be presented here for the analysis of the potential of Li-H. The variational calculations of Docken and Hinze[120] can be refined on the basis of Eq. (4.19) in the following way.

Let us analyze the results obtained by Docken and Hinze[120] for the triplet state ($s = 1$). These results are evidence of the absence of a van der Waals potential well up to large distances ($\sim 12a_0$). The method used by Docken and Hinze[120] to calculate the potential did not allow them to accurately determine the long-range interaction. Their method describes essentially the exchange part of the potential. This conclusion is substantiated* in Table 4.10, in which the results of calculation of the triplet exchange interaction on the basis of Eq. (4.17) for $J^1 = 1.94$ are compared with the results of Docken and Hinze.[120] As can be seen from this comparison, the analytic form of potential [Eq. (4.17)] is sufficiently flexible to describe the exchange interaction.

There is another indirect confirmation of the validity of Eq. (4.19). The distance σ at which the exchange interaction is equal to the van der Waals interaction is $7a_0$ for H-H and $10.4\ a_0$ for Li-Li (Ref. 57). If the interaction potential is thought of as a sum of the exchange term and the van der Waals term, then the meaning of the parameter σ will be the same as for the Lenard-

*The number of calculated data points in the asymptotic region provided by Docken and Hinze[120] is inadequate for a meaningful comparison. The results of their study have therefore been approximated (with a relative error of 2%) in the region of a_0 (8 – 12) by a smooth curve.

Table 4.11. Singlet potential of Li-H (a.u.) [comparison of the results obtained in Refs. 93 and 120 and calculated on the basis of Eq. (4.19)].

R, a_0	Eq. (4.19)	Ref. 93	Ref. 120	R, a_0	Eq. (4.19)	Ref. 93	Ref. 120
6	2.8×10^{-2}	2.14×10^{-2}	2.3×10^{-2}	8.5	1.84×10^{-3}	1.84×10^{-3}	1.94×10^{-3}
6.5	1.65×10^{-2}	1.4×10^{-2}	1.5×10^{-2}	9	1.05×10^{-3}	1.06×10^{-3}	—
7	9.62×10^{-3}	8.78×10^{-3}	—	9.5	5.09×10^{-4}	6.17×10^{-4}	—
7.5	5.58×10^{-3}	5.32×10^{-3}	5.74×10^{-3}	10	3.4×10^{-4}	3.62×10^{-4}	3.55×10^{-4}
8	3.2×10^{-3}	3.15×10^{-3}	—	10.5	1.92×10^{-4}	2.14×10^{-4}	—

Table 4.12. Parameters which determine the interaction potentials of Li-H and Li-Li.

System	J^0	J^1	β	C_6	C_8
Li-H	-11.96	1.84	0.63	66.5	312.7
Li-Li	2.71×10^{-2}	2.71×10^{-2}	0.63	1400	724×10^2

Jones potential, $4\epsilon\,[(\sigma/r)^{12} - (\sigma/r)^6]$. According to known empirical combination rules, the parameter σ for Li-H is therefore $(7\text{–}10.4)/2 \simeq 9$. Incorporating the term $-C_6/R^6$ into Eq. (4.19), we find that the triplet potential of Li-H in this case does in fact change sign at $R \approx 9a_0$. Accordingly, at $R \gtrsim 7a_0$, the preferable results are those for the triplet potential, which are given by Eq. (4.19).

For the singlet state, the long-range interaction in Eq. (4.19) is considerably weaker than the exchange interaction at distances $(6\text{–}10)a_0$. Table 4.11 compares the results of calculations based on Eq. (4.19) for $J^0 = -11.96$ with the results of Refs. 93 and 120. This comparison shows that asymptotic expression [Eq. (4.19)] accurately describes the change in the singlet potential at these distances. It is more appropriate to use Eq. (4.19) for large distances, where the variational method cannot in principle give an asymptotically exact expression.[141]

The expression for an interaction potential between lithium atoms found by an asymptotic method[56] is

$$V^{s}(R) = \mp (J/2)\, R^{7/2\beta - 1} e^{-2\beta R} + V_g, \tag{4.20}$$

where the $\mp$ superscript corresponds to the singlet state and the subscript corresponds to the triplet state, respectively. Interestingly, in contrast with the interaction between hydrogen and lithium atoms, in the case of lithium atoms the exchange interaction for singlet and triplet terms is the same in absolute value.

In summary, the parameter values which determine the interaction potentials of Li-H and Li-Li are given (in atomic units) in Table 4.12 for reference.

The interaction potentials of Li atoms with Li_2, H_2, LiH, Li_2H, and Li_3 molecules

For atoms with a single outer *s* electron, the interaction potential between a triatomic molecule and an atom, which correlates with the ground state of the interacting particles at infinity, is[17]

$$U=\frac{1}{2}\sum_{a\neq b} Q_{ab}-\frac{1}{2}\left[\frac{1}{2}(J_{ab}+J_{cd}-J_{ac}-J_{bd})^2\right.$$
$$\left.+\frac{1}{2}(J_{ac}+J_{bd}-J_{bc}-J_{ad})^2+\frac{1}{2}(J_{bc}+J_{ad}-J_{ab}-J_{cd})^2\right]^{1/2};$$
$$Q=(U^{(1)}+U^{(3)})/2;\quad J=U^{(1)}-U^{(3)}, \tag{4.21}$$

where the subscripts *a*, *b*, *c*, and *d* denote different atoms, and $U^{(1)}$ and $U^{(3)}$ are the lowest singlet and triplet terms of diatomic systems; the equation for the interaction potential surface of the *a* atom with the *bc* molecule can be derived from Eq. (4.21) by eliminating all quantities that contain the subscript *d*.

The quantities $J(R)$ and $Q(R)$ in Eq. (4.21) can be determined by various methods. One possible method involves the use of asymptotic potentials. An approximation of this sort can legitimately be used at distances greater than the size of the interacting particles. The asymptotic equations can easily be used because of their simplicity. As asymptotic potentials we have used the results of Smirnov[56] for the interaction of lithium atoms and the results of Shpil'rain and Polishchuk[80] for lithium and hydrogen atoms.

In the calculation of the atom-molecule interaction potentials, the distances in the molecule were assumed equal to the equilibrium distances. For potentials of binary interaction between atoms in the molecule, we have therefore used semiempirical potentials of Li-Li, taken from Ref. 202, and the results of variational calculations for H-H and Li-H (see Refs. 120 and 202), which are accurate at intermediate and small distances.

An important refinement is required in the calculation of the potential of LiH-Li, because Eq. (4.21) ignores the dipole moment μ of the LiH molecule. In this case, an appropriate term $-\mu^2\alpha(3\cos^2\theta+1)/2R^6$, where α is the polarizability of the lithium atom—should be added to Eq. (4.21); $\mu=6.0D$ and $\alpha=165$ a.u. If the atom–molecule potential of interaction at intermediate distances must be determined, the semiempirical and variational results mentioned above can be used instead of the asymptotic potentials.

Let us analyze the accuracy of the semiempirical and variational atom–atom curves that are used below. For the singlet curves of Li_2, H_2, and LiH, the well depth D_e and the equilibrium distance r_e for all pairs under consideration are in good agreement with the experimental data. For the triplet potentials, molecular-beam experiments give values for the triplet wells of various alkali atoms, *M* and *M'*, which are approximately equal for all pairs and which gradually increase from Na to Cs. These values are[118,205] 0.025–0.027 eV for Na-K and 0.029–0.033 eV for K-Cs. The value of D_e, used for the triplet curve for Li_2, is approximately equal to the experimental values presented above.

There are no data on the triplet potential of Li-H. However, the accuracy of the results of a variational calculation by Docken and Hinze[120] increases with decreasing distance, as indicated by the good agreement between the

Table 4.13. Parameters ϵ'(K) and r_{min} (nm) of the effective spherically symmetric Lennard–Jones potential.

System	ϵ'	r_{min}	System	ϵ'	r_{min}
Li-Li_2	6 500	0.37	Li-Li_2H	9500	0.34
Li-H_2	11 600	0.21	Li-Li_3	5300	0.50
Li-LiH	4 900	0.34			

experimental and calculated results for singlet potentials at intermediate distances. Since Docken and Hinze[120] used the same set of basis functions for the triplet and singlet distances, the accuracy of the triplet potential is expected to be good at intermediate distances where these results are actually used. The variational calculations of Kolos and Volnevich for the H-H interaction, which were approximated by Pickup,[202] are precision calculations for a broad range of distances.

In the calculations of collision integrals, we use both the atom–molecule asymptotic potentials at large distances and the "semiempirical" results at intermediate distances. The calculated potentials are attractive and anisotropic in nature. These potentials are consistent with the existence of stable trimers and more complex alkali-metal clusters observed experimentally.[141] A comparison of the calculated and experimental data on the dissociation energy of Li_3, Li_2H, and LiH_2 molecules has shown[154,257,258] that the method of diatomic complexes in molecules gives generally lower absolute values of the potential at intermediate distances. For example, the energy of the dissociation reaction $Li_3 \rightarrow Li_2 + Li$ has the calculated value 0.4 eV, which is slightly lower[257] than the experimental value, 0.64 eV.

The following characteristics of the atom–molecule potentials at intermediate distances are used below: the potential well depth ϵ, and the position of the minimum, for the potentials which are averaged over all atom–molecule orientations.

The availability of experimental data on the dissociation energy of Li_3, Li_2H, and LiH_2 molecules makes it possible to refine the initial working parameters ϵ for the potentials of Li-Li_2, Li-LiH, and Li-H_2. A more accurate depth of a spherically symmetric potential well ϵ', can be determined from the equation

$$\epsilon' = \epsilon \, (D_{exp} / D_{calc}), \tag{4.22}$$

where D_{exp} is the experimental value of the dissociation energy and D_{calc} is the calculated value which is determined by the method of diatomic complexes in molecules. The values of ϵ' (ϵ without an experiment) and r_m for the potentials considered above are listed in Table 4.13.

4.3 Calculation methods for the atomic-particle collision integrals in lithium-hydrogen gas mixtures

Calculation of the cross sections for scattering of atomic particles is a formidable computational problem even in the very simple case of atom–atom scattering. The situation becomes even more complicated in the calculation of scattering processes involving molecules, where the anisotropy of the interaction potential and the internal degrees of freedom of molecules are important. At the prevailing accuracy of calculation of potential surfaces, the use of numerical methods (the method of classical trajectories or exact quantum-mechanical analysis) is not justifiable, and approximate calculation methods can instead be used. There are generally no clear-cut guidelines, however, for the use of the available approximate calculation methods. In the more difficult cases, we have therefore used various physically justifiable calculation methods in parallel. As we shall see below, the results of various approximate methods are nearly the same.

Collision integrals for hydrogen and lithium atoms with lithium atoms

We begin our analysis of the calculation of the collision integrals with the simple case of atom-atom scattering. As a candidate for calculation, we consider the collision integrals for hydrogen and lithium atoms. The calculations are based on the asymptotic potentials obtained by Shpil'rain and Polishchuk.[80]

The basic equations which express the collision integrals in terms of the differential scattering cross section[34] $d\sigma(\theta)$ are

$$\sigma^2\Omega^{(l,m)*} = \left[(m+1)!\,\frac{1}{2}\left(1-\frac{1}{2}\,\frac{1+(-1)^l}{1+l}\right)\pi\right]^{-1}\int_0^\infty e^{-\gamma^2}\gamma^{2m+3}Q^{(l)}(g)d\gamma,$$

$$Q^{(l)} = \int_0^\infty (1-\cos^l\theta)d\sigma(\theta), \qquad (4.23)$$

where $\gamma = \frac{1}{2}\mu g^2/T$, μ is the reduced mass of the colliding particles, g is the relative velocity, and T is the temperature. Since hydrogen and lithium atoms in the ground state have one valence s electron each, they can interact along two potential curves corresponding to a total electron spin 0 or 1. Using accordingly s and a to denote the collision integrals and allowing for the statistical weight of each state, we find

$$\sigma^2\Omega^{(l,m)*} = \tfrac{3}{4}\,\sigma^2\Omega_a^{(l,m)*} + \tfrac{1}{4}\,\sigma^2\Omega_s^{(l,m)*}. \qquad (4.24)$$

Let us first consider the scattering at a triplet branch. In the energy range on the order of thermal energies and higher, the triplet potential, a repulsive potential, decreases markedly. A rigid-sphere model, proposed by Palkina *et al.*,[44] can therefore be used in the calculations. This rather simple method proved to be quite effective in calculating the collision integrals for monatomic alkali-metal vapors. The collision integrals are given by simple formulas[19,58]

$$\sigma^2\Omega_a^{(1,1)*} = R_{1-1}^2, \quad U^{(3)}(R_{1-1}) = 2.25T;$$
$$\sigma^2\Omega_a^{(2,2)} = R_{2-2}^2, \quad U^{(3)}(R_{2-2}) = 0.83T, \qquad (4.25)$$

where $U^{(3)}$ is a triplet potential. The difference between the numerical factors in front of T on the right-hand sides of Eqs. (4.25), which determine the colli-

Table 4.14. Collision integrals for hydrogen atoms and lithium atoms, Å².

	Li-H		Li-Li	
T,K	$\sigma^2\Omega^{(1,1)*}$	$\sigma^2\Omega^{(2,2)*}$	$\sigma^2\Omega^{(1,1)*}$	$\sigma^2\Omega^{(2,2)*}$
700	13.9	15.9	17.6	20.1
900	13.0	14.9	16.5	18.9
1100	12.3	14.1	15.6	17.9
1300	11.7	13.5	14.9	17.0
1500	11.2	12.9	14.2	16.3

sion parameters R_{1-1} and R_{2-2}, stems from the determination of the scattering angle as a function of the impact parameter.

In the singlet state, the interaction between hydrogen atoms and alkali metals corresponds to an attraction (the depth of the potential well is ~2 eV). A "capture" model, according to which the scattering is accompanied by the formation of a long-lived complex which subsequently decays isotropically in all directions, can therefore be used. The cross section $\sigma_0(E)$ for the formation of a complex during scattering, with a relative kinetic energy E, is given by the relations[29]

$$\sigma_0(E) = \pi R_0^2 [1 + |U^{(1)}(R_0)|/E];$$
$$U^{(1)}(R_0)/E = 1 - R_0 U^{(1)\prime}(R_0)/2E. \tag{4.26}$$

It was shown in Refs. 19, 44, and 58 that if the slow change in $\sigma_0(E)$ as a function of energy is taken into account, the averaging based on Eq. (4.23) leads to the following results for the collision integrals:

$$\sigma^2\Omega_s^{(1,1)*} = \frac{1}{\pi}\sigma_0(2.52T);$$
$$\sigma^2\Omega_s^{(2,2)*} = \frac{1}{\pi}\sigma_0(3.51T). \tag{4.27}$$

The collision integrals calculated from Eqs. (4.24)–(4.27) are presented in Table 4.14.

The collision integrals for Li-H found in the literature have been calculated previously. The data of Krupenie *et al.*[173] and Belov and Klyuchnikov[7] are 50%–60% higher than the present results, consistent with the error of the collision-integral calculations estimated by the authors. The reason for the discrepancy is that the values of the interaction energy of Li-H atoms in the triplet state are nearly twice as large as they should be. It also stems from the use of a very crude method by Krupenie *et al.*[173] for obtaining the triplet curve.

Unfortunately, a direct determination of the accuracy of the present calculations is more difficult because of the absence of experimental studies. An indirect determination is, however, possible. A comparison of the collision integrals for atomic alkali-metal vapors, calculated by Smirnov and Chibisov[58] by the asymptotic method, and the calculations of Krupenie *et al.*[173] for Li-Li shows that the results of the latter study are, as in our case, 50%–60% higher than those of the former study. A comparison of the data of Smirnov and Chibisov[58] with the experimental results reported by Yargin *et al.*[89] shows a 15%–20% discrepancy. Since the method used by Smirnov and Chibisov[58] to calculate the collision integrals is the same as the method used in this section, the error of the results for Li-H presented in Table 4.14 is estimated to be 15%–20%.

The results obtained by Smirnov and Chibisov[58] are, in our view, the best results of all those found in the literature and are in good agreement with the experimental data for Na, K, Rb, and Cs. There are no experimental data available for lithium. The discrepancy between the theoretical and experimental data for the atom–atom collision integrals is slight (only 15%).

The errors of the results for the collision integrals of alkali-metal atoms are attributed to the approximations that were used in the calculations. In the calculations of singlet collision integrals, Smirnov and Chibisov[58] ignored the repulsive part of the potential. In the case of triplet scattering, the presence of the van der Waals well, which is on the order of 400 K for Li-Li, was apparently ignored in the calculation. The availability of experimental data on the atom–atom collision integrals for alkali metals has made it possible to refine the results of Smirnov and Chibisov[58] for Li-Li scattering. The refinement was possible because the collision integrals of the atoms in the temperature interval up to 1500 K are determined by the asymptotic behavior of the interaction potentials. The asymptotic behavior of the potential curves of Li-Li and Na-Na is very similar.[60] The results of a calculation by Smirnov and Chibisov[58] for Na-Na are 15% higher than the experimental data with a 2%–3% error (see Ref. 89). The results obtained by Smirnov and Chibisov[58] for the collision integrals of Li atoms, which were reduced by 15%, are therefore expected to be very close to the true results. These more accurate values are given in Table 4.14.

Collision integrals for atoms and molecules in a lithium–hydrogen gas mixture

In the absence of experimental data on the transport properties and collision integrals for pure lithium vapors and the lithium–hydrogen system, it is difficult to carry out calculations and to estimate their error. Polishchuk *et al.*[48] suggested a method for calculating the atom–molecule collision integrals for alkali-metal vapors. A comparison with known experimental results for alkali-metal vapor, which were reported in a review by Yargin *et al.*,[89] shows that collision integrals can be determined within 10% by using this method.

Polishchuk *et al.*[49] have shown that the results obtained for the atom-dimer collision integrals for alkali metals have made it possible to theoretically describe for the first time, in a systematic way, the decrease in the viscosity of alkali-metal vapors with increasing pressure, and to eliminate the discrepancy between theoretical and experimental values of the heat flux resulting from the energy transfer in the reaction leading to the formation and destruction of dimers. They have also obtained the first reliable data on the heat conductivity and viscosity of lithium vapor, whose error is within 10%.

It was already pointed out that the general nature of the potential surfaces for atom-molecule interactions makes it possible to use a single procedure for calculating the collision integrals both for a lithium-hydrogen gas mixture and for alkali-metal vapor. This procedure, suggested by Polishchuk *et al.*,[48] will be described below for the particular case of the atom-lithium dimer system.

We shall first discuss the method which is based on the assumption that an intermediate triatomic complex is formed during scattering (the "capture" model). At the energy of relative motion, E, a complex is formed when the colliding particles close in on each other to a distance R, at which point the interaction potential $|U(R)| \gg E$. If the parameter $1/\alpha$ is a typical distance at which a sharply decreasing potential changes appreciably, the colliding particles will converge in a time $\tau \simeq 1/\alpha v$ from a distance R, where $|U(R)| \ll E$, to a distance at which $|U(R)| \gg E$. In this time, the molecule turns through an angle $\Delta\theta \sim v\tau/r_e \sim 1/\alpha r_e$. For abruptly changing, strongly anisotropic potentials $r_e \gg 1/\alpha$, so that the orientation angle changes by $\Delta\theta \ll 1$. During the formation of a complex, the angle θ may therefore be assumed constant. If the angle θ is constant, the "capture" cross section $\sigma_0(\theta,E)$ can be calculated[29] on the basis of the following equations:

$$\sigma_0(\theta,E) = \pi b_0^2(\theta,E), \quad U(r_0,\theta) + r_0 U'(r_0,\theta)/2 = E,$$
$$b_0 = \sqrt{n_0/(n_0-2)}, \quad n_0 = -r_0 U'(r_0,\theta)/U(r_0,\theta). \tag{4.28}$$

If we now assume that all orientations are equally probable for an average cross section, we find

$$\sigma_0(E) = \tfrac{1}{2}\int_0^\pi \sigma_0(\theta,E)\sin\theta\, d\theta. \tag{4.29}$$

If the exact equations of the kinetic theory, in which the inelastic processes are taken into account,[34] are used and if the scattering in the "capture" model is assumed isotropic, we can use the following expressions for the collision integrals:

$$\sigma^2\Omega^{(1,1)*}(T) = \frac{1}{\pi}\int d\gamma\, \gamma^5 \exp(-\gamma^2)\sigma_0(\gamma^2 T);$$
$$\sigma^2\Omega^{(2,2)*}(T) = \frac{1}{3\pi}\int d\gamma\, \gamma^7 \exp(-\gamma^2)\sigma_0(\gamma^2 T), \tag{4.30}$$

where $\gamma = \sqrt{E/T}$. In view of the slight energy dependence of σ_0, it follows from Eq. (4.30) that the energies, which constitute the principal contribution to the transport scattering cross sections, range from 0 to $4T$ for $\sigma^2\Omega^{(1,1)*}$ and from 0 to $6T$ for $\sigma^2\Omega^{(2,2)*}$.

Let us now discuss the feasibility of using the capture model for the scattering in the atom-alkali-metal dimer system. Experimental studies of the exchange reaction $M' + M_2 \to MM' + M$ in molecular beams of alkali metals have shown that at energies of the relative motion of up to $\sim$0.2 eV, the distribution of products is isotropic in the c.m. system.[253] This behavior can be explained in terms of the formation of the collisional complex during the scattering, caused by the attractive nature of the $M-M_2$ interaction. For energies that constitute the principal contribution to the collision integrals, the mechanism for the scattering through the formation of an intermediate complex can occur only near the lower boundary of the temperature interval under consideration, 700–1500 K. At higher temperatures, the number of captured atoms decreases and the repulsive part of the potential becomes dominant.

In view of this circumstance, a different approach to this problem should also be considered. This approach involves the use of an effective, spherically symmetric potential for which the collision integrals have already been tabulated. We shall use the Lennard–Jones potential $\bar{\epsilon}[(r_{min}/r)^{12}-2(r_{min}/r)^6]$. The advantage of this approach is that it has no upper bounds in terms of the energies of the relative motion, E. The repulsive region of the potential is important in the determination of the parameters $\bar{\epsilon}$ and r_{min}. To estimate these parameters, we have therefore used the potential surfaces M_2-M which were found by the method of diatomic complexes in molecules. Since the M_2-M potential is strongly anisotropic, the parameters $\bar{\epsilon}$ and r_{min} cannot be chosen *a priori*. A more reasonable way to determine these parameters is by averaging the M_2-M potential over all orientations. This procedure was carried out in Sec. 4.2, where we have also introduced appropriate refinements in which the experimental data on the dissociation energy of trimers were taken into account.

The monomer–dimer collision integrals for lithium were calculated by two methods which were discussed above. The asymptotic potentials and the capture model were used in the first case and the effective, spherically symmetric Lennard–Jones potential was used in the second case. The calculated cross sections which were obtained by different methods differ by 20%. According to Polishchuk *et al.*,[48] the average of the two theoretical values may be used for the integral $\sigma^2\Omega^{(1,1)*}$, and the scatter between these values may be viewed as an approximate measure of the accuracy of the calculation, which is 10%.

The principal disadvantage of using the capture model in our case is the assumption that the scattering is isotropic and the disregard of the repulsive part of the potential. As a result, the quantity $A^*_{1-2}=\sigma^2\Omega^{(2,2)*}_{1-2}/\sigma^2\Omega^{(1,1)*}_{1-2}$ calculated in this model is smaller than unity. For the Lennard-Jones potential, $A^*_{1-2}>1$, but in the energy interval under consideration it is smaller than the experimental value, $A^*_{1-2}=1.2$, found for all alkali metals.[89] A more accurate calculation of the collision integrals would require a correct calculation of the repulsive part of the potential and the angular dependence of the differential scattering cross section. To calculate the collision integrals $\sigma^2\Omega^{(2,2)*}_{1-2}$, we used the experimental value $A^*_{1-2}=1.2$ and the values of $\sigma^2\Omega^{(1,1)*}_{1-2}$ obtained by us. The procedure presented above for calculating the collision integrals can be used in full measure not only for atom–alkali-metal dimer collisions but also for collisions of an Li atom with H_2, LiH, Li_2H, and Li_3 molecules in the temperature interval under study.

The collision integrals $\sigma^2\Omega^{(1,1)*}$ for the colliding pairs indicated above,

Table 4.15. Collision integrals $\sigma^2\Omega^{(1,1)*}$ (Å²) for the lithium–hydrogen components in the gas mixture.

	Li-H_2			Li-LiH			Li-Li_2H			Li-Li_3		
T,K	*A*	*L*	*M*	*A*	*L*	*M*	*A*	*L*	*M*	*A*	*L*	*M*
700	22.4	17.9	20.2	42.5	34.3	38.4	57.2	48.3	52.8	82.7	67.7	75.2
800	21.6	16.6	19.1	40.8	32.6	36.7	55.3	46.2	50.8	79.0	65.0	72.0
900	21.0	16.0	18.5	39.4	31.0	35.2	53.9	44.5	49.2	75.8	62.2	69.0
1000	20.4	15.4	17.9	38.2	29.4	33.8	52.3	43.6	48.0	73.3	60.3	66.8
1100	19.9	14.9	17.4	37.1	28.5	32.8	51.1	42.2	46.7	71.9	56.9	64.4
1200	19.5	14.5	17.0	36.2	27.6	31.9	50.0	41.2	45.6	69.0	56.6	62.8
1300	19.1	14.1	16.6	35.3	26.5	30.9	49.0	40.1	44.6	67.1	55.1	61.1
1400	18.7	13.9	16.3	34.5	25.5	30.0	48.1	39.1	43.6	65.6	53.8	59.7
1500	18.4	13.6	16.0	33.8	24.8	29.3	47.3	38.1	42.7	64.1	52.5	58.3

Note: A—Asymptotic potentials; *L*—the Lennard–Jones potential; *M*—mean value.

which were found by the two methods mentioned above, are given in Table 4.15. As can be seen from this table, the results obtained on the basis of these two methods are approximately the same, differing by no more than 20%. Polishchuk *et al.*[48] suggested that the average of the two calculated values should be used as the final result and the scatter between these two values can be regarded as the calculation error.

The collision integrals $\sigma^2\Omega^{(2,2)*}$ for the H_2–H_2 collision can be reliably determined from the experimental data on hydrogen viscosity[33] by using the standard relation

$$\sigma^2\Omega^{(2,2)*} = 266.93\times10^{-8}\sqrt{MT}\,\mu^{-1}, \tag{4.31}$$

where $\sigma^2\Omega^{(2,2)*}$ (Å²), M is the reduced molecular mass, T is the temperature (K), and μ is the viscosity (Pa s). The energy range extending from very low energies to $6T$, which is the dominant contribution to the collision integral $\sigma^2\,\Omega^{(2,2)*}$, is well above the corresponding energy interval $(0$–$4)T$ for $\sigma^2\Omega^{(1,1)*}$. This difference makes it more difficult to reliably determine the collision integrals $\sigma^2\Omega^{(2,2)*}$ which must be known in order to calculate $A^* = \sigma^2\Omega^{(2,2)*}/\sigma^2\Omega^{(1,1)*}$.

Using the experimental results for alkali-metal vapor, we have found above that $A^* = 1.2$ for a Li-Li_2 collision. The value $A^* = 1.2$ was also found for a H_2–H_2 collision from the best agreement between the viscosity and the thermal conductivity obtained from an analysis of experimental data on molecular hydrogen based on the use of the equations of the kinetic theory of a low-density gas. Since ordinarily $A^* > 1$, we set $A^* = 1.2$ for all binary collisions of the various components of the gas mixture. The quantity $B^* = (5\sigma^2\Omega^{(1,2)*} - 4\sigma^2\Omega^{(1,3)*})/\sigma^2\Omega^{(1,1)*}$ has virtually no effect on the results and, for definiteness, we set, as it was done by Yargin *et al.*,[89] $B^* = A^*$ in all cases.

It is very difficult to accurately calculate molecule-molecule collision integrals. A relatively low concentration of molecules in the vapor accounts for the relatively low molecule–molecule collision frequency and hence for the slight effect this type of collision has on the transport coefficients. The following equations can therefore be used to estimate the collision integrals. We list the components of the vapor in the following order: 1)H_2, 2) Li, 3) Li_2, 4) LiH, 5) Li_2H, and 6) Li_3. We can then use the following combination relations to estimate the collision integral $\Omega(i,j)$:

$$\Omega(l,j)=[\Omega(1,2)+\Omega(2,j)]/2, \quad j\geqslant 3;$$
$$\Omega(i,j)=[-\Omega(2,2)+\Omega(2,i)+\Omega(2,j)], \quad i,j\geqslant 3. \tag{4.32}$$

Clearly, the relationship between the quantities found from this estimate is consistent with the relationship between the geometric dimensions of the molecules.

4.4. The procedure used for calculating the transport coefficients in a lithium–hydrogen (deuterium, tritium) gas mixture

The computer calculations of the diffusion, thermal-conductivity, and viscosity coefficients of a gas mixture of hydrogen and lithium vapor were based on exact equations of the kinetic theory of gases in the approximation of local chemical equilibrium. The vapor composition corresponds to the region in which it is in equilibrium with the solution of hydrogen in lithium. This vapor composition was calculated in Chap. 3.

We have used the following equations[12,34]:

$$\mu=-\frac{\begin{vmatrix} H_{11}H_{12}\ldots H_{1\nu}x_1 \\ \ldots\ldots\ldots\ldots \\ H_{1\nu}H_{2\nu}\ldots H_{\nu\nu}x_\nu \\ x_1x_2\ldots x_\nu \quad 0 \end{vmatrix}}{\begin{vmatrix} H_{11}H_{12}\ldots H_{1\nu} \\ \ldots\ldots\ldots\ldots \\ H_{1\nu}H_{2\nu}\ldots H_{\nu\nu} \end{vmatrix}}; \tag{4.33}$$

$$H_{ii}=\frac{x_i^2}{\mu_i}+\sum_{k\neq i}\frac{2x_ix_k}{\mu_{ik}}\frac{M_iM_k}{(M_i+M_k)^2}\left(\frac{5}{3A_{ik}^*}+\frac{M_k}{M_i}\right);$$

$$H_{ij}=-\frac{2x_ix_j}{\mu_{ij}}\frac{M_iM_j}{(M_i+M_j)^2}\left(\frac{5}{3A_{ij}^*}=1\right) \quad i\neq j;$$

$$\mu_{ik}=26.693\,\frac{\sqrt{2M_iM_k/(M_i+M_k)T}}{\sigma^2\Omega_{ik}^{(2,2)*}}\,10^{-7},$$

where x_i is the partial molar fraction of the component i, M_i is its molecular mass, μ_i is the viscosity (Pa s), $\sigma^2\Omega_{ik}^{(l,m)*}$ is the collision integral for i and k particles (Å^2), and $A_{ik}^*=\Omega_{ik}^{(2,2)}/\Omega_{ik}^{(1,1)}$.

If the particle collisions are inelastic and the probability for a chemical reaction is high enough to establish local equilibrium and for an average local temperature to be maintained, we can write[34] the equation for thermal conductivity λ[W/(m K)] as a sum of three terms:

$$\lambda = \lambda_{\text{transl}} + \lambda_{\text{int}} + \lambda_r, \tag{4.34}$$

where λ_{transl} and λ_{int} are thermal conductivities which result from the translational motion of the molecule and from the internal degrees of freedom in the absence of chemical reaction (frozen thermal conductivity), and λ_r is the thermal conductivity caused by chemical reaction. The value of λ_{transl} is given by[19,34]

$$\lambda_{\text{transl}} = 4 \frac{\begin{vmatrix} L_{11}L_{12}...L_{1\nu}x_1 \\ .... \\ L_{1\nu}L_{2\nu}...L_{\nu\nu}x_\nu \\ x_1x_2...\, x_\nu \quad 0 \end{vmatrix}}{\begin{vmatrix} L_{11}L_{12}...L_{1\nu} \\ .... \\ L_{1\nu}L_{2\nu}...L_{\nu\nu} \end{vmatrix}}; \tag{4.35}$$

$$L_{ii} = -\frac{4x_i^2}{\lambda_i} - \sum_{k \neq i} \frac{2x_i x_k \left[(15/2)\, M_i^2 + (25/4)\, M_k^2 - 3M_k^2 B_{ik}^* + 4M_i M_k A_{ik}^* \right]}{(M_i + M_k)^2 A_{ik}^* \lambda_{ik}};$$

$$L_{ij} = \frac{2x_i x_j M_i M_j}{(M_i + M_j)^2 A_{ij}^* \lambda_{ij}} \left(\frac{55}{4} - 3B_{ij}^* - 4A_{ij}^* \right) \quad i \neq j;$$

$$\lambda_{ij} = 832.2 \times 10^{-4}\, \frac{\sqrt{T(M_i + M_j)/2M_i M_j}}{\sigma_{ij}^2 \Omega_{ij}^{(2,2)*}}.$$

The thermal conductivity λ_r is produced as a result of thermal flux due to chemical reactions[112]:

$$\lambda_r = -\frac{1}{RT^2} \frac{\begin{vmatrix} A_{11}...\, A_{1r} \quad \Delta H_1 \\ .... \\ A_{r1}...\, A_{rr} \quad \Delta H_r \\ \Delta H_1 ... \Delta H_r \quad 0 \end{vmatrix}}{\begin{vmatrix} A_{11}...\, A_{1r} \\ .... \\ A_{r1}...\, A_{rr} \end{vmatrix}} \tag{4.36}$$

$$A_{ij} = A_{ji} = \sum_{k=1}^{N-1} \sum_{l=k+1}^{N} \frac{RT}{pD_{kl}} x_k x_l \left(\frac{\nu_k^i}{x_k} - \frac{\nu_l^i}{x_l} \right) \left(\frac{\nu_k^j}{x_k} - \frac{\nu_l^j}{x_l} \right).$$

Here all quantities can be expressed in any consistent units: N is the number of components in the gas mixture, r is the number of reactions, ν_k^i is the stoichiometric coefficient of the kth component in the equation of the ith reaction, and ΔH_i is the heat of the ith reaction, as given by

$$\Delta H_i = \sum_{k=1}^{N} \nu_k^i H_k, \tag{4.37}$$

where H_k is the absolute enthalpy of the kth component.

In Eq. (4.36), R is the universal gas constant and D_{ik} is the diffusion coefficient (cm^2/s) given by

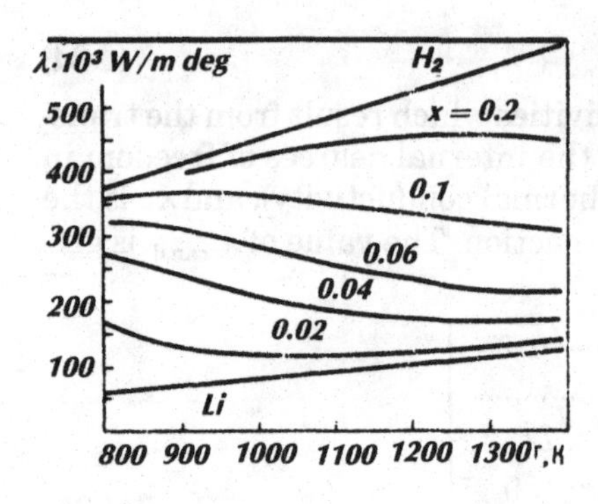

Figure 4.5. Thermal conductivity of vapor above the Li-LiH solution as a function of temperature for various mole fractions (x) of LiH in the solution.

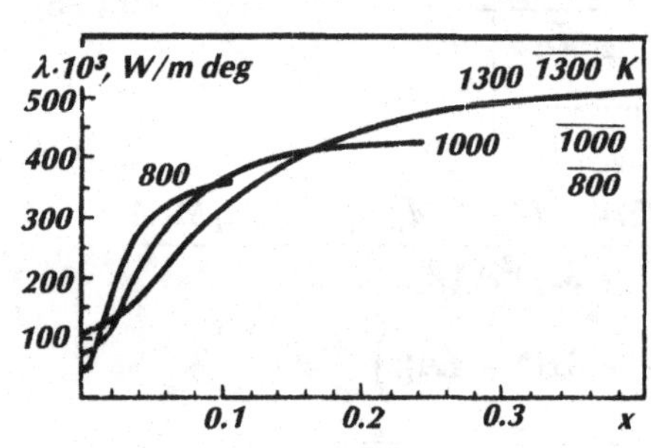

Figure 4.6. Thermal conductivity of vapor above the Li-LiH solution as a function of the composition of the solution (x—mole fraction of LiH in the solution at various temperatures; the horizontal bars show the thermal conductivity λ of pure hydrogen).

$$D_{ik} = 0.002\,628 \frac{\sqrt{T^3(M_i + M_k)/2M_iM_k}}{P\sigma^2\Omega_{ik}^{(1,1)*}}, \tag{4.38}$$

where P is the pressure (atm). The thermal conductivity λ_{int} results from the transfer of energy by means of the internal degrees of freedom of molecules[195]:

$$\lambda_{\text{int}} = \sum_i \frac{6}{5} A_{ii}^* \eta_i \, c_{\text{int}}^{-i} \left(1 + \sum_{j \neq i} \frac{x_j}{x_i} \frac{D_{ii}}{D_{ij}}\right)^{-1}, \tag{4.39}$$

where all quantities can be expressed in terms of one system of units, and c_{int}^i is the specific heat determined by the internal degrees of freedom of the ith component.

The results of calculations of the thermal conductivity and viscosity of vapor above a solution of hydrogen, deuterium, and tritium in lithium are given in Table 5.18. To grasp the total picture, consider the results for hydrogen as illustrated in Figs. 4.5–4.8. Figure 4.5 is a plot of the temperature dependence of the thermal conductivity at various concentrations of hydrogen in the solution, and Fig. 4.6 is a plot of the thermal conductivity as a function of the concentration of hydrogen in the solution at various temperatures. Figures 4.7 and 4.8 are plots of the same functional dependences for the viscosity of vapor.

As can be seen from Fig. 4.5, with increase of the temperature, the thermal conductivity of the mixture tends to change in the direction of the thermal conductivity of lithium, because the concentration of hydrogen in the vapor decreases with increasing temperature and constant concentration of hydrogen in the solution. This behavior is manifested particularly clearly on

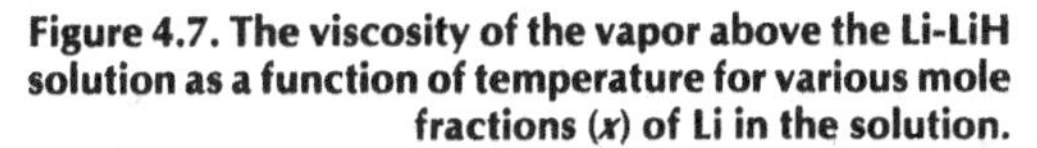
Figure 4.7. The viscosity of the vapor above the Li-LiH solution as a function of temperature for various mole fractions (x) of Li in the solution.

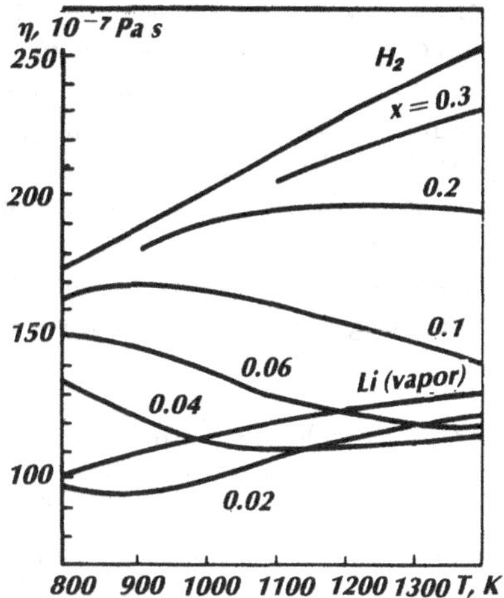

Figure 4.8. The viscosity of the vapor above the Li-LiH solution as a function of the composition of the solution (x—mole fraction of LiH in the solution) at various temperatures. The horizontal bars show the viscosity (μ) of pure hydrogen.

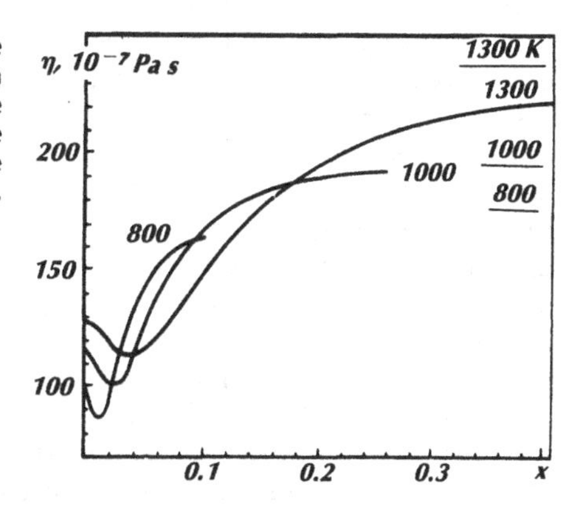

the curve for $x = 0.02$, which asymptotically approaches the curve for pure lithium vapor. At lower temperatures, and high concentrations (0.1 and 0.2), the thermal conductivity of vapor changes in accordance with the increase in the thermal conductivity of hydrogen with temperature. As the temperature is raised, however, the behavior of the thermal conductivity of vapor changes, manifesting the tendency indicated above.

As is shown in Sec. 3.1, at certain concentrations of hydrogen in the solution the liquid phase separates into two phases: a solution of hydrogen in lithium and a solution of lithium in lithium hydride (the immiscibility zone). At these concentrations of hydrogen in the solution, the vapor above the solution consists almost entirely of molecular hydrogen, because the thermal conductivity of vapor, as can be seen in Fig. 4.5, is rather close to the thermal conductivity of hydrogen. The curves in Fig. 4.6 break off at concentrations corresponding to the beginning of the immiscibility zone.

It is interesting to note the curious way in which the vapor viscosity changes with temperature (see Figs. 4.7 and 4.8). Two factors determine the general way in which the viscosity changes. One factor, which was discussed above, has to do with the decrease in the fraction of hydrogen in the vapor as the temperature is raised, provided that the composition of the solution remains constant. The other factor, which is more important, is related to the particular features of the two principal components of the vapor: Li and H_2. The point we wish to make is that the collisional integral $\sigma^2\Omega^{(2,2)*}$ for $Li-Li$ is approximately equal to the collision integral for H_2-H_2, but the reduced masses m_{Li-Li} and m_{Li-H_2} of these pairs differ by a factor of more than 2. The viscosity governed by the $Li-H_2$ collision

$$\eta_{Li-H_2} = \tfrac{5}{16}\sqrt{\pi m_{Li-H_2}T}/\pi\sigma^2\Omega^{(2,2)*}_{Li-H_2},$$

is therefore lower than the viscosity of lithium

$$\eta_{Li} = \tfrac{5}{16}\sqrt{\pi m_{Li-Li}T}/\pi\sigma^2\Omega^{(2,2)*}_{Li-Li}.$$

This circumstance accounts for the fact that the viscosity of the mixture may be lower than the viscosity of lithium, as is evident in Fig. 4.7, where the curves for the concentrations $x = 0.02$, 0.04, and 0.06 go below the region of the curve for pure lithium vapor.

The particular feature under consideration occurs at low concentrations of hydrogen, when the H_2–H_2 collision has only a slight effect on the viscosity (second-order terms in hydrogen). As the hydrogen concentration in the vapor is raised, the effect vanishes, since the viscosity of pure hydrogen is considerably higher than the viscosity of lithium, so that addition of an appreciably amount of hydrogen to the vapor increases its viscosity (in the limiting case, the viscosity of vapor is the same as that of hydrogen).

The results obtained for the interaction potentials and collision integrals for a lithium-hydrogen gas mixture remain the same when the components of the gas mixture are replaced by their isotopes. These results can therefore be used directly for the calculation of the transport coefficients of lithium-deuterium and lithium-tritium gas mixtures. Since the diffusion coefficients [Eq. (4.38)] depend just on the mass, we present here only the results for a lithium-hydrogen mixture (see Table 5.19). The results for isotopic mixtures can be obtained by a simple conversion.

Chapter 5
Tables of the thermophysical properties of lithium hydride, deuteride, and tritide and their solutions with lithium

The tables on the thermophysical properties of lithium hydride, deuteride, and tritide contained in this chapter were compiled on the basis of a critical analysis and statistical evaluation of experimental data, and on the basis of the developed theoretical and calculation methods for determining the thermophysical characteristics of the isotopic modifications of lithium hydride in the condensed and vapor phases and of their lithium-containing solutions when the liquid phase is at equilibrium with the vapor. The relative errors of the tabulated data presented in this chapter were estimated in the relevant sections of the preceding chapters. Let us summarize the results of this analysis. The caloric properties of LiH, LiD, and LiT in the solid state—the heat capacity c_P, enthalpy $H(T) - H(0)$, entropy S, Planck's function $\Phi = S - [H(\tau) - H(0)]/T$—have a 2% error; the caloric properties of these compounds in the liquid state (aside from the enthalpy of liquid lithium hydride) have a 3% error; and the enthalpy of liquid lithium hydride has a 2% error. The tabulated data on the density (ρ) of LiH, LiD, and LiT in the solid state have a 0.2% error. These data have a 2% error for LiH in the liquid state and a 3% error for LiD and LiT in the liquid state. The data on the thermal expansion coefficients (α) of LiH, LiD, and LiT in the solid phase have a 2% error; the data on the thermal conductivity (λ) of solid and liquid LiH have a 15% error; and the data on the electrical conductivity ($j\sigma$) have a 10% error. The data on the surface tension (σ) of liquid lithium hydride have a 1% error, and the data on the kinematic viscosity (ν) and dynamic viscosity (μ) of liquid LiH have a 5% and 7% error, respectively. The data on the partial pressure of H_2 (D_2) when the Li-LiH (LiD) systems are at phase equilibrium have a 2% error to the left of the immiscibility zone and in the immiscibility zone, and the data on the partial pressure of T_2 in the Li-LiT system have a 5% error. To the right of the immiscibility zone, the data on the partial pressure of H_2 (D_2, T_2) have a 5% error. All kinetic coefficients in the vapor phase have a 10% error.

Table 5.1. Caloric properties of lithium hydride.

T,K	c_P, J/(mole deg)	*H*(*T*) − *H*(0), J/mole	*S*, J/(mole deg)	Φ, J/(mole deg)
50	0.81	8.5	0.23	0.06
100	6.37	172.3	2.28	0.56
150	13.5	668.7	6.21	1.75
200	19.0	1 486	10.9	3.44
250	24.0	2 560	15.6	5.40
298	29.0	3 826	20.3	7.44
300	29.2	3 880	20.4	7.51
350	33.5	5 451	25.3	9.69
400	36.9	7 214	30.0	11.9
450	39.6	9 130	34.5	14.2
500	41.9	11 170	38.8	16.4
550	44.0	13 320	42.9	18.6
600	46.0	15 570	46.8	20.8
650	48.1	17 920	50.5	23.0
700	50.3	20 380	54.2	25.1
750	52.8	22 950	57.7	27.1
800	55.6	25 660	61.2	29.1
850	58.6	28 520	64.7	31.1
900	62.1	31 530	68.1	33.1
950	65.7	34 730	71.6	35.0
965 (S)	66.9	35 720	72.6	35.6
965 (l)	61.8	57 060	94.7	35.6
1000	58.3	59 150	96.9	37.7
1050	54.5	61 970	99.6	40.6
1100	52.0	64 630	102.1	43.3
1150	50.7	67 190	104.4	45.9
1200	50.7	69 720	106.5	48.4
1250	52.0	72 280	108.6	50.8
1300	54.5	74 920	110.7	53.0

Table 5.2. Caloric properties of lithium deuteride.

T,K	c_p, J/(mole deg)	$H(T) - H(0)$, J/mole	S, J/(mole deg)	Φ, J/(mole deg)
50	0.84	10.1	0.26	0.06
100	6.69	177.6	2.35	0.58
150	15.1	720.6	6.64	1.84
200	23.0	1 677	12.1	3.70
250	29.5	2 988	17.9	5.96
298	34.6	4 534	23.6	8.41
300	34.8	4 598	23.8	8.47
350	39.3	6 458	29.5	11.1
400	42.5	8 510	35.0	13.7
450	44.9	10 700	40.1	16.4
500	46.7	12 990	45.0	19.0
550	48.4	15 370	49.5	21.5
600	50.0	17 820	53.7	24.0
650	51.8	20 370	57.8	26.5
700	53.8	23 000	61.7	28.9
750	56.0	25 750	65.5	31.2
800	58.5	28 610	69.2	33.4
850	61.4	31 600	72.8	35.6
900	64.6	34 750	76.4	37.8
950	68.2	38 070	80.0	39.9
967 (S)	69.5	39 240	81.2	40.6
967 (l)	62.4	59 730	102.4	40.6
1000	59.0	61 730	104.3	42.6
1050	55.2	64 590	107.1	45.6
1100	52.9	67 280	109.7	48.5
1150	51.5	69 890	112.0	51.2
1200	51.6	72 460	114.1	53.7
1250	53.1	75 070	116.2	56.1
1300	55.5	77 770	118.2	58.4

Table 5.3. Caloric properties of lithium tritide.

T,K	C_P, J/(mole deg)	$H(T) - H(0)$, J/mole	S, J/(mole deg)	Φ, J/(mole deg)
50	0.84	10.1	0.26	0.06
100	7.17	183	2.40	0.57
150	17.3	803	7.28	1.93
200	25.6	1 885	13.4	4.01
250	32.1	3 336	19.9	6.5
298	37.5	5 013	26.0	9.2
300	37.7	5 082	26.2	9.3
350	41.8	7 074	32.3	12.1
400	44.6	9 238	38.1	15.0
450	46.8	11 570	43.5	17.9
500	48.4	13 970	48.5	20.7
550	49.9	16 360	53.2	23.4
600	51.4	18 900	57.6	26.1
650	53.0	21 510	61.7	28.6
700	54.9	24 200	65.7	31.1
750	57.0	27 000	69.6	33.6
800	59.5	29 910	73.3	35.9
850	62.3	32 960	77.0	38.2
900	65.5	36 150	80.7	40.5
950	69.1	39 510	84.3	42.7
968 (S)	70.5	40 770	85.6	43.6
968 (l)	62.9	62 890	108.6	43.6
1000	59.5	64 850	110.8	45.9
1050	55.8	67 640	113.6	49.1
1100	53.4	70 460	116.1	52.0
1150	52.1	73 090	118.4	54.8
1200	52.2	75 690	120.6	57.5
1250	53.5	78 330	122.7	60.0
1300	56.0	81 060	124.8	62.4

Table 5.4. Densities of lithium hydride, deuteride, and tritide (kg/m³).

T,K	LiH	LiD	LiT	T,K	LiH	LiD	LiT
100	785.2	896.4	1002.1	850	710.0	801.7	892.0
150	783.8	894.6	1000.0	900	702.4	792.8	882.0
200	781.5	891.6	996.1	950	694.5	783.5	871.5
250	778.5	887.5	991.0	965 (S)	692.1	—	—
298	775.0	882.6	985.2	965 (l)	585.0	—	—
300	774.9	882.4	984.8	967 (S)	—	780.3	—
350	770.6	876.6	977.9	967 (l)	—	659.0	—
400	765.8	870.3	970.4	968 (S)	—	—	867.8
450	760.6	863.4	962.5	968 (l)	—	—	733.0
500	755.1	856.5	954.4	1000	577.0	651.0	724.0
550	749.4	849.2	946.1	1050	567.0	639.0	711.0
600	743.4	841.9	937.7	1100	557.0	626.0	698.0
650	737.2	834.3	929.0	1150	546.0	616.0	685.0
700	730.8	826.6	920.2	1200	536.0	604.0	672.0
750	724.1	818.5	911.1	1250	525.0	592.0	659.0
800	717.2	810.3	901.8	1300	515.0	580.0	646.0

Table 5.5. Coefficients of linear thermal expansion of lithium hydride, deuteride, and tritide in the solid state (10^{-6} K^{-1}).

T,K	LiH	LiD	LiT	T,K	LiH	LiD	LiT
50	0.9	1.0	1.0	550	52.1	57.3	59.1
100	7.6	7.9	8.5	600	54.5	59.2	60.8
150	16.0	17.9	20.5	650	56.9	61.3	62.7
200	22.5	27.2	30.3	700	59.5	63.7	65.0
250	28.4	34.9	38.0	750	62.5	66.3	67.5
298	34.3	41.0	44.4	800	65.8	69.2	70.4
300	34.6	41.2	44.6	850	69.4	72.7	73.7
350	39.7	46.5	49.5	900	73.5	76.5	77.5
400	43.7	50.3	52.8	950	77.8	80.7	81.8
450	46.9	53.1	55.4	Melting point*	79.2	82.3	83.5
500	49.6	55.3	57.3				

*The melting points of LiH, LiD, and LiT are 965, 967, and 968 K, respectively.

Table 5.6. Surface tension of lithium hydride (N/m).

T,K	$^{\sigma}P_{H_2} = 2\times10^5$ Pa	$^{\sigma}P_{H_2} = 10^6$ Pa	σ	T,K	$^{\sigma}P_{H_2} = 2\times10^5$ Pa	$^{\sigma}P_{H_2} = 10^6$ Pa	σ
965	0.213	0.213	0.213	1150	0.201	0.197	0.196
1000	0.211	0.210	0.209	1200	0.197	0.193	0.192
1050	0.208	0.206	0.205	1250	0.194	0.189	0.187
1100	0.204	0.201	0.201	1300	0.191	0.184	0.183

Table 5.7. Partial pressures of the components and total vapor pressure in the Li-LiH system (*x* is the mole fraction of lithium hydride in the liquid phase).

T,K	*x*	Partial pressure, Pa					Total pressure, Pa
		Li	LiH	H_2	Li_2	Li_2H	
800	0.000 01	1.076	9.232×10^{-7}	7.56×10^{-7}	8.336×10^{-3}	4.934×10^{-7}	1.084
	0.000 02	1.076	1.846×10^{-6}	3.02×10^{-6}	8.366×10^{-3}	9.867×10^{-7}	1.084
	0.000 05	1.076	4.616×10^{-6}	1.89×10^{-5}	8.335×10^{-3}	2.467×10^{-6}	1.084
	0.000 07	1.076	6.462×10^{-6}	3.71×10^{-5}	8.335×10^{-3}	3.453×10^{-6}	1.084
	0.000 10	1.076	9.232×10^{-6}	7.56×10^{-5}	8.334×10^{-3}	4.933×10^{-6}	1.084
	0.000 50	1.075	4.615×10^{-5}	1.89×10^{-3}	8.328×10^{-3}	2.465×10^{-5}	1.085
	0.001 00	1.075	9.227×10^{-5}	7.57×10^{-3}	8.319×10^{-3}	4.926×10^{-5}	1.091
	0.005 00	1.070	4.599×10^{-4}	1.90×10^{-1}	8.253×10^{-3}	2.446×10^{-4}	1.269
	0.010 00	1.065	9.155×10^{-4}	7.59×10^{-1}	8.171×10^{-3}	4.844×10^{-4}	1.833
	0.050 00	1.024	4.340×10^{-3}	18.5	7.550×10^{-3}	2.207×10^{-3}	19.49
	0.100 00	0.977	7.913×10^{-3}	67.3	6.881×10^{-3}	3.842×10^{-3}	68.30
900	0.000 01	12.71	1.950×10^{-5}	4.44×10^{-6}	1.808×10^{-1}	1.227×10^{-5}	12.89
	0.000 02	12.71	3.900×10^{-5}	1.78×10^{-5}	1.808×10^{-1}	2.455×10^{-5}	12.89
	0.000 05	12.71	9.751×10^{-5}	1.11×10^{-4}	1.808×10^{-1}	6.137×10^{-5}	12.89
	0.000 07	12.71	1.365×10^{-4}	2.18×10^{-4}	1.808×10^{-1}	8.592×10^{-5}	12.89
	0.000 10	12.71	1.950×10^{-4}	4.44×10^{-4}	1.808×10^{-1}	1.277×10^{-4}	12.89
	0.000 50	12.71	9.749×10^{-4}	1.11×10^{-2}	1.807×10^{-1}	6.133×10^{-4}	12.90
	0.001 00	12.70	1.949×10^{-3}	4.45×10^{-2}	1.805×10^{-1}	1.226×10^{-3}	12.93
	0.005 00	12.56	9.713×10^{-3}	1.11	1.790×10^{-1}	6.083×10^{-3}	13.96
	0.010 00	12.59	1.933×10^{-2}	4.45	1.773×10^{-1}	1.205×10^{-2}	17.25
	0.050 00	12.10	9.152×10^{-2}	1.08×10^{2}	1.638×10^{-1}	5.482×10^{-2}	1.205×10^{2}
	0.100 00	11.55	1.674×10^{-1}	3.97×10^{2}	1.492×10^{-1}	9.569×10^{-2}	4.088×10^{2}
	0.150 00	11.08	2.255×10^{-1}	7.82×10^{2}	1.373×10^{-1}	1.236×10^{-1}	7.93×10^{2}
	0.200 00	10.71	2.652×10^{-1}	1.16×10^{3}	1.283×10^{-1}	1.406×10^{-1}	1.170×10^{3}
1000	0.000 10	96.92	2.155×10^{-4}	1.83×10^{-5}	2.359	1.634×10^{-4}	99.28
	0.000 02	96.92	4.309×10^{-4}	7.33×10^{-5}	2.359	3.267×10^{-4}	99.28
	0.000 05	96.91	1.077×10^{-3}	4.58×10^{-4}	2.359	8.167×10^{-4}	99.27
	0.000 07	96.91	1.508×10^{-3}	8.98×10^{-4}	2.359	1.143×10^{-3}	99.27
	0.000 01	96.91	2.154×10^{-3}	1.83×10^{-3}	2.358	1.633×10^{-3}	99.27
	0.000 50	96.87	1.077×10^{-2}	4.58×10^{-2}	2.357	8.162×10^{-3}	99.29
	0.001 00	96.82	2.153×10^{-2}	1.83×10^{-1}	2.354	1.631×10^{-2}	99.40
	0.005 00	96.44	1.073×10^{-1}	4.59	2.335	8.094×10^{-2}	1.036×10^{2}

	0.010 00	95.95	2.135×10^{-1}	18.4	2.312	1.802×10^{-1}	1.170×10^{2}
	0.050 00	92.24	1.010	4.44×10^{2}	2.137	7.286×10^{-1}	5.405×10^{2}
	0.100 00	88.01	1.851	1.64×10^{3}	1.945	1.275	1.734×10^{3}
	0.150 00	84.33	2.511	3.29×10^{3}	1.786	1.657	$3.379\,03\times10^{3}$
	0.200 00	81.28	2.996	5.04×10^{3}	1.659	1.905	5.126×10^{3}
	0.250 00	78.92	3.320	6.56×10^{3}	1.564	2.050	6.651×10^{3}
	0.255 53	78.70	3.347	6.71×10^{3}	1.555	2.061	6.794×10^{3}
	0.983 00	6.249	2.984×10^{-1}	8.46×10^{3}	9.807×10^{-3}	1.459×10^{-2}	8.463×10^{3}
	0.988 00	4.527	2.783×10^{-1}	1.40×10^{4}	5.145×10^{-3}	9.856×10^{-3}	1.403×10^{4}
	0.993 00	2.703	2.610×10^{-1}	3.46×10^{4}	1.834×10^{-3}	5.519×10^{-3}	3.461×10^{4}
	0.998 00	7.874×10^{-1}	2.472×10^{-1}	3.66×10^{5}	1.557×10^{-4}	1.523×10^{-3}	3.656×10^{4}
1100	0.000 01	5.058×10^{2}	1.383×10^{-3}	5.84×10^{-5}	18.84	1.388×10^{-3}	5.246×10^{2}
	0.000 02	5.058×10^{2}	3.167×10^{-3}	2.34×10^{-4}	18.84	2.776×10^{-3}	5.246×10^{2}
	0.000 05	5.058×10^{2}	7.917×10^{-3}	1.46×10^{-3}	18.84	6.939×10^{-3}	5.246×10^{2}
	0.000 07	5.058×10^{2}	1.108×10^{-2}	2.86×10^{-3}	18.84	9.715×10^{-3}	5.246×10^{2}
	0.000 10	5.057×10^{2}	1.583×10^{-2}	5.84×10^{-3}	18.84	1.388×10^{-2}	5.246×10^{2}
	0.000 50	5.055×10^{2}	7.915×10^{-2}	1.46×10^{-1}	18.82	6.934×10^{-2}	5.247×10^{2}
	0.001 00	5.053×10^{2}	1.582×10^{-1}	5.85×10^{-1}	18.80	1.386×10^{-1}	5.250×10^{2}
	0.005 00	5.033×10^{2}	7.884×10^{-1}	14.6	18.65	6.876×10^{-1}	5.380×10^{2}
	0.010 00	5.008×10^{2}	1.568	58.5	18.47	1.361	5.806×10^{2}
	0.050 00	4.814×10^{2}	7.413	1.41×10^{3}	17.07	6.184	1.925×10^{2}
	0.100 00	4.592×10^{2}	13.62	5.24×10^{3}	15.53	10.84	5.739×10^{3}
	0.150 00	4.396×10^{2}	18.59	1.07×10^{4}	14.24	14.16	1.114×10^{4}
	0.200 00	4.227×10^{2}	22.42	1.68×10^{4}	13.16	16.42	1.724×10^{4}
	0.250 00	4.085×10^{2}	25.24	2.28×10^{4}	12.29	17.87	2.322×10^{4}
	0.300 00	3.972×10^{2}	27.20	2.80×10^{4}	11.62	18.72	2.841×10^{4}
	0.350 00	3.88×10^{2}	28.45	3.19×10^{4}	11.13	19.17	3.237×10^{4}
	0.360 32	3.874×10^{2}	28.63	3.28×10^{4}	11.05	19.22	3.301×10^{4}
	0.940 00	76.94	6.357	4.07×10^{4}	4.360×10^{-1}	8.476×10^{-1}	4.078×10^{4}
	0.945 00	72.72	6.061	4.14×10^{4}	3.894×10^{-1}	7.638×10^{-1}	4.149×10^{4}
	0.950 00	68.07	5.780	4.30×10^{4}	3.413×10^{-1}	6.819×10^{-1}	4.306×10^{4}
	0.955 00	63.01	5.514	4.57×10^{4}	2.924×10^{-1}	6.021×10^{-1}	4.573×10^{4}
	0.960 00	57.53	5.263	4.99×10^{4}	2.438×10^{-1}	5.248×10^{-1}	4.996×10^{4}
	0.965 00	51.65	5.027	5.65×10^{4}	1.964×10^{-1}	4.500×10^{-1}	5.654×10^{4}
	0.970 00	45.36	4.607	6.69×10^{4}	1.516×10^{-1}	3.788×10^{-1}	6.698×10^{4}
	0.975 00	38.69	4.601	8.43×10^{4}	1.102×10^{-1}	3.085×10^{-1}	8.434×10^{4}
	0.980 00	31.63	4.410	1.16×10^{5}	7.371×10^{-2}	2.418×10^{-1}	1.159×10^{5}
	0.985 00	24.22	4.236	1.82×10^{5}	4.319×10^{-2}	1.778×10^{-1}	1.824×10^{5}
	0.990 00	16.45	4.080	3.67×10^{5}	1.993×10^{-2}	1.163×10^{-1}	3.667×10^{5}

Table 5.7. (con't.).

T,K	x	Partial pressure, Pa					Total pressure, Pa
		Li	LiH	H_2	Li_2	Li_2H	
	0.995 00	8.365	3.945	1.33×10^{6}	5.154×10^{-3}	5.719×10^{-2}	1.326×10^{6}
1200	0.000 01	1.988×10^{3}	8.293×10^{-3}	1.54×10^{-4}	1.048×10^{2}	8.175×10^{-3}	2.093×10^{3}
	0.000 02	1.988×10^{3}	1.659×10^{-2}	6.14×10^{-4}	1.048×10^{2}	1.635×10^{-2}	2.093×10^{3}
	0.000 05	1.988×10^{3}	4.147×10^{-2}	3.84×10^{-3}	1.048×10^{2}	4.087×10^{-2}	2.093×10^{3}
	0.000 07	1.988×10^{3}	5.805×10^{-2}	7.52×10^{-3}	1.048×10^{2}	5.722×10^{-2}	2.093×10^{3}
	0.000 10	1.988×10^{3}	8.293×10^{-2}	1.54×10^{-2}	1.048×10^{2}	8.174×10^{-2}	2.093×10^{3}
	0.000 50	1.987×10^{3}	4.146×10^{-1}	3.84×10^{-1}	1.047×10^{2}	4.085×10^{-1}	2.093×10^{3}
	0.001 00	1.986×10^{3}	8.288×10^{-1}	1.54	1.046×10^{2}	8.162×10^{-1}	2.094×10^{3}
	0.005 00	1.978×10^{3}	4.129	38.4	1.038×10^{2}	4.050	2.128×10^{3}
	0.010 00	1.968×10^{3}	8.212	1.54×10^{2}	1.028×10^{2}	8.015	2.241×10^{3}
	0.050 00	1.892×10^{3}	38.79	3.71×10^{3}	94.98	36.39	5.769×10^{3}
	0.100 00	1.805×10^{3}	71.36	1.38×10^{4}	86.41	63.87	1.582×10^{4}
	0.150 00	1.726×10^{3}	97.94	2.84×10^{4}	79.07	83.85	3.038×10^{4}
	0.200 00	1.657×10^{3}	1.192×10^{2}	4.57×10^{4}	72.81	97.93	4.763×10^{4}
	0.250 00	1.595×10^{3}	1.360×10^{2}	6.41×10^{4}	67.49	1.076×10^{2}	6.602×10^{4}
	0.300 00	1.541×10^{3}	1.490×10^{2}	8.25×10^{5}	62.98	1.139×10^{2}	8.439×10^{5}
	0.350 00	1.493×10^{3}	1.591×10^{2}	1.00×10^{5}	59.16	1.178×10^{2}	1.019×10^{5}
	0.400 00	1.452×10^{3}	1.668×10^{2}	1.16×10^{5}	55.90	1.201×10^{2}	1.183×10^{5}
	0.840 00	4.823×10^{2}	59.11	1.32×10^{5}	6.172	14.14	1.330×10^{5}
	0.850 00	4.691×10^{2}	58.03	1.35×10^{5}	5.838	13.50	1.355×10^{5}
	0.860 00	4.535×10^{2}	56.93	1.39×10^{5}	5.457	12.80	1.395×10^{5}
	0.870 00	4.357×10^{2}	55.82	1.45×10^{5}	5.037	12.06	1.452×10^{5}
	0.880 00	4.158×10^{2}	54.72	1.53×10^{5}	4.583	11.28	1.534×10^{5}
	0.890 00	3.932×10^{3}	53.62	1.64×10^{5}	4.102	10.45	1.645×10^{5}
	0.900 00	3.685×10^{2}	52.53	1.79×10^{5}	3.602	9.599	1.797×10^{5}
	0.910 00	3.415×10^{2}	51.47	2.00×10^{5}	3.093	8.715	2.008×10^{5}
	0.920 00	3.122×10^{2}	50.43	2.30×10^{5}	2.585	7.807	2.305×10^{5}
	0.930 00	2.806×10^{2}	49.42	2.74×10^{5}	2.089	6.877	2.739×10^{5}
	0.940 00	2.468×10^{2}	48.45	3.40×10^{5}	1.616	5.930	3.402×10^{5}
	0.950 00	2.108×10^{2}	47.52	4.48×10^{5}	1.179	4.967	4.485×10^{5}
	0.960 00	1.726×10^{2}	46.65	6.44×10^{5}	7.904×10^{-1}	3.993	6.444×10^{5}
	0.970 00	1.323×10^{2}	45.84	1.06×10^{6}	4.645×10^{-1}	3.008	1.058×10^{6}

	0.980 00	9.001×10^{2}	45.11	2.22×10^{6}	2.150×10^{-1}	2.014	2.215×10^{6}
1300	0.000 01	6.252×10^{3}	3.329×10^{-2}	3.48×10^{-4}	4.364×10^{2}	3.575×10^{-2}	6.688×10^{3}
	0.000 02	6.252×10^{3}	6.658×10^{-2}	1.39×10^{-3}	4.364×10^{2}	7.150×10^{-2}	6.688×10^{3}
	0.000 05	6.252×10^{3}	1.685×10^{-1}	8.69×10^{-3}	4.363×10^{2}	1.787×10^{-1}	6.688×10^{3}
	0.000 07	6.251×10^{3}	2.330×10^{-1}	1.70×10^{-2}	4.363×10^{2}	2.502×10^{-1}	6.688×10^{3}
	0.000 10	6.251×10^{3}	3.329×10^{-1}	3.48×10^{-2}	4.383×10^{2}	3.574×10^{-1}	6.688×10^{3}
	0.000 50	6.249×10^{3}	1.864	8.70×10^{-1}	4.359×10^{2}	1.786	6.689×10^{3}
	0.001 00	6.248×10^{3}	3.327	3.48	4.355×10^{2}	3.569	6.692×10^{3}
	0.010 00	6.190×10^{3}	32.96	3.48×10^{2}	4.277×10^{2}	35.04	7.033×10^{3}
	0.050 00	5.951×10^{3}	1.556×10^{2}	8.38×10^{3}	3.953×10^{2}	1.590×10^{2}	1.504×10^{4}
	0.100 00	5.675×10^{3}	2.866×10^{2}	3.13×10^{4}	3.596×10^{2}	2.794×10^{2}	3.788×10^{4}
	0.150 00	5.425×10^{3}	3.952×10^{2}	6.50×10^{4}	3.286×10^{2}	3.682×10^{2}	7.156×10^{4}
	0.200 00	5.197×10^{3}	4.847×10^{2}	1.07×10^{5}	3.016×10^{2}	4.327×10^{2}	1.131×10^{5}
	0.250 00	4.988×10^{3}	5.591×10^{2}	1.54×10^{5}	2.777×10^{2}	4.789×10^{2}	1.603×10^{5}
	0.300 00	4.792×10^{3}	6.216×10^{2}	2.06×10^{5}	2.564×10^{2}	5.116×10^{2}	2.125×10^{5}
	0.350 00	4.606×10^{3}	6.751×10^{2}	2.63×10^{5}	2.368×10^{2}	5.340×10^{2}	2.694×10^{5}
	0.400 00	4.424×10^{3}	7.220×10^{2}	3.27×10^{5}	2.185×10^{2}	5.486×10^{2}	3.325×10^{5}
	0.450 00	4.241×10^{3}	7.645×10^{2}	3.98×10^{5}	2.008×10^{2}	5.569×10^{2}	4.042×10^{5}
	0.895 00	1.150×10^{3}	2.183×10^{2}	4.42×10^{5}	14.75	43.10	4.435×10^{5}
	0.900 00	1.074×10^{3}	2.234×10^{2}	5.31×10^{5}	12.88	41.23	5.319×10^{5}
	0.905 00	1.001×10^{3}	2.286×10^{2}	6.39×10^{5}	11.19	39.30	6.405×10^{5}
	0.910 00	9.299×10^{2}	2.336×10^{2}	7.74×10^{5}	9.653	37.32	7.753×10^{5}
	0.915 00	8.609×10^{2}	2.387×10^{2}	9.42×10^{5}	8.275	35.29	9.433×10^{5}
	0.920 00	7.941×10^{2}	2.436×10^{2}	1.15×10^{6}	7.040	33.23	1.155×10^{6}
	0.925 00	7.294×10^{2}	2.485×10^{2}	1.42×10^{6}	5.940	31.13	1.424×10^{6}
	0.930 00	6.669×10^{2}	2.533×10^{2}	1.77×10^{6}	4.965	29.01	1.769×10^{6}
	0.935 00	6.064×10^{2}	2.579×10^{2}	2.22×10^{6}	4.105	26.86	2.219×10^{6}
	0.940 00	5.480×10^{2}	2.626×10^{2}	2.81×10^{6}	3.353	24.71	2.813×10^{6}

Table 5.8. Partial pressures of the components and total vapor pressure in the Li-LiD system (*x* is the mole fraction of lithium deuteride in the liquid phase).

T,K	*x*	Partial pressure, Pa					Total pressure, Pa
		Li	LiD	D_2	Li_2	Li_2D	
800	0.000 01	1.076	9.204×10^{-7}	1.43×10^{-6}	8.366×10^{-3}	5.095×10^{-7}	1.084
	0.000 05	1.076	4.602×10^{-6}	3.57×10^{-5}	8.335×10^{-3}	2.547×10^{-6}	1.084
	0.000 10	1.076	9.204×10^{-6}	1.43×10^{-4}	8.334×10^{-3}	5.094×10^{-6}	1.084
	0.000 50	1.075	4.601×10^{-5}	3.58×10^{-3}	8.328×10^{-3}	2.545×10^{-5}	1.087
	0.001 00	1.075	9.199×10^{-5}	1.43×10^{-2}	8.319×10^{-3}	5.087×10^{-5}	1.097
	0.005 00	1.070	4.585×10^{-4}	3.58×10^{-1}	8.255×10^{-3}	2.525×10^{-4}	1.438
	0.010 00	1.065	9.127×10^{-4}	1.43	8.171×10^{-3}	5.002×10^{-4}	2.509
	0.050 00	1.024	4.327×10^{-3}	34.9	7.550×10^{-3}	2.279×10^{-3}	35.92
	0.100 00	0.977	7.890×10^{-3}	127	6.881×10^{-3}	3.968×10^{-3}	1.283×10^{2}
900	0.000 01	12.71	1.797×10^{-5}	7.48×10^{-6}	1.808×10^{-1}	1.178×10^{-5}	12.89
	0.000 02	12.71	3.593×10^{-5}	2.99×10^{-5}	1.808×10^{-1}	2.355×10^{-5}	12.89
	0.000 05	12.71	8.983×10^{-5}	1.87×10^{-4}	1.808×10^{-1}	5.888×10^{-5}	12.89
	0.000 07	12.71	1.258×10^{-4}	3.87×10^{-4}	1.808×10^{-1}	8.243×10^{-4}	12.89
	0.000 10	12.71	1.797×10^{-4}	7.48×10^{-4}	1.808×10^{-1}	1.178×10^{-4}	12.89
	0.000 50	12.71	8.981×10^{-4}	1.87×10^{-2}	1.807×10^{-1}	5.884×10^{-4}	12.91
	0.001 00	12.70	1.796×10^{-3}	7.49×10^{-2}	1.805×10^{-1}	1.176×10^{-3}	12.96
	0.005 00	12.65	8.949×10^{-3}	1.87	1.790×10^{-1}	5.837×10^{-3}	14.72
	0.010 00	12.59	1.781×10^{-2}	7.50	1.773×10^{-1}	1.156×10^{-2}	20.29
	0.050 00	12.10	8.452×10^{-2}	1.82×10^{2}	1.638×10^{-1}	5.260×10^{-2}	1.943×10^{2}
	0.100 00	11.55	1.542×10^{-1}	6.68×10^{2}	1.492×10^{-1}	9.181×10^{-2}	6.800×10^{2}
	0.150 00	11.08	2.078×10^{-1}	1.32×10^{3}	1.373×10^{-1}	1.186×10^{-1}	1.327×10^{3}
	0.200 00	10.71	2.443×10^{-1}	1.95×10^{3}	1.283×10^{-1}	1.349×10^{-1}	1.962×10^{3}
1000	0.000 01	96.92	2.044×10^{-4}	2.81×10^{-5}	2.359	1.631×10^{-4}	99.28
	0.000 05	96.91	1.022×10^{-3}	7.03×10^{-4}	2.359	8.154×10^{-4}	99.27
	0.000 10	96.91	2.044×10^{-3}	2.81×10^{-3}	2.358	1.631×10^{-3}	99.27
	0.000 50	96.87	1.022×10^{-2}	7.03×10^{-2}	2.357	8.148×10^{-3}	99.32
	0.001 00	96.82	2.043×10^{-2}	2.81×10^{-1}	2.354	1.628×10^{-2}	99.49
	0.005 00	96.44	1.081×10^{-1}	7.04	2.355	8.081×10^{-2}	1.060×10^{2}
	0.010 00	95.96	2.025×10^{-1}	28.2	2.312	1.600×10^{-1}	1.268×10^{2}
	0.050 00	92.24	9.578×10^{-1}	6.82×10^{2}	2.137	7.274×10^{-1}	7.779×10^{2}
	0.100 00	88.01	1.756	2.52×10^{3}	1.945	1.272	2.610×10^{3}
	0.200 00	81.28	2.842	7.73×10^{3}	1.659	1.902	7.818×10^{3}

	0.255 53	78.70	3.175	1.03×10^{4}	1.555	2.057	1.038×10^{4}
	0.983 00	6.249	2.849×10^{-1}	1.31×10^{4}	9.807×10^{-3}	1.466×10^{-2}	1.315×10^{4}
	0.988 00	4.527	2.657×10^{-1}	2.18×10^{4}	5.145×10^{-3}	9.903×10^{-3}	2.179×10^{4}
	0.993 00	2.703	2.492×10^{-1}	5.38×10^{4}	1.834×10^{-3}	5.545×10^{-3}	5.378×10^{4}
	0.998 00	7.874×10^{-1}	2.360×10^{-1}	5.68×10^{5}	1.557×10^{-4}	1.530×10^{-3}	5.682×10^{5}
1100	0.000 01	505.8	1.478×10^{-3}	8.31×10^{-5}	18.84	1.376×10^{-3}	5.246×10^{2}
	0.000 05	505.8	7.389×10^{-3}	2.08×10^{-3}	18.84	6.881×10^{-3}	5.246×10^{2}
	0.000 10	505.7	1.478×10^{-2}	8.31×10^{-3}	18.84	1.376×10^{-2}	5.246×10^{2}
	0.000 50	505.5	7.387×10^{-2}	2.08×10^{-1}	18.82	6.876×10^{-2}	5.247×10^{2}
	0.001 00	505.3	1.477×10^{-1}	8.31×10^{-1}	18.80	1.374×10^{-1}	5.252×10^{2}
	0.005 00	503.3	7.358×10^{-1}	70.8	18.65	6.818×10^{-1}	5.441×10^{2}
	0.010 00	500.8	1.464	83.1	18.47	1.350	6.052×10^{2}
	0.050 00	481.4	6.918	2.01×10^{3}	17.07	6.132	2.521×10^{3}
	0.100 00	459.2	12.71	7.45×10^{3}	15.53	10.75	7.949×10^{3}
	0.200 00	422.7	20.92	2.38×10^{4}	13.16	16.29	2.432×10^{4}
	0.300 00	397.2	25.39	3.97×10^{4}	11.62	18.57	4.020×10^{4}
	0.360 32	387.4	26.72	4.63×10^{4}	11.05	19.06	4.675×10^{4}
	0.940 00	76.94	5.410	4.81×10^{4}	4.360×10^{-1}	7.664×10^{-1}	4.820×10^{4}
	0.950 00	68.07	4.919	5.08×10^{4}	3.413×10^{-1}	6.166×10^{-1}	5.089×10^{4}
	0.960 00	57.53	4.479	5.90×10^{4}	2.438×10^{-1}	4.745×10^{-1}	5.905×10^{4}
	0.970 00	45.36	4.090	7.91×10^{4}	1.516×10^{-1}	3.416×10^{-1}	7.918×10^{4}
	0.980 00	31.63	3.753	1.37×10^{5}	7.371×10^{-2}	2.186×10^{-1}	1.370×10^{5}
	0.990 00	16.45	3.472	4.33×10^{5}	1.993×10^{-2}	1.052×10^{-1}	4.335×10^{5}
	0.995 00	8.365	3.357	1.57×10^{6}	5.154×10^{-3}	5.171×10^{-2}	1.568×10^{6}
1200	0.000 01	1.988×10^{3}	7.640×10^{-3}	2.05×10^{-4}	1.048×10^{2}	8.059×10^{-3}	2.093×10^{3}
	0.000 05	1.988×10^{3}	3.820×10^{-2}	5.12×10^{-3}	1.048×10^{2}	4.029×10^{-2}	2.093×10^{3}
	0.000 10	1.988×10^{3}	7.640×10^{-2}	2.05×10^{-2}	1.048×10^{2}	8.058×10^{-2}	2.093×10^{3}
	0.000 50	1.987×10^{3}	3.819×10^{-1}	5.12×10^{-1}	1.047×10^{2}	4.026×10^{-1}	2.093×10^{3}
	0.001 00	1.986×10^{3}	7.635×10^{-1}	2.05	1.046×10^{2}	8.046×10^{-1}	2.094×10^{3}
	0.005 00	1.978×10^{3}	3.804	51.3	1.038×10^{2}	3.992	2.141×10^{3}
	0.010 00	1.968×10^{3}	7.565	7.05×10^{2}	1.028×10^{2}	7.900	2.291×10^{3}
	0.050 00	1.892×10^{3}	35.73	4.95×10^{3}	94.98	35.87	7.006×10^{3}
	0.100 00	1.805×10^{3}	65.74	1.84×10^{4}	86.41	62.96	2.043×10^{4}
	0.200 00	1.657×10^{3}	1.098×10^{2}	6.10×10^{4}	72.81	96.54	6.290×10^{4}
	0.300 00	1.541×10^{3}	1.373×10^{2}	1.10×10^{5}	62.98	1.122×10^{2}	1.120×10^{5}
	0.400 00	1.452×10^{3}	1.537×10^{2}	1.55×10^{5}	55.90	1.184×10^{2}	1.573×10^{5}
	0.840 00	4.823×10^{2}	51.60	1.59×10^{5}	6.172	13.21	1.593×10^{5}
	0.860 00	4.535×10^{2}	49.70	1.67×10^{5}	5.457	11.96	1.670×10^{5}
	0.880 00	4.156×10^{2}	47.77	1.83×10^{5}	4.583	10.53	1.837×10^{5}

Table 5.8. (con't.).

T,K	x	Partial pressure, Pa					Total pressure, Pa
		Li	LiD	D_2	Li_2	Li_2D	
	0.900 00	3.685×10^{2}	45.86	2.15×10^{5}	3.602	8.966	2.153×10^{5}
	0.920 00	3.122×10^{2}	44.02	2.76×10^{5}	2.585	7.292	2.762×10^{5}
	0.940 00	2.468×10^{2}	42.29	4.07×10^{5}	1.616	5.539	4.076×10^{5}
	0.960 00	1.726×10^{2}	40.72	7.72×10^{5}	7.904×10^{-1}	3.730	7.723×10^{5}
	0.980 00	90.01	39.38	2.65×10^{6}	2.150×10^{-1}	1.881	2.655×10^{6}
1300	0.000 01	6.252×10^{3}	3.029×10^{-2}	4.40×10^{-4}	4.364×10^{2}	3.494×10^{-2}	6.688×10^{3}
	0.000 05	6.252×10^{3}	1.514×10^{-1}	1.10×10^{-2}	4.363×10^{2}	1.747×10^{-1}	6.688×10^{3}
	0.000 10	6.251×10^{3}	3.028×10^{-1}	4.40×10^{-2}	4.363×10^{2}	3.493×10^{-1}	6.688×10^{3}
	0.000 50	6.249×10^{3}	1.514	1.10	4.359×10^{2}	1.746	6.689×10^{3}
	0.001 00	6.246×10^{3}	3.027	4.40	4.355×10^{2}	3.488	6.692×10^{3}
	0.005 00	6.221×10^{3}	15.08	1.10×10^{2}	4.320×10^{2}	17.31	6.795×10^{3}
	0.010 00	6.190×10^{3}	29.98	4.40×10^{2}	4.277×10^{2}	34.25	7.121×10^{3}
	0.050 00	5.951×10^{3}	1.415×10^{2}	1.06×10^{4}	3.953×10^{2}	1.554×10^{2}	1.724×10^{4}
	0.100 00	5.675×10^{3}	2.608×10^{2}	3.96×10^{4}	3.596×10^{2}	2.731×10^{2}	4.613×10^{4}
	0.200 00	5.197×10^{3}	4.410×10^{2}	1.35×10^{5}	3.016×10^{2}	4.229×10^{2}	1.413×10^{5}
	0.300 00	4.792×10^{3}	5.654×10^{2}	2.61×10^{5}	2.564×10^{2}	5.000×10^{2}	2.671×10^{5}
	0.400 00	4.424×10^{3}	6.568×10^{2}	4.13×10^{5}	2.185×10^{2}	5.362×10^{2}	4.190×10^{5}
	0.450 00	4.241×10^{3}	6.955×10^{2}	5.04×10^{5}	2.008×10^{2}	5.443×10^{2}	5.097×10^{5}
	0.900 00	1.074×10^{3}	2.204×10^{2}	7.89×10^{5}	12.88	43.69	7.904×10^{5}
	0.910 00	9.299×10^{2}	2.505×10^{2}	1.15×10^{6}	9.653	39.55	1.152×10^{6}
	0.920 00	7.941×10^{2}	2.403×10^{2}	1.72×10^{6}	7.040	35.21	1.717×10^{6}
	0.930 00	6.669×10^{2}	2.498×10^{2}	2.63×10^{6}	4.965	30.74	2.631×10^{6}
	0.940 00	5.480×10^{2}	2.589×10^{2}	4.18×10^{6}	3.353	26.18	4.184×10^{6}

Table 5.9. Partial pressures of the components and total vapor pressure in the Li-LiT system (*x* is the mole fraction of lithium tritide in the liquid phase).

T,K	*x*	Partial pressure, Pa					Total pressure, Pa
		Li	LiT	T_2	Li_2	Li_2T	
800	0.000 01	1.076	8.376×10^{-7}	1.51×10^{-6}	8.336×10^{-3}	4.636×10^{-7}	1.084
	0.000 05	1.076	4.188×10^{-6}	3.78×10^{-5}	8.335×10^{-3}	2.318×10^{-6}	1.084
	0.000 10	1.076	8.375×10^{-6}	1.51×10^{-4}	8.334×10^{-3}	4.635×10^{-6}	1.084
	0.000 50	1.075	4.187×10^{-5}	3.78×10^{-3}	8.328×10^{-3}	2.316×10^{-5}	1.087
	0.001 00	1.075	8.371×10^{-5}	1.51×10^{-2}	8.319×10^{-3}	4.629×10^{-5}	1.098
	0.005 00	1.070	4.173×10^{-4}	3.79×10^{-1}	8.253×10^{-3}	2.298×10^{-4}	1.458
	0.010 00	1.065	8.306×10^{-4}	1.52	8.171×10^{-3}	4.552×10^{-4}	2.591
	0.050 00	1.024	3.938×10^{-3}	36.9	7.550×10^{-3}	2.074×10^{-3}	37.91
	0.100 00	0.998	7.179×10^{-3}	135	6.881×10^{-3}	3.610×10^{-3}	1.355×10^{2}
900	0.000 01	12.71	1.660×10^{-5}	7.85×10^{-6}	1.808×10^{-1}	1.093×10^{-5}	12.89
	0.000 05	12.71	8.300×10^{-5}	1.96×10^{-4}	1.808×10^{-1}	5.466×10^{-5}	12.89
	0.000 10	12.71	1.660×10^{-4}	7.85×10^{-4}	1.808×10^{-1}	1.093×10^{-4}	12.8931
	0.000 50	12.71	8.298×10^{-4}	1.96×10^{-2}	1.807×10^{-1}	5.462×10^{-4}	12.91
	0.001 00	12.70	1.659×10^{-3}	7.86×10^{-2}	1.805×10^{-1}	1.092×10^{-3}	12.96
	0.005 00	12.65	8.268×10^{-3}	1.97	1.790×10^{-1}	5.418×10^{-3}	14.81
	0.010 00	12.59	1.645×10^{-2}	7.87	1.773×10^{-1}	1.073×10^{-2}	20.66
	0.050 00	12.10	7.791×10^{-2}	1.91×10^{2}	1.638×10^{-1}	4.883×10^{-2}	2.053×10^{2}
	0.100 00	11.55	1.425×10^{-1}	7.01×10^{2}	1.492×10^{-1}	8.523×10^{-2}	7.128×10^{2}
	0.200 00	10.71	2.258×10^{-1}	2.05×10^{3}	1.283×10^{-1}	1.252×10^{-1}	2.058×10^{3}
1000	0.000 01	96.92	1.917×10^{-4}	2.92×10^{-5}	2.359	1.555×10^{-4}	99.28
	0.000 05	96.91	9.586×10^{-4}	7.33×10^{-4}	2.359	7.777×10^{-4}	99.27
	0.000 10	96.91	1.917×10^{-3}	2.93×10^{-3}	2.358	1.555×10^{-3}	99.27
	0.000 50	96.87	9.584×10^{-3}	7.33×10^{-2}	2.357	7.771×10^{-3}	99.32
	0.001 00	96.82	1.916×10^{-2}	2.93×10^{-1}	2.356	1.553×10^{-2}	99.50
	0.005 00	96.44	9.547×10^{-2}	7.35	2.335	7.707×10^{-2}	1.063×10^{2}
	0.010 00	95.95	1.899×10^{-1}	29.4	2.312	1.526×10^{-1}	1.280×10^{2}
	0.050 00	92.24	8.985×10^{-1}	7.11×10^{2}	2.137	6.938×10^{-1}	8.070×10^{2}
	0.100 00	88.01	1.647	2.62×10^{3}	1.945	1.214	2.718×10^{3}
	0.200 00	81.28	2.666	8.06×10^{3}	1.659	1.814	8.149×10^{3}
	0.300 00	77.32	3.120	1.22×10^{4}	1.501	2.019	1.228×10^{4}
	0.983 00	6.249	2.655×10^{-1}	1.35×10^{4}	9.807×10^{-3}	1.389×10^{-2}	1.334×10^{4}
	0.988 00	4.527	2.477×10^{-1}	2.24×10^{4}	5.145×10^{-3}	9.384×10^{-3}	2.246×10^{4}
	0.993 00	2.703	2.323×10^{-1}	5.54×10^{4}	1.634×10^{-3}	5.255×10^{-3}	5.557×10^{4}

Table 5.9. (con't.).

T,K	x	Partial pressure, Pa					Total pressure, Pa
		Li	LiT	T_2	Li_2	Li_2T	
	0.998 00	7.874×10^{-1}	2.200×10^{-1}	5.85×10^{5}	1.557×10^{-4}	1.450×10^{-3}	5.850×10^{5}
1100	0.000 01	5.058×10^{2}	1.354×10^{-3}	8.62×10^{-5}	18.84	1.310×10^{-3}	5.246×10^{2}
	0.000 05	5.058×10^{2}	6.767×10^{-3}	2.16×10^{-3}	18.84	6.548×10^{-3}	5.246×10^{2}
	0.000 10	5.057×10^{2}	1.353×10^{-2}	8.62×10^{-3}	1.884×10	1.309×10^{-2}	5.246×10^{2}
	0.000 50	5.055×10^{2}	8.766×10^{-2}	2.16×10^{-1}	1.882×10	6.543×10^{-2}	5.247×10^{2}
	0.001 00	5.053×10^{2}	1.353×10^{-1}	8.63×10^{-1}	1.880×10	1.308×10^{-1}	5.252×10^{2}
	0.005 00	5.033×10^{2}	6.739×10^{-1}	21.6	1.865×10	6.488×10^{-1}	5.448×10^{2}
	0.010 00	5.008×10^{2}	1.341	86.3	1.847×10	1.284	6.081×10^{2}
	0.050 00	4.814×10^{2}	6.336	2.09×10^{3}	1.707×10	5.835	2.596×10^{3}
	0.100 00	4.592×10^{2}	11.64	7.73×10^{3}	1.553×10	10.23	8.229×10^{3}
	0.200 00	4.227×10^{2}	19.16	2.47×10^{4}	1.316×10	15.50	2.522×10^{4}
	0.300 00	3.972×10^{2}	23.25	4.12×10^{4}	1.182×10	17.67	4.170×10^{4}
	0.350 00	3.888×10^{2}	24.32	4.71×10^{4}	1.113×10	18.09	4.755×10^{4}
	0.950 00	68.07	4.941	6.34×10^{4}	3.413×10^{-1}	6.434×10^{-1}	6.350×10^{4}
	0.960 00	57.53	4.499	7.36×10^{4}	2.438×10^{-1}	4.952×10^{-1}	7.369×10^{4}
	0.970 00	45.36	4.109	9.88×10^{4}	1.516×10^{-1}	3.565×10^{-1}	9.881×10^{4}
	0.980 00	31.63	3.770	1.71×10^{5}	7.371×10^{-2}	2.282×10^{-1}	1.710×10^{5}
	0.990 00	16.45	3.488	5.41×10^{5}	1.993×10^{-2}	1.098×10^{-1}	5.410×10^{5}
	0.995 00	8.365	3.372	1.96×10^{6}	5.154×10^{-3}	5.397×10^{-2}	1.957×10^{6}
1200	0.000 01	1.988×10^{3}	7.270×10^{-3}	2.12×10^{-4}	1.048×10^{2}	2.668×10^{-3}	2.093×10^{3}
	0.000 05	1.988×10^{3}	3.635×10^{-2}	5.29×10^{-3}	1.048×10^{2}	3.834×10^{-2}	2.093×10^{3}
	0.000 10	1.988×10^{3}	7.270×10^{-2}	2.12×10^{-2}	1.048×10^{2}	7.668×10^{-2}	2.093×10^{3}
	0.000 50	1.987×10^{3}	3.634×10^{-1}	5.30×10^{-1}	1.047×10^{2}	3.831×10^{-1}	2.093×10^{3}
	0.001 00	1.986×10^{3}	7.266×10^{-1}	2.12	1.046×10^{2}	7.656×10^{-1}	2.094×10^{3}
	0.005 00	1.978×10^{3}	3.620	53.0	1.038×10^{2}	3.799	2.142×10^{3}
	0.010 00	1.968×10^{3}	7.199	2.12×10^{2}	1.026×10^{2}	7.518	2.297×10^{3}
	0.050 00	1.892×10^{3}	34.00	5.11×10^{3}	94.98	34.14	7.168×10^{3}
	0.100 00	1.805×10^{3}	62.56	1.90×10^{4}	86.41	59.91	2.104×10^{4}
	0.200 00	1.657×10^{3}	1.045×10^{2}	6.30×10^{4}	72.81	91.86	6.494×10^{4}
	0.300 00	1.541×10^{3}	1.307×10^{2}	1.14×10^{5}	62.98	1.068×10^{2}	1.157×10^{5}
	0.400 00	1.452×10^{3}	1.462×10^{2}	1.61×10^{5}	55.90	1.126×10^{2}	1.625×10^{5}
	0.880 00	4.156×10^{2}	47.97	2.11×10^{5}	4.583	10.58	2.114×10^{5}

	0.900 00	3.685×10^{2}	46.05	2.47×10^{5}	3.602	9.004	2.477×10^{5}
	0.920 00	3.122×10^{2}	44.21	3.17×10^{5}	2.585	7.323	3.179×10^{5}
	0.940 00	2.468×10^{2}	42.47	4.69×10^{5}	1.616	5.562	4.691×10^{5}
	0.960 00	1.726×10^{2}	40.89	8.89×10^{5}	7.904×10^{-1}	3.745	8.889×10^{5}
	0.980 00	90.01	39.55	3.06×10^{6}	2.150×10^{-1}	1.889	3.056×10^{6}
1300	0.000 01	6.252×10^{3}	2.903×10^{-2}	4.53×10^{-4}	4.364×10^{2}	3.360×10^{-2}	6.688×10^{3}
	0.000 05	6.252×10^{3}	1.451×10^{-1}	1.13×10^{-2}	4.363×10^{2}	1.680×10^{-1}	6.688×10^{3}
	0.000 10	6.251×10^{3}	2.903×10^{-1}	4.53×10^{-2}	4.363×10^{2}	3.360×10^{-1}	6.688×10^{3}
	0.000 50	0.249×10^{3}	1.451	1.13	4.359×10^{2}	1.679	6.689×10^{3}
	0.001 00	6.246×10^{3}	2.901	4.53	4.355×10^{2}	3.355	6.692×10^{3}
	0.005 00	6.221×10^{3}	14.45	1.13×10^{2}	4.320×10^{2}	16.64	6.797×10^{3}
	0.010 00	6.190×10^{3}	28.74	4.53×10^{2}	4.277×10^{2}	32.93	7.132×10^{3}
	0.050 00	5.951×10^{3}	1.357×10^{2}	1.09×10^{4}	3.953×10^{2}	1.495×10^{2}	1.755×10^{4}
	0.100 00	5.675×10^{3}	2.499×10^{2}	4.08×10^{4}	3.596×10^{2}	2.626×10^{2}	4.730×10^{4}
	0.200 00	5.197×10^{3}	4.227×10^{2}	1.39×10^{5}	3.016×10^{2}	4.067×10^{2}	1.453×10^{5}
	0.300 00	4.792×10^{3}	5.420×10^{2}	2.69×10^{5}	2.564×10^{2}	4.808×10^{2}	2.748×10^{5}
	0.400 00	4.424×10^{3}	6.296×10^{2}	4.26×10^{5}	2.185×10^{2}	5.157×10^{2}	4.314×10^{5}
	0.450 00	4.241×10^{3}	6.666×10^{2}	5.19×10^{5}	2.008×10^{2}	5.234×10^{2}	5.248×10^{5}
	0.900 00	1.074×10^{3}	1.948×10^{2}	6.91×10^{5}	12.88	38.75	6.925×10^{5}
	0.910 00	9.299×10^{2}	2.037×10^{2}	1.01×10^{6}	9.653	35.07	1.010×10^{6}
	0.920 00	7.941×10^{2}	2.124×10^{2}	1.50×10^{6}	7.040	31.23	1.504×10^{6}
	0.930 00	6.669×10^{2}	2.208×10^{6}	2.30×10^{6}	4.965	27.26	2.305×10^{6}
	0.940 00	5.480×10^{2}	2.289×10^{2}	3.66×10^{6}	3.353	23.22	3.665×10^{6}

Table 5.10. Composition of the vapor phase in the Li-LiH system when the liquid is in equilibrium with the vapor (x is the mole fraction of lithium hydride in the liquid phase).

T,K	x	Li	LiH	H_2	Li_2	Li_2H
800	0.000 01	9.9231×10^{-1}	8.516×10^{-7}	6.9750×10^{-8}	7.6895×10^{-3}	4.551×10^{-7}
	0.000 05	9.9229×10^{-1}	4.258×10^{-6}	1.7439×10^{-5}	7.6891×10^{-3}	2.276×10^{-6}
	0.000 10	9.9223×10^{-1}	8.516×10^{-6}	6.9759×10^{-5}	7.6882×10^{-3}	4.551×10^{-6}
	0.000 50	9.9052×10^{-1}	4.252×10^{-5}	1.7424×10^{-3}	7.6719×10^{-3}	2.271×10^{-5}
	0.001 00	9.8530×10^{-1}	8.460×10^{-5}	6.9391×10^{-3}	7.6277×10^{-3}	4.517×10^{-5}
	0.005 00	8.4356×10^{-1}	3.625×10^{-4}	1.4938×10^{-1}	6.5043×10^{-3}	1.927×10^{-4}
	0.010 00	5.8098×10^{-1}	4.994×10^{-4}	4.1380×10^{-1}	4.4573×10^{-3}	2.642×10^{-4}
	0.050 00	5.2532×10^{-2}	2.227×10^{-4}	9.4674×10^{-1}	3.8741×10^{-4}	1.133×10^{-4}
	0.100 00	1.4309×10^{-2}	1.159×10^{-4}	9.8542×10^{-1}	1.0074×10^{-4}	5.625×10^{-5}
900	0.000 01	9.8597×10^{-1}	1.513×10^{-6}	3.4468×10^{-7}	1.4025×10^{-2}	9.520×10^{-7}
	0.000 05	9.8595×10^{-1}	7.563×10^{-6}	8.6176×10^{-6}	1.4024×10^{-2}	4.760×10^{-6}
	0.000 10	9.8592×10^{-1}	1.513×10^{-5}	3.4473×10^{-5}	1.4023×10^{-2}	9.519×10^{-6}
	0.000 50	9.8501×10^{-1}	7.557×10^{-5}	8.6170×10^{-4}	1.4005×10^{-2}	4.754×10^{-5}
	0.001 00	9.8235×10^{-1}	1.508×10^{-4}	3.4405×10^{-3}	1.3960×10^{-2}	9.480×10^{-5}
	0.005 00	9.0626×10^{-1}	6.959×10^{-4}	7.9781×10^{-2}	1.2827×10^{-2}	4.358×10^{-4}
	0.010 00	7.2969×10^{-1}	1.121×10^{-3}	2.5822×10^{-1}	1.0276×10^{-2}	6.983×10^{-4}
	0.050 00	1.0043×10^{-1}	7.597×10^{-4}	8.9699×10^{-1}	1.3597×10^{-3}	4.551×10^{-4}
	0.100 00	2.8246×10^{-2}	4.095×10^{-4}	9.7074×10^{-1}	3.6499×10^{-4}	2.341×10^{-4}
	0.200 00	9.1494×10^{-3}	2.267×10^{-4}	9.9039×10^{-1}	1.0951×10^{-4}	1.201×10^{-4}
1000	0.000 01	9.7624×10^{-1}	2.170×10^{-6}	1.8462×10^{-7}	2.3760×10^{-2}	1.645×10^{-6}
	0.000 05	9.7622×10^{-1}	1.085×10^{-5}	4.6158×10^{-6}	2.3759×10^{-2}	8.227×10^{-6}
	0.000 10	9.7619×10^{-1}	2.176×10^{-5}	1.8465×10^{-5}	2.3757×10^{-2}	1.645×10^{-5}
	0.000 50	9.7561×10^{-1}	1.085×10^{-4}	4.6170×10^{-4}	2.3733×10^{-2}	8.220×10^{-5}
	0.001 00	9.7409×10^{-1}	2.168×10^{-4}	1.8454×10^{-3}	2.3684×10^{-2}	1.641×10^{-4}
	0.005 00	9.3129×10^{-1}	1.036×10^{-3}	4.4335×10^{-2}	2.2553×10^{-2}	7.817×10^{-4}
	0.010 00	8.2016×10^{-1}	1.825×10^{-3}	1.5688×10^{-1}	1.9783×10^{-2}	1.370×10^{-3}
	0.050 00	1.7065×10^{-1}	1.868×10^{-3}	8.2218×10^{-1}	3.9531×10^{-3}	1.348×10^{-3}
	0.100 00	5.0770×10^{-2}	1.068×10^{-3}	9.4631×10^{-1}	1.1221×10^{-3}	7.352×10^{-4}
	0.200 00	1.5855×10^{-2}	5.844×10^{-4}	9.8287×10^{-1}	3.2361×10^{-4}	3.716×10^{-4}
	0.255 53	1.1585×10^{-2}	4.927×10^{-4}	9.8739×10^{-1}	2.2896×10^{-4}	3.033×10^{-4}
	0.983 00	7.3841×10^{-4}	3.526×10^{-5}	9.9922×10^{-1}	1.1588×10^{-6}	1.724×10^{-6}
	0.988 00	3.2275×10^{-4}	1.984×10^{-5}	9.9966×10^{-1}	3.6687×10^{-7}	9.027×10^{-7}
	0.993 00	7.8096×10^{-5}	7.543×10^{-6}	9.9991×10^{-1}	5.3002×10^{-8}	1.595×10^{-7}
	0.998 00	2.1536×10^{-6}	6.762×10^{-7}	1.0000	4.2582×10^{-10}	4.165×10^{-9}

1100	0.000 01	9.6408×10^{-1}	3.018×10^{-6}	1.1136×10^{-7}	3.5913×10^{-2}	2.646×10^{-6}
	0.000 10	9.6402×10^{-1}	3.018×10^{-5}	1.1138×10^{-5}	3.5908×10^{-2}	2.645×10^{-5}
	0.000 50	9.6356×10^{-1}	1.509×10^{-4}	2.7852×10^{-4}	3.5877×10^{-2}	1.322×10^{-4}
	0.001 00	9.6250×10^{-1}	3.014×10^{-4}	1.1138×10^{-3}	3.5819×10^{-2}	2.640×10^{-4}
	0.005 00	9.3539×10^{-1}	1.465×10^{-3}	2.7191×10^{-2}	3.4671×10^{-2}	1.278×10^{-3}
	0.010 00	8.6244×10^{-1}	2.701×10^{-3}	1.0070×10^{-1}	3.1808×10^{-2}	2.344×10^{-3}
	0.050 00	2.5002×10^{-1}	3.850×10^{-3}	7.3405×10^{-1}	8.8646×10^{-3}	3.212×10^{-3}
	0.100 00	8.0021×10^{-2}	2.373×10^{-3}	9.1301×10^{-1}	2.7066×10^{-3}	1.888×10^{-3}
	0.200 00	2.4512×10^{-2}	1.300×10^{-3}	9.7247×10^{-1}	7.6312×10^{-4}	9.524×10^{-4}
	0.300 00	1.3983×10^{-2}	9.575×10^{-4}	9.8399×10^{-1}	4.0906×10^{-4}	6.591×10^{-4}
	0.360 32	1.1736×10^{-2}	8.674×10^{-4}	9.8648×10^{-1}	3.3486×10^{-4}	5.824×10^{-4}
	0.940 00	1.8866×10^{-3}	1.559×10^{-4}	9.9793×10^{-1}	1.0691×10^{-5}	2.078×10^{-5}
	0.950 00	1.5810×10^{-3}	1.343×10^{-4}	9.9826×10^{-1}	7.9264×10^{-6}	1.584×10^{-5}
	0.960 00	1.1516×10^{-3}	1.054×10^{-4}	9.9873×10^{-1}	4.8794×10^{-6}	1.050×10^{-5}
	0.970 00	6.7729×10^{-4}	7.177×10^{-5}	9.9924×10^{-1}	2.2628×10^{-6}	5.642×10^{-6}
	0.980 00	2.7295×10^{-4}	3.805×10^{-5}	9.9969×10^{-1}	6.3595×10^{-7}	2.088×10^{-6}
	0.990 00	4.4869×10^{-5}	1.113×10^{-5}	9.9994×10^{-1}	5.4367×10^{-8}	3.173×10^{-7}
	0.995 00	6.3085×10^{-6}	2.975×10^{-6}	9.9999×10^{-1}	3.8866×10^{-9}	4.313×10^{-8}
1200	0.000 01	9.4990×10^{-1}	3.963×10^{-6}	7.3354×10^{-8}	5.0097×10^{-2}	3.906×10^{-6}
	0.000 05	9.4987×10^{-1}	7.981×10^{-5}	1.8340×10^{-6}	5.0093×10^{-2}	1.953×10^{-5}
	0.000 10	9.4983×10^{-1}	3.963×10^{-5}	7.3363×10^{-6}	5.0089×10^{-2}	3.906×10^{-5}
	0.000 50	9.4938×10^{-1}	1.981×10^{-4}	1.8346×10^{-4}	5.0045×10^{-2}	1.952×10^{-4}
	0.001 00	9.4851×10^{-1}	3.959×10^{-4}	7.3373×10^{-4}	4.9974×10^{-2}	3.898×10^{-4}
	0.005 00	9.2933×10^{-1}	1.940×10^{-3}	1.8056×10^{-2}	4.8768×10^{-2}	1.903×10^{-3}
	0.010 00	8.7836×10^{-1}	3.665×10^{-3}	6.8529×10^{-2}	4.5864×10^{-2}	3.577×10^{-3}
	0.050 00	3.2799×10^{-1}	6.724×10^{-3}	6.4252×10^{-1}	1.6464×10^{-2}	6.308×10^{-3}
	0.100 00	1.1410×10^{-1}	4.512×10^{-3}	8.7189×10^{-1}	5.4631×10^{-3}	4.038×10^{-3}
	0.200 00	3.4783×10^{-2}	2.503×10^{-3}	9.5913×10^{-1}	1.5287×10^{-3}	2.056×10^{-3}
	0.300 00	1.8257×10^{-2}	1.766×10^{-3}	9.7786×10^{-1}	7.4627×10^{-4}	1.349×10^{-3}
	0.400 00	1.2271×10^{-2}	1.410×10^{-3}	9.8483×10^{-1}	4.7256×10^{-4}	1.015×10^{-3}
	0.840 00	3.6259×10^{-3}	4.444×10^{-4}	9.9578×10^{-1}	4.6399×10^{-5}	1.063×10^{-4}
	0.860 00	3.2517×10^{-3}	4.081×10^{-4}	9.9621×10^{-1}	3.9127×10^{-5}	9.179×10^{-5}
	0.880 00	2.7100×10^{-3}	3.568×10^{-4}	9.9683×10^{-1}	2.9881×10^{-5}	7.353×10^{-5}
	0.900 00	2.0504×10^{-3}	2.923×10^{-4}	9.9758×10^{-1}	2.0044×10^{-5}	5.341×10^{-5}
	0.920 00	1.3541×10^{-3}	2.187×10^{-4}	9.9838×10^{-1}	1.1215×10^{-5}	3.386×10^{-5}
	0.940 00	7.2552×10^{-4}	1.424×10^{-4}	9.9911×10^{-1}	4.7506×10^{-6}	1.743×10^{-5}
	0.960 00	2.6783×10^{-4}	7.258×10^{-5}	9.9965×10^{-1}	1.2265×10^{-6}	6.196×10^{-6}
	0.980 00	4.0631×10^{-5}	2.036×10^{-5}	9.9994×10^{-1}	9.7028×10^{-8}	9.089×10^{-7}

Table 5.10. (con't.).

T,K	x	Li	LiH	H_2	Li_2	Li_2H
1300	0.000 01	9.3475×10^{-1}	4.978×10^{-6}	5.1980×10^{-8}	6.5243×10^{-2}	5.345×10^{-6}
	0.000 05	9.3471×10^{-1}	2.489×10^{-5}	1.2996×10^{-6}	6.5237×10^{-2}	2.672×10^{-5}
	0.000 10	9.3466×10^{-1}	4.977×10^{-5}	5.1986×10^{-6}	6.5231×10^{-2}	5.344×10^{-5}
	0.000 50	9.3418×10^{-1}	2.488×10^{-4}	1.2999×10^{-4}	6.5171×10^{-2}	2.670×10^{-4}
	0.001 00	9.3337×10^{-1}	4.972×10^{-4}	5.1992×10^{-4}	6.5082×10^{-2}	5.344×10^{-4}
	0.005 00	9.1832×10^{-1}	2.446×10^{-3}	1.2846×10^{-2}	6.3777×10^{-2}	2.614×10^{-3}
	0.010 00	8.8009×10^{-1}	4.686×10^{-3}	4.9423×10^{-2}	6.0818×10^{-2}	4.982×10^{-3}
	0.050 00	3.9562×10^{-1}	1.034×10^{-2}	5.5718×10^{-1}	2.6283×10^{-2}	1.057×10^{-2}
	0.100 00	1.4983×10^{-1}	7.568×10^{-3}	8.2573×10^{-1}	9.4931×10^{-3}	7.377×10^{-3}
	0.200 00	4.5967×10^{-2}	4.287×10^{-3}	9.4325×10^{-1}	2.6671×10^{-3}	3.827×10^{-3}
	0.300 00	2.2555×10^{-2}	2.926×10^{-3}	9.7090×10^{-1}	1.2067×10^{-3}	2.408×10^{-3}
	0.400 00	1.3303×10^{-2}	2.171×10^{-3}	9.8222×10^{-1}	6.5702×10^{-4}	1.650×10^{-3}
	0.450 00	1.0492×10^{-2}	1.691×10^{-3}	9.8574×10^{-1}	4.9679×10^{-4}	1.378×10^{-3}
	0.900 00	2.0197×10^{-3}	4.201×10^{-4}	9.9746×10^{-1}	2.4222×10^{-5}	7.752×10^{-5}
	0.910 00	1.1994×10^{-3}	3.014×10^{-4}	9.9844×10^{-1}	1.2452×10^{-5}	4.813×10^{-5}
	0.920 00	6.8759×10^{-4}	2.109×10^{-4}	9.9907×10^{-1}	6.0959×10^{-6}	2.877×10^{-5}
	0.930 00	3.7690×10^{-4}	1.431×10^{-4}	9.9946×10^{-1}	2.8060×10^{-6}	1.639×10^{-5}
	0.940 00	1.9478×10^{-4}	9.330×10^{-5}	9.9970×10^{-1}	1.1917×10^{-6}	8.782×10^{-6}

Table 5.11. Composition of the vapor phase in the Li-LiD system when the liquid is in equilibrium with the vapor (*x* is the mole fraction of lithium deuteride in the liquid phase).

T,K	x	Li	LiD	D_2	Li_2	Li_2D
800	0.000 01	9.9231×10^{-1}	8.491×10^{-7}	1.3188×10^{-6}	7.6895×10^{-3}	4.700×10^{-7}
	0.000 05	9.9227×10^{-1}	4.245×10^{-6}	3.2973×10^{-5}	7.6889×10^{-3}	2.350×10^{-6}
	0.000 10	9.9217×10^{-1}	8.490×10^{-6}	1.3189×10^{-4}	7.6878×10^{-3}	4.699×10^{-6}
	0.000 50	9.8898×10^{-1}	4.232×10^{-5}	3.2894×10^{-3}	7.6600×10^{-3}	2.341×10^{-5}
	0.001 00	9.7925×10^{-1}	8.383×10^{-5}	1.3040×10^{-2}	7.5808×10^{-3}	4.635×10^{-5}
	0.005 00	7.4448×10^{-1}	3.189×10^{-4}	2.4928×10^{-1}	5.7404×10^{-3}	1.757×10^{-4}
	0.010 00	4.2450×10^{-1}	3.638×10^{-4}	5.7168×10^{-1}	3.2568×10^{-3}	1.994×10^{-4}
	0.050 00	2.8498×10^{-2}	1.205×10^{-4}	9.7111×10^{-1}	2.1017×10^{-4}	6.345×10^{-5}
	0.100 00	7.6200×10^{-3}	6.151×10^{-5}	9.9223×10^{-1}	5.3647×10^{-5}	3.093×10^{-5}
900	0.000 01	9.8597×10^{-1}	1.393×10^{-6}	5.8027×10^{-7}	1.4025×10^{-2}	9.134×10^{-7}
	0.000 05	9.8595×10^{-1}	6.967×10^{-6}	1.4508×10^{-5}	1.4024×10^{-2}	4.567×10^{-6}
	0.000 10	9.8590×10^{-1}	1.393×10^{-5}	5.8035×10^{-5}	1.4023×10^{-2}	9.133×10^{-6}
	0.000 50	9.8444×10^{-1}	6.958×10^{-5}	1.4499×10^{-3}	1.3996×10^{-2}	4.559×10^{-5}
	0.001 00	9.8006×10^{-1}	1.386×10^{-4}	5.7787×10^{-3}	1.3927×10^{-2}	9.075×10^{-5}
	0.005 00	8.5946×10^{-1}	6.080×10^{-4}	1.2738×10^{-1}	1.2165×10^{-2}	3.966×10^{-4}
	0.010 00	6.2028×10^{-1}	8.766×10^{-4}	3.6954×10^{-1}	8.7356×10^{-3}	5.696×10^{-4}
	0.050 00	6.2263×10^{-2}	4.339×10^{-4}	9.3619×10^{-1}	8.4292×10^{-4}	2.707×10^{-4}
	0.100 00	1.6982×10^{-2}	2.268×10^{-4}	9.8244×10^{-1}	2.1941×10^{-4}	1.350×10^{-4}
	0.200 00	5.4560×10^{-3}	1.245×10^{-4}	9.9429×10^{-1}	6.5361×10^{-5}	6.874×10^{-5}
1000	0.000 01	9.7624×10^{-1}	2.059×10^{-6}	2.8324×10^{-7}	2.3760×10^{-2}	1.643×10^{-6}
	0.000 05	9.7622×10^{-1}	1.029×10^{-5}	7.0814×10^{-6}	2.3759×10^{-2}	8.213×10^{-6}
	0.000 10	9.7618×10^{-1}	2.059×10^{-5}	2.8328×10^{-5}	2.3757×10^{-2}	1.643×10^{-5}
	0.000 50	9.7538×10^{-1}	1.029×10^{-4}	7.0816×10^{-4}	2.3728×10^{-2}	8.204×10^{-5}
	0.001 00	9.7314×10^{-1}	2.053×10^{-4}	2.8285×10^{-3}	2.3661×10^{-2}	1.637×10^{-4}
	0.005 00	9.0980×10^{-1}	9.602×10^{-4}	6.6448×10^{-2}	2.2033×10^{-2}	7.624×10^{-4}
	0.010 00	7.5681×10^{-1}	1.597×10^{-3}	2.2210×10^{-1}	1.8236×10^{-2}	1.262×10^{-3}
	0.050 00	1.1858×10^{-1}	1.231×10^{-3}	8.7650×10^{-1}	2.7469×10^{-3}	9.352×10^{-4}
	0.100 00	3.3724×10^{-2}	6.728×10^{-4}	9.6437×10^{-1}	7.4539×10^{-4}	4.876×10^{-4}
	0.200 00	1.0396×10^{-2}	3.635×10^{-4}	9.8878×10^{-1}	2.1220×10^{-4}	2.433×10^{-4}
	0.255 53	7.5845×10^{-3}	3.060×10^{-4}	9.9176×10^{-1}	1.4990×10^{-4}	1.983×10^{-4}
	0.983 00	4.7526×10^{-4}	2.167×10^{-5}	9.9950×10^{-1}	7.4584×10^{-7}	1.115×10^{-6}
	0.988 00	2.0769×10^{-4}	1.219×10^{-5}	9.9978×10^{-1}	2.3609×10^{-7}	4.544×10^{-7}
	0.993 00	5.0252×10^{-5}	4.634×10^{-6}	9.9995×10^{-1}	3.4105×10^{-8}	1.031×10^{-7}

Table 5.11. (con't.).

T,K	x	Li	LiD	D_2	Li_2	Li_2D
	0.998 00	1.3857×10^{-6}	4.154×10^{-7}	1.0000	2.7399×10^{-10}	2.693×10^{-9}
1100	0.000 01	9.6408×10^{-1}	2.817×10^{-6}	1.5835×10^{-7}	3.5913×10^{-2}	2.623×10^{-6}
	0.000 05	9.6406×10^{-1}	1.508×10^{-5}	3.9589×10^{-6}	3.5911×10^{-2}	1.312×10^{-5}
	0.000 10	9.6402×10^{-1}	2.817×10^{-5}	1.5837×10^{-5}	3.5908×10^{-2}	2.623×10^{-5}
	0.000 50	9.6346×10^{-1}	1.408×10^{-4}	3.9598×10^{-4}	3.5873×10^{-2}	1.310×10^{-4}
	0.001 00	9.6207×10^{-1}	2.812×10^{-4}	1.5829×10^{-3}	3.5803×10^{-2}	2.616×10^{-4}
	0.005 00	9.2488×10^{-1}	1.352×10^{-3}	3.8229×10^{-2}	3.4282×10^{-2}	1.253×10^{-3}
	0.010 00	8.2745×10^{-1}	2.419×10^{-3}	1.3738×10^{-1}	3.0518×10^{-2}	2.230×10^{-3}
	0.050 00	1.9094×10^{-1}	2.744×10^{-3}	7.9711×10^{-1}	6.7699×10^{-3}	2.432×10^{-3}
	0.100 00	5.7775×10^{-2}	1.599×10^{-3}	9.3732×10^{-1}	1.9542×10^{-3}	1.352×10^{-3}
	0.200 00	1.7382×10^{-2}	8.604×10^{-4}	9.8050×10^{-1}	5.4114×10^{-4}	6.697×10^{-4}
	0.300 00	9.8811×10^{-3}	6.315×10^{-4}	9.8874×10^{-1}	2.8907×10^{-4}	4.619×10^{-4}
	0.360 32	8.2872×10^{-3}	5.717×10^{-4}	9.9050×10^{-1}	2.3646×10^{-4}	4.078×10^{-4}
	0.940 00	1.5963×10^{-3}	1.122×10^{-4}	9.9827×10^{-1}	9.0453×10^{-6}	1.590×10^{-5}
	0.950 00	1.3376×10^{-3}	9.666×10^{-5}	9.9855×10^{-1}	6.7060×10^{-6}	1.212×10^{-5}
	0.960 00	9.7419×10^{-4}	7.585×10^{-5}	9.9894×10^{-1}	4.1278×10^{-6}	8.035×10^{-6}
	0.970 00	5.7292×10^{-4}	5.166×10^{-5}	9.9937×10^{-1}	1.9141×10^{-6}	4.315×10^{-6}
	0.980 00	2.3087×10^{-4}	2.739×10^{-5}	9.9974×10^{-1}	5.3791×10^{-7}	1.596×10^{-6}
	0.990 00	3.7951×10^{-5}	8.010×10^{-6}	9.9995×10^{-1}	4.5984×10^{-8}	2.426×10^{-7}
	0.995 00	5.3357×10^{-5}	3.142×10^{-6}	9.9999×10^{-1}	3.2872×10^{-9}	3.299×10^{-8}
1200	0.000 01	9.4990×10^{-1}	3.651×10^{-6}	9.7902×10^{-8}	5.0097×10^{-2}	3.851×10^{-6}
	0.000 05	9.4987×10^{-1}	1.825×10^{-5}	2.4477×10^{-6}	5.0093×10^{-2}	1.925×10^{-5}
	0.000 10	9.4983×10^{-1}	3.651×10^{-5}	9.7914×10^{-6}	5.0089×10^{-2}	3.850×10^{-5}
	0.000 50	9.4934×10^{-1}	1.825×10^{-4}	2.4484×10^{-4}	5.0043×10^{-2}	1.924×10^{-4}
	0.001 00	9.4831×10^{-1}	3.646×10^{-4}	9.7907×10^{-4}	4.9964×10^{-2}	3.842×10^{-4}
	0.005 00	9.2392×10^{-1}	1.777×10^{-3}	2.3958×10^{-2}	4.8484×10^{-2}	1.864×10^{-3}
	0.010 00	8.5896×10^{-1}	3.302×10^{-3}	8.9441×10^{-2}	4.4851×10^{-2}	3.448×10^{-3}
	0.050 00	2.7008×10^{-1}	5.101×10^{-3}	7.0614×10^{-1}	1.3557×10^{-2}	5.121×10^{-3}
	0.100 00	8.8354×10^{-2}	3.219×10^{-3}	9.0111×10^{-1}	4.2305×10^{-3}	3.082×10^{-3}
	0.200 00	2.6336×10^{-2}	1.746×10^{-3}	9.6923×10^{-1}	1.1575×10^{-3}	1.535×10^{-3}
	0.300 00	1.3757×10^{-2}	1.226×10^{-3}	$9.,8345\times10^{-1}$	5.6233×10^{-4}	1.002×10^{-3}
	0.400 00	9.2301×10^{-13}	9.773×10^{-4}	9.8868×10^{-1}	3.5545×10^{-4}	7.527×10^{-4}
	0.840 00	3.0278×10^{-3}	3.239×10^{-4}	9.9653×10^{-1}	3.8745×10^{-5}	8.290×10^{-5}
	0.860 00	2.7151×10^{-3}	2.975×10^{-4}	9.9688×10^{-1}	3.2670×10^{-5}	7.159×10^{-5}

	0.880 00	2.2625×10^{-3}	2.600×10^{-4}	9.9740×10^{-1}	2.4947×10^{-5}	5.734×10^{-5}
	0.900 00	1.7116×10^{-3}	2.130×10^{-4}	9.9802×10^{-1}	1.6732×10^{-5}	4.165×10^{-5}
	0.920 00	1.1302×10^{-3}	1.594×10^{-4}	9.9867×10^{-1}	9.3608×10^{-6}	2.640×10^{-5}
	0.940 00	6.0548×10^{-4}	1.038×10^{-4}	9.9927×10^{-1}	3.9646×10^{-6}	1.359×10^{-5}
	0.960 00	2.2350×10^{-4}	5.273×10^{-5}	9.9972×10^{-1}	1.0234×10^{-6}	4.829×10^{-6}
	0.980 00	3.3903×10^{-5}	1.483×10^{-5}	9.9995×10^{-1}	8.0962×10^{-8}	7.084×10^{-7}
1300	0.000 01	9.3475×10^{-1}	4.528×10^{-6}	6.5755×10^{-8}	6.5243×10^{-2}	5.224×10^{-6}
	0.000 05	9.3471×10^{-1}	2.264×10^{-5}	1.6440×10^{-6}	6.5238×10^{-2}	2.612×10^{-5}
	0.000 10	9.3467×10^{-1}	4.528×10^{-5}	6.5762×10^{-6}	6.5231×10^{-2}	5.223×10^{-5}
	0.000 50	9.3418×10^{-1}	2.263×10^{-4}	1.6444×10^{-4}	6.5171×10^{-2}	2.610×10^{-4}
	0.001 00	9.3329×10^{-1}	4.523×10^{-4}	6.5764×10^{-4}	6.5077×10^{-2}	5.213×10^{-4}
	0.005 00	9.1546×10^{-1}	2.219×10^{-3}	1.6199×10^{-2}	6.3578×10^{-2}	2.547×10^{-3}
	0.010 00	8.6917×10^{-1}	4.210×10^{-3}	6.1745×10^{-2}	6.0063×10^{-2}	4.809×10^{-3}
	0.050 00	3.4508×10^{-1}	8.207×10^{-3}	6.1478×10^{-1}	2.2925×10^{-2}	9.012×10^{-3}
	0.100 00	1.2301×10^{-1}	5.652×10^{-3}	6.5762×10^{-1}	7.7942×10^{-3}	5.919×10^{-3}
	0.200 00	3.6788×10^{-2}	3.121×10^{-3}	9.5496×10^{-1}	2.1346×10^{-3}	2.994×10^{-3}
	0.300 00	1.7944×10^{-2}	2.117×10^{-3}	9.7711×10^{-1}	9.5997×10^{-4}	1.872×10^{-3}
	0.400 00	1.0558×10^{-2}	1.568×10^{-3}	9.8607×10^{-1}	5.2142×10^{-4}	1.280×10^{-3}
	0.450 00	8.3206×10^{-3}	1.365×10^{-3}	9.8885×10^{-1}	3.9396×10^{-4}	1.068×10^{-3}
	0.900 00	1.3591×10^{-3}	2.789×10^{-4}	9.9829×10^{-1}	1.6299×10^{-5}	5.528×10^{-5}
	0.910 00	8.0685×10^{-4}	2.000×10^{-4}	9.9895×10^{-1}	8.3762×10^{-6}	3.431×10^{-5}
	0.920 00	4.6244×10^{-4}	1.399×10^{-4}	9.9937×10^{-1}	4.0998×10^{-6}	2.051×10^{-5}
	0.930 00	2.5345×10^{-4}	9.495×10^{-5}	9.9964×10^{-1}	1.8869×10^{-6}	1.168×10^{-5}
	0.940 00	1.3098×10^{-4}	6.188×10^{-5}	8.9980×10^{-1}	8.0135×10^{-7}	6.258×10^{-6}

Table 5.12. Composition of the vapor phase in the Li-LiT system when the liquid is in equilibrium with the vapor (*x* is the mole fraction of lithium tritide in the liquid phase).

T,K	*x*	Li	LiT	T_2	Li_2	Li_2T
800	0.000 01	9.9231×10^{-1}	7.726×10^{-7}	1.3941×10^{-6}	7.6895×10^{-3}	4.277×10^{-7}
	0.000 02	9.9230×10^{-1}	1.545×10^{-6}	5.5765×10^{-6}	7.6894×10^{-3}	8.553×10^{-7}
	0.000 05	9.9227×10^{-1}	3.863×10^{-6}	3.4855×10^{-5}	7.6889×10^{-3}	2.138×10^{-6}
	0.000 07	9.9223×10^{-1}	5.408×10^{-6}	6.8316×10^{-5}	7.6885×10^{-3}	2.993×10^{-6}
	0.000 10	9.9216×10^{-1}	7.726×10^{-6}	1.3942×10^{-4}	7.6877×10^{-3}	4.276×10^{-6}
	0.000 50	9.8880×10^{-1}	3.851×10^{-5}	3.4765×10^{-3}	7.6588×10^{-3}	2.130×10^{-5}
	0.001 00	9.7853×10^{-1}	7.623×10^{-5}	1.3774×10^{-2}	7.5753×10^{-3}	4.215×10^{-5}
	0.005 00	7.3407×10^{-1}	2.862×10^{-4}	2.5982×10^{-1}	5.6601×10^{-3}	1.576×10^{-4}
	0.010 00	4.1111×10^{-1}	3.206×10^{-4}	5.8524×10^{-1}	3.1541×10^{-3}	1.757×10^{-4}
	0.050 00	2.7002×10^{-2}	1.039×10^{-4}	9.7264×10^{-1}	1.9913×10^{-4}	5.471×10^{-5}
	0.100 00	7.2117×10^{-3}	5.298×10^{-5}	9.9266×10^{-1}	5.0773×10^{-5}	2.664×10^{-5}
900	0.000 01	9.8597×10^{-1}	1.288×10^{-6}	6.0880×10^{-7}	1.4025×10^{-2}	8.479×10^{-7}
	0.000 05	9.8595×10^{-1}	6.438×10^{-6}	1.5221×10^{-5}	1.4024×10^{-2}	4.239×10^{-6}
	0.000 10	9.8590×10^{-1}	1.288×10^{-5}	6.0888×10^{-5}	1.4023×10^{-2}	8.478×10^{-6}
	0.000 50	9.8438×10^{-1}	6.429×10^{-5}	1.5210×10^{-3}	1.3996×10^{-2}	4.232×10^{-5}
	0.001 00	9.7980×10^{-1}	1.280×10^{-4}	6.0612×10^{-3}	1.3924×10^{-2}	8.421×10^{-5}
	0.005 00	8.5417×10^{-1}	5.584×10^{-4}	1.3282×10^{-1}	1.2090×10^{-2}	3.659×10^{-4}
	0.010 00	6.0928×10^{-1}	7.965×10^{-4}	3.8083×10^{-1}	8.5806×10^{-3}	5.193×10^{-4}
	0.050 00	5.9526×10^{-2}	3.833×10^{-4}	9.3904×10^{-1}	8.0587×10^{-4}	2.402×10^{-4}
	0.100 00	1.6200×10^{-2}	1.999×10^{-4}	9.8327×10^{-1}	2.0931×10^{-4}	1.196×10^{-4}
	0.200 00	5.2018×10^{-3}	1.097×10^{-4}	9.9457×10^{-1}	6.2316×10^{-5}	6.084×10^{-5}
1000	0.000 01	9.7624×10^{-1}	1.931×10^{-6}	2.9539×10^{-7}	2.3760×10^{-2}	1.567×10^{-6}
	0.000 05	9.7622×10^{-1}	9.656×10^{-6}	7.3852×10^{-6}	2.3759×10^{-2}	7.833×10^{-6}
	0.000 10	9.7618×10^{-1}	1.931×10^{-5}	2.9543×10^{-5}	2.3757×10^{-2}	1.567×10^{-5}
	0.000 50	9.7536×10^{-1}	9.649×10^{-5}	7.3852×10^{-4}	2.3727×10^{-2}	7.825×10^{-5}
	0.001 00	9.7304×10^{-1}	1.926×10^{-4}	2.9495×10^{-3}	2.3659×10^{-2}	1.561×10^{-4}
	0.005 00	9.0730×10^{-1}	8.983×10^{-4}	6.9108×10^{-2}	2.1972×10^{-2}	7.251×10^{-4}
	0.010 00	7.4978×10^{-1}	1.484×10^{-3}	2.2947×10^{-1}	1.8067×10^{-2}	1.192×10^{-3}
	0.050 00	1.1430×10^{-1}	1.113×10^{-3}	8.8108×10^{-1}	2.6477×10^{-3}	8.597×10^{-4}
	0.100 00	3.2386×10^{-2}	6.061×10^{-4}	9.6585×10^{-1}	7.1583×10^{-4}	4.466×10^{-4}
	0.200 00	9.9737×10^{-3}	3.271×10^{-4}	9.8927×10^{-1}	2.0357×10^{-4}	2.226×10^{-4}
	0.300 00	6.2952×10^{-3}	2.540×10^{-4}	9.9316×10^{-1}	1.2224×10^{-4}	1.644×10^{-4}
	0.983 00	4.6164×10^{-4}	1.962×10^{-5}	9.9952×10^{-1}	1.2448×10^{-7}	1.026×10^{-6}

	0.988 00	2.0174×10^{-4}	1.104×10^{-5}	9.9979×10^{-1}	2.2933×10^{-7}	4.182×10^{-7}
	0.993 00	4.8812×10^{-5}	4.195×10^{-6}	9.9995×10^{-1}	3.3127×10^{-8}	9.491×10^{-8}
	0.998 00	1.3460×10^{-6}	3.761×10^{-7}	1.0000	2.6614×10^{-10}	2.479×10^{-9}
1100	0.000 01	9.6408×10^{-1}	2.580×10^{-6}	1.6433×10^{-7}	3.5913×10^{-2}	2.496×10^{-6}
	0.000 05	9.6406×10^{-1}	1.290×10^{-5}	4.1085×10^{-6}	3.5911×10^{-2}	1.248×10^{-5}
	0.000 10	9.6402×10^{-1}	2.580×10^{-5}	1.6435×10^{-5}	3.5908×10^{-2}	2.496×10^{-5}
	0.000 50	9.6346×10^{-1}	1.289×10^{-4}	4.1095×10^{-4}	3.5873×10^{-2}	1.247×10^{-4}
	0.001 00	9.6205×10^{-1}	2.576×10^{-4}	1.6427×10^{-3}	3.5802×10^{-2}	2.490×10^{-4}
	0.005 00	9.2371×10^{-1}	1.237×10^{-3}	3.9623×10^{-2}	3.4238×10^{-2}	1.191×10^{-3}
	0.010 00	8.2344×10^{-1}	2.204×10^{-3}	1.4188×10^{-1}	3.0369×10^{-2}	2.112×10^{-3}
	0.050 00	1.8542×10^{-1}	2.441×10^{-3}	8.0332×10^{-1}	6.5742×10^{-3}	2.248×10^{-3}
	0.100 00	5.5810×10^{-2}	1.414×10^{-3}	9.3965×10^{-1}	1.8877×10^{-3}	1.243×10^{-3}
	0.200 00	1.6763×10^{-2}	7.600×10^{-4}	9.8134×10^{-1}	5.2186×10^{-4}	6.146×10^{-4}
	0.300 00	9.5259×10^{-3}	5.576×10^{-4}	9.8921×10^{-1}	2.7868×10^{-4}	4.237×10^{-4}
	0.350 00	8.1768×10^{-3}	5.115×10^{-4}	9.9070×10^{-1}	2.3414×10^{-4}	3.804×10^{-4}
	0.950 00	1.0720×10^{-3}	7.781×10^{-5}	9.9883×10^{-1}	5.3746×10^{-6}	1.013×10^{-5}
	0.960 00	7.8071×10^{-4}	6.106×10^{-5}	9.9915×10^{-1}	3.3080×10^{-6}	6.720×10^{-6}
	0.970 00	4.5909×10^{-4}	4.158×10^{-5}	9.9949×10^{-1}	1.5338×10^{-6}	3.608×10^{-6}
	0.980 00	1.8499×10^{-4}	2.205×10^{-5}	9.9979×10^{-1}	4.3101×10^{-7}	1.334×10^{-6}
	0.990 00	3.0407×10^{-5}	6.446×10^{-6}	9.9996×10^{-1}	3.6844×10^{-8}	2.029×10^{-7}
	0.995 00	4.2751×10^{-6}	1.724×10^{-6}	9.9999×10^{-1}	2.6338×10^{-9}	2.758×10^{-8}
1200	0.000 01	9.4990×10^{-1}	3.474×10^{-6}	1.0119×10^{-7}	5.0097×10^{-2}	3.664×10^{-6}
	0.000 05	9.4987×10^{-1}	1.737×10^{-5}	2.5298×10^{-6}	5.0093×10^{-2}	1.852×10^{-5}
	0.000 10	9.4983×10^{-1}	3.474×10^{-5}	1.0120×10^{-5}	5.0089×10^{-2}	3.664×10^{-5}
	0.000 50	9.4935×10^{-1}	1.736×10^{-4}	2.5306×10^{-4}	5.0043×10^{-2}	1.831×10^{-4}
	0.001 00	9.4831×10^{-1}	3.469×10^{-4}	1.0119×10^{-3}	4.9964×10^{-2}	3.656×10^{-4}
	0.005 00	9.2334×10^{-1}	1.690×10^{-3}	2.4747×10^{-2}	4.8454×10^{-2}	1.773×10^{-3}
	0.010 00	8.5667×10^{-1}	3.133×10^{-3}	9.2196×10^{-2}	4.4731×10^{-2}	3.277×10^{-3}
	0.050 00	2.6396×10^{-1}	4.744×10^{-3}	7.1329×10^{-1}	1.3250×10^{-2}	4.762×10^{-3}
	0.100 00	8.5787×10^{-2}	2.974×10^{-3}	9.0428×10^{-1}	4.1075×10^{-3}	2.848×10^{-3}
	0.200 00	2.5511×10^{-2}	1.609×10^{-3}	9.7034×10^{-1}	1.1212×10^{-3}	1.415×10^{-3}
	0.300 00	1.3319×10^{-2}	1.129×10^{-3}	9.8408×10^{-1}	5.4443×10^{-4}	9.233×10^{-4}
	0.400 00	8.9345×10^{-3}	9.001×10^{-4}	9.8913×10^{-1}	3.4407×10^{-4}	6.933×10^{-4}
	0.880 00	1.9664×10^{-3}	2.269×10^{-4}	9.9773×10^{-1}	2.1681×10^{-5}	5.005×10^{-5}
	0.900 00	1.4874×10^{-3}	1.859×10^{-4}	9.9828×10^{-1}	1.4541×10^{-5}	3.635×10^{-5}
	0.920 00	9.8213×10^{-4}	1.391×10^{-4}	9.9885×10^{-1}	8.1341×10^{-6}	2.304×10^{-5}
	0.940 00	5.2609×10^{-4}	9.053×10^{-5}	9.9937×10^{-1}	3.4448×10^{-6}	1.186×10^{-5}
	0.960 00	1.9418×10^{-4}	4.601×10^{-5}	9.9975×10^{-1}	8.8921×10^{-7}	4.213×10^{-6}
	0.980 00	2.9455×10^{-5}	1.294×10^{-5}	9.9996×10^{-1}	7.0341×10^{-8}	6.181×10^{-7}

Table 5.12. (con't.).

T,K	x	Li	LiT	T_2	Li_2	Li_2T
1300	0.000 01	9.3475×10^{-1}	4.340×10^{-6}	6.7726×10^{-8}	6.5243×10^{-2}	5.024×10^{-6}
	0.000 05	9.3471×10^{-1}	2.170×10^{-5}	1.6933×10^{-6}	6.5238×10^{-2}	2.512×10^{-5}
	0.000 10	9.3467×10^{-1}	4.340×10^{-5}	6.7734×10^{-6}	6.5231×10^{-2}	5.023×10^{-5}
	0.000 50	9.3419×10^{-1}	2.169×10^{-4}	1.6937×10^{-4}	6.5172×10^{-2}	2.510×10^{-4}
	0.001 00	9.3331×10^{-1}	4.335×10^{-4}	6.7737×10^{-4}	6.5078×10^{-2}	5.013×10^{-4}
	0.005 00	9.1519×10^{-1}	2.126×10^{-3}	1.6680×10^{-2}	6.3559×10^{-2}	2.449×10^{-3}
	0.010 00	8.6788×10^{-1}	4.029×10^{-3}	6.3501×10^{-2}	5.9974×10^{-2}	4.618×10^{-3}
	0.050 00	3.3906×10^{-1}	7.729×10^{-3}	6.2217×10^{-1}	2.2525×10^{-2}	8.516×10^{-3}
	0.100 00	1.1998×10^{-1}	5.284×10^{-3}	8.6158×10^{-1}	7.6022×10^{-3}	5.552×10^{-3}
	0.200 00	3.5773×10^{-2}	2.909×10^{-3}	9.5644×10^{-1}	2.0757×10^{-3}	2.799×10^{-3}
	0.300 00	1.7436×10^{-2}	1.972×10^{-3}	9.7791×10^{-1}	9.3279×10^{-4}	1.750×10^{-3}
	0.400 00	1.0256×10^{-2}	1.460×10^{-3}	9.8658×10^{-1}	5.0650×10^{-4}	1.195×10^{-3}
	0.450 00	8.0818×10^{-3}	1.270×10^{-3}	9.8927×10^{-1}	3.8265×10^{-4}	9.975×10^{-4}
	0.900 00	1.5511×10^{-3}	2.813×10^{-4}	9.9809×10^{-1}	1.8602×10^{-5}	5.595×10^{-5}
	0.910 00	9.2093×10^{-4}	2.018×10^{-4}	9.9883×10^{-1}	9.5604×10^{-6}	3.474×10^{-5}
	0.920 00	5.2785×10^{-4}	1.412×10^{-4}	9.9931×10^{-1}	4.6797×10^{-6}	2.076×10^{-5}
	0.930 00	2.8931×10^{-4}	9.580×10^{-5}	9.9960×10^{-1}	2.1539×10^{-6}	1.183×10^{-5}
	0.940 00	1.4951×10^{-4}	6.244×10^{-5}	9.9978×10^{-1}	9.1474×10^{-7}	6.335×10^{-6}

Table 5.13. Partial pressures of hydrogen, deuterium, and tritium in the two-dimensional region of the Li-LiH (LiD, LiT) systems (Pa).

T,K	P_{H_2}	P_{D_2}	P_{T_2}
800	23.69	34.21	42.4
850	132.2	193.8	234
900	609.2	905.9	1.07×10^3
950	2.39×10^3	3.60×10^3	4.14×10^3
961	3.24×10^3	—	—
963	—	5.03×10^3	—
964	—	—	5.84×10^3
1000	6.70×10^3	9.605×10^3	1.072×10^4
1050	1.536×10^4	2.145×10^4	2.356×10^4
1100	3.265×10^4	4.452×10^4	4.818×10^4
1150	6.500×10^4	8.672×10^4	9.260×10^4
1200	1.222×10^5	1.598×10^5	1.686×10^5

Table 5.14. Thermal conductivity of crystalline lithium hydride λ_{cryst}, and pressed lithium hydride, λ_{pr} [W/(m deg)].

T,K	λ_{cryst}	λ_{pr}	T,K	λ_{cryst}	λ_{pr}
300	14.3	7.9	700	4.7	4.2
350	11.0	7.0	750	4.4	4.1
400	9.0	6.3	800	4.1	3.9
450	7.7	5.7	850	3.9	3.8
500	6.8	5.3	900	3.7	3.6
550	6.0	5.0	950	3.6	3.5
600	5.5	4.7	965	3.5	3.5
650	5.1	4.4			

Table 5.15. Thermal conductivity of liquid lithium hydride, λ [W/(m deg)].

T,K	λ	T,K	λ
965	1.2	1150	1.9
1000	1.3	1200	2.1
1050	1.5	1250	2.3
1100	1.7	1300	2.5

Table 5.16. Dynamic viscosity μ and kinematic viscosity ν of liquid lithium hydride.

T,K	μ,Pa s	ν,m²/s	T,K	μ,Pa s	ν,m²/s
965	7.77×10^{-10}	1.33×10^{-6}	1150	5.39×10^{-10}	9.88×10^{-7}
1000	7.38×10^{-10}	1.27×10^{-6}	1200	4.79×10^{-10}	8.98×10^{-7}
1050	6.71×10^{-10}	1.18×10^{-6}	1250	4.25×10^{-10}	8.14×10^{-7}
1100	6.04×10^{-10}	1.08×10^{-6}	1300	3.76×10^{-10}	7.31×10^{-7}

Table 5.17. Electrical conductivity of lithium hydride.

T,K	$\bar{\sigma}$,S/m	T,K	$\bar{\sigma}$,S/m
700	9.75×10^{-4}	900	0.783
750	7.25×10^{-3}	950	2.68
800	4.20×10^{-2}	965 (S)	3.79
850	1.98×10^{-1}	965 (l)	30×10^{2}

Table 5.18. Viscosities μ and thermal conductivities λ of lithium–hydrogen (deuterium, tritium) gas mixtures.*

		μ,10^{-7} Pa s			λ,10^{-3} W/m deg)		
T,K	x	LiH	LiD	LiT	LiH	LiD	LiT
800	0.00	99	99	99	59	59	59
	0.01	87	102	119	86	93	85
	0.02	97	153	189	160	163	143
	0.03	117	187	234	227	208	179
	0.04	133	209	258	223	232	198
	0.06	15	227	279	321	254	215
	0.08	159	234	288	343	263	222
	0.10	164	238	292	354	267	225
900	0.00	108	108	108	69	69	69
	0.01	98	102	110	83	83	79
	0.02	95	128	154	130	133	119
	0.03	108	165	204	188	182	159
	0.04	123	193	239	239	217	188
	0.06	147	225	278	307	256	219
	0.08	161	240	296	344	274	232
	0.10	169	248	305	364	283	240
	0.20	181	259	318	396	297	250
1000	0.00	115	115	115	80	80	80
	0.01	108	108	112	89	87	84
	0.02	100	116	132	116	115	106
	0.03	104	142	171	159	154	138
	0.04	114	169	208	204	194	168
	0.06	136	211	261	279	243	210
	0.08	155	235	291	329	273	234
	0.10	167	244	307	361	290	247
	0.20	188	272	334	416	317	268
	0.26	192	276	338	434	321	271
1100	0.00	121	121	121	91	91	91
	0.01	115	116	118	98	96	94
	0.02	107	116	125	116	112	105
	0.03	106	129	149	146	139	127
	0.04	111	150	180	182	170	152
	0.06	129	191	236	252	224	196
	0.08	148	222	275	308	262	227
	0.10	162	243	230	348	287	247
	0.20	143	281	345	430	333	283
	0.30	201	290	356	451	343	291
	0.35	203	292	358	455	345	292
1200	0.00	125	125	125	101	101	101
	0.01	121	121	123	107	105	103

Table 5.18. (con't.).

T,K	x	μ,10^{-7} Pa s			λ,10^{-3} W/m deg)		
		LiH	LiD	LiT	LiH	LiD	LiT
	0.02	114	118	125	120	116	110
	0.03	110	124	139	142	134	124
	0.04	111	138	161	170	157	143
	0.06	123	173	211	231	206	183
	0.08	140	205	253	286	247	216
	0.10	155	231	285	331	278	241
	0.20	195	285	351	438	344	293
	0.30	208	300	369	471	362	307
	0.40	214	306	376	484	369	313
1300	0.00	128	128	128	108	108	108
	0.01	125	125	126	114	112	111
	0.02	119	122	127	125	120	116
	0.03	115	124	134	142	134	126
	0.04	114	132	149	165	152	139
	0.06	121	159	190	216	193	173
	0.08	134	189	231	268	233	205
	0.10	148	216	265	313	266	232
	0.20	195	285	351	441	349	299
	0.30	213	308	370	488	377	321
	0.40	222	319	392	511	390	331
	0.45	226	323	396	518	394	334

*The values are given for the state at which the liquid is in a thermodynamic equilibrium with the gas; x is the mole fraction of lithium hydride (deuteride, tritide) in the liquid solution.

Table 5.19. Coefficients of binary diffusion for a lithium–hydrogen gas mixture.*

T,K	i	j					
		1	2	3	4	5	6
800	1	820	180	100	120	91	69
	2	180	140	47	61	38	26
	3	104	47	24	30	21	15
	4	120	61	30	38	26	18
	5	91	38	26	26	18	14
	6	69	26	15	19	14	10
1000	1	1200	270	160	180	140	100
	2	270	210	70	92	57	39
	3	160	70	35	45	31	23
	4	180	92	45	57	39	29
	5	140	57	31	39	27	20
	6	100	39	23	29	20	16
1300	1	1800	430	250	290	220	170
	2	430	340	110	150	91	63
	3	250	110	56	73	49	36
	4	290	150	73	92	63	47
	5	220	91	49	63	43	33
	6	170	53	36	47	33	25

*D_{ij} (10^{-4} m^2/s) are scaled to a pressure of 10^3 Pa [see Eq. (4.38)]. The components of the vapor are numbered in the following order: 1) H_2; 2) Li; 3) Li_2; 4) LiH; 5) Li_2H; 6) Li_3.

Appendix Calculation of the temperature dependence of the adiabatic elastic moduli

Let us consider the temperature dependence of the adiabatic moduli c_{ij}^{ad} in the quasi-harmonic approximation. We have[32]

$$vc_{11}^{\text{ad}}(T)=\frac{\partial^2 U}{\partial V_{11}^2}-\frac{\partial}{\partial V_{11}}(\bar{\gamma}_{11}\epsilon_{\text{vib}})_s,$$

$$vc_{44}^{\text{ad}}(T)=\frac{\partial^2 U}{\partial V_{12}^2}-\frac{\partial}{\partial V_{12}}(\bar{\gamma}_{12}\epsilon_{\text{vib}})_s,$$

$$vc_{12}^{\text{ad}}(T)=\frac{\partial^2 U}{\partial V_{11}\partial V_{22}}-\frac{\partial}{\partial V_{11}}(\bar{\gamma}_{22}\epsilon_{\text{vib}})_s.$$

Here V_{ij} is the elastic-strain tensor; the derivatives are used for the unstrained state [for $v=v(T)$, where $v(T)$ is the equilibrium volume of the cell at the temperature T]; the tensor $\bar{\gamma}_{ij}$ is a generalized Grüneisen constant

$$\bar{\gamma}_{ij}=\sum_k \gamma_{ij}^k\epsilon_k \quad \sum_k \epsilon_k,$$

$$\gamma_{ij}^k=-\frac{1}{2}\frac{\partial \ln \omega_k^2}{\partial V_{ij}}, \quad \epsilon_k=\frac{\hbar\omega_k}{2}+\frac{\hbar\omega_k}{e^{\hbar\omega_k/kT}-1}$$

is the energy of the kth oscillator, and u is the potential energy of the crystal per cell.

We assume that the components of the tensor $\bar{\gamma}_{ij}$, just as the Grüneisen constant γ considered above, depend only slightly on the volume, the temperature and the isotopic composition. In the case of lithium hydride, γ is virtually independent of the temperature,[160] beginning at $T\approx 50$ K. All calculations (of l, α, c_P-c_V, and other thermodynamic quantities for LiH, LiD, and LiT) carried out under this assumption yield values that are in good agreement with the experiment. Since in the quasiharmonic approximation[32] $(\partial\epsilon_{\text{vib}}/\partial V_{ij})_s=-\bar{\gamma}_{ij}\epsilon_{\text{vib}}$, we find

$$c_{11}^{\mathrm{ad}}(T)=\frac{1}{v}\frac{\partial^2 U}{\partial V_{11}^2}+\bar{\gamma}_{11}^2\frac{\epsilon_{\mathrm{vib}}}{v}=\frac{1}{v}\frac{\partial^2 U}{\partial V_{11}^2}+\frac{\bar{\gamma}_{11}^2}{\gamma^2}\left(P+\frac{dU}{dv}\right),$$
$$c_{44}^{\mathrm{ad}}(T)=\frac{1}{v}\frac{\partial^2 U}{\partial V_{12}}+\bar{\gamma}_{12}\frac{\epsilon_{\mathrm{vib}}}{v}=\frac{1}{v}\frac{\partial^2 U}{\partial V_{12}^2}+\frac{\bar{\gamma}_{12}^2}{\gamma^2}\left(P+\frac{dU}{dv}\right), \tag{A.1}$$
$$c_{12}^{\mathrm{ad}}(T)=\frac{1}{v}\frac{\partial^2 U}{\partial V_{11}\partial V_{22}}+\bar{\gamma}_{11}\bar{\gamma}_{22}\frac{\epsilon_{\mathrm{vib}}}{v}=\frac{1}{v}\frac{\partial^2 U}{\partial V_{11}\partial V_{22}}+\frac{\bar{\gamma}_{11}\bar{\gamma}_{22}}{\gamma^2}\left(P+\frac{dU}{dv}\right).$$

It is nevertheless difficult to theoretically calculate the adiabatic moduli on the basis of Eqs. (A.1), since $\bar{\gamma}_{ij}$ are not known. It is also difficult to calculate the potential of interaction between the atoms in the crystal. For the typically ionic crystals such as alkali-metal halides, a calculation of the elastic moduli from Eqs. (A.1), with the use of the Born–Mayer potential and under the assumption that the interionic forces are of the nature of the central forces, gives only a qualitative agreement with the experiment (see Ref. 32, for example). (For certain salts such as[32] LiF, there is not even any qualitative agreement.)

In our case, the central-force model is even less suitable for calculating c_{ij}^{ad}, since the derivatives dc_{ij}^{ad}/dT in this model are always less than zero, while experiments[125,138,225] show that $dc_{12}^{\mathrm{ad}}/dT>0$ for LiH.

Calculations based on the Born–Mayer model potential, in which the lattice vibrations are ignored and the elastic moduli c_{ij}^{ad} for LiH are assumed to have noncentral forces, also give unsatisfactory results. Since the approximations obtained from different moduli are crude, and since they are unsuitable for lithium hydride, we shall use here another approach to calculate c_{ij}^{ad}. We find the derivatives $(dc_{ij}^{\mathrm{ad}}/dT)_P$ by working from Eq. (A.1) under the assumption that the components of the tensor $\bar{\gamma}_{ij}$ are, as in the previous case of γ, independent of the temperature and the isotopic composition. We find

$$\left(\frac{dc_{11}^{\mathrm{ad}}}{dT}\right)_P=3\alpha v\frac{d}{dv}\left(\frac{1}{v}\frac{\partial^2 U}{\partial V_{11}^2}+\frac{\bar{\gamma}_{11}^2}{\gamma}\frac{dU}{dv}\right),$$
$$\left(\frac{dc_{44}^{\mathrm{ad}}}{dT}\right)_P=3\alpha v\frac{d}{dv}\left(\frac{1}{v}\cdot\frac{\partial^2 U}{\partial V_{12}^2}+\frac{\bar{\gamma}_{12}^2}{\gamma}\frac{dU}{dv}\right), \tag{A.2}$$
$$\left(\frac{dc_{12}^{\mathrm{ad}}}{dT}\right)_P=3\alpha v\frac{d}{dv}\left(\frac{1}{v}\frac{\partial^2 U}{\partial V_{11}\partial V_{22}}+\frac{\bar{\gamma}_{11}\bar{\gamma}_{12}}{\gamma}\frac{dU}{dv}\right).$$

Retaining in the derivatives $(dc_{ij}^{\mathrm{ad}}/dT)_P$ only the lowest-order terms in anharmonicity, we can use the coefficients of α in the harmonic approximation. We then find

$$\left(\frac{dc_{11}^{\mathrm{ad}}}{dT}\right)_P\cong 3\alpha v\frac{d}{dv}\left[\frac{1}{v}\frac{\partial^2 U}{\partial V_{11}^2}+\frac{\bar{\gamma}_{11}^2}{\gamma}\frac{dU}{dv}\right]_{v=\bar{v}}\equiv 3\Gamma_{11}\alpha$$

and, correspondingly, we find

$$\left(\frac{dc_{12}^{\mathrm{ad}}}{dT}\right)_P=3\Gamma_{12}\alpha,\quad \left(\frac{dc_{44}^{\mathrm{ad}}}{dT}\right)_P=3\Gamma_{44}\alpha. \tag{A.3}$$

Since more exactly $(dc_{ij}^{\mathrm{ad}}/dT)_P=3\Gamma_{ij}\alpha[1+0(\delta v/v)]$, we will drop the terms $\sim\delta v/\tilde{v}$ and higher-order terms. (We note that the thermal coefficient of

Table A.I. Adiabatic elastic moduli of ^{7}LiH, ^{7}LiD, and ^{7}LiT, 10^{10} N/m².

	^{7}LiH			^{7}LiD			^{7}LiT		
T,K	c_{11}^{ad}	c_{12}^{ad}	c_{44}^{ad}	c_{11}^{ad}	c_{12}^{ad}	c_{44}^{ad}	c_{11}^{ad}	c_{12}^{ad}	c_{44}^{ad}
0	7.20	1.38	4.83	7.70	1.25	5.03	7.91	1.20	5.11
100	7.18	1.38	4.82	7.67	1.26	5.02	7.89	1.21	5.20
200	7.00	1.42	4.75	7.47	1.31	4.94	7.67	1.26	5.02
298	6.72[1]	1.49[1]	4.34[1]	7.10	1.40	4.79	7.26	1.37	4.85
400	6.23	1.62	4.45	6.56	1.54	4.58	6.70	1.51	4.63
500	5.70	1.75	4.24	5.96	1.69	4.35	6.07	1.67	4.39
600	5.11	1.90	4.00	5.30	1.86	4.09	5.38	1.84	4.12
700	4.46	2.06	3.75	4.59	2.04	4.08	4.66	2.02	3.83
800	3.75	2.24	3.47	3.83	2.23	3.51	3.89	2.22	3.53
900	2.96	2.44	3.16	3.03	2.43	3.20	3.06	2.42	3.21
T_m	2.40	2.59	2.94	2.40	2.59	2.95	2.41	2.60	2.95

[1]The values were taken from Ref. 125.

linear expansion α was calculated in the same approximation, yielding results that are in good agreement with the available experimental results.) The coefficients Γ_{11}, Γ_{12}, and Γ_{44} can be determined from the experimental data. Gerlich and Smith[125] and Haussühl and Skorczyk[138] measured the thermoelastic constants $\gamma_{ij} = dc_{ij}^{ad}/dT$ at $T = 273$ K (Ref. 138) and at $T = 298$ K (Ref. 125). Using the data on λ_{ij} of the more accurate measurements (those of Gerlich and Smith[125]) and our calculated value of the coefficient α (see Chap. 2, Sec. 2.2), we find

$$\Gamma_{11} = -113.63\times10^{11}\ \text{dyn/cm}^2, \quad \Gamma_{12} = 28.728\times10^{11}\ \text{dyn/cm}^2,$$
$$\Gamma_{44} = -44.613\times10^{11}\ \text{dyn/cm}^2.$$

Integrating Eqs. (A.3), we find the temperature dependence of the adiabatic moduli

$$c_{ij}^{ad}(T) - c_{ij}^{ad}(0) = 3\Gamma_{ij}\ \ln\frac{l(T)}{l(0)}. \tag{A.4}$$

The values of $c_{11}^{ad}(0)$, $c_{12}^{ad}(0)$, and $c_{44}^{ad}(0)$ can be determined from the data on c_{ij}^{ad} at a temperature of 298 K (Ref. 125) and the data on l at temperatures of 0 K and 298 K (Ref. 96). The values of the adiabatic elastic moduli calculated on the basis of Eq. (A.4) are presented in Table A.1. The lattice constant $l(T)$ is taken from Table 2.39.

The values of $c_{ij}^{ad}(0)$ for ^{7}LiH calculated by us are in good agreement with the experimental data obtained by Terras[225] at a temperature of 0 K. According to Terra's data,

$$c_{11}^{ad}(0) = 7.16\times10^{11}\ \text{dyn/cm}^2, \quad c_{12}^{ad}(0) = 1.39\times10^{11}\ \text{dyn/cm}^2,$$
$$c_{44}^{ad}(0) = 4.72\times10^{11}\ \text{dyn/cm}^2,$$

whereas our calculations yield

$$c_{11}^{ad}(0) = 7.20\times10^{11}\ \text{dyn/cm}^2, \quad c_{12}^{ad}(0) = 1.38\times10^{11}\ \text{dyn/cm}^2,$$
$$c_{44}^{ad}(0) = 4.83\times10^{11}\ \text{dyn/cm}^2.$$

The values of c_{11}^{ad}, c_{12}^{ad}, and c_{44}^{ad} were not measured at temperatures above room temperature.

References

1. N. I. Akulichev and Yu. A. Klyachko, in *Research Communications of the Scientific Staff of the D. I. Mendeleev Chemistry Department, No. 3* (Academy of Sciences of the USSR, Moscow, 1954), pp. 26–30.
2. A. A. Aleksandrov and M. S. Trakhtengerts, Teploenergetika **11**, 86 (1970).
3. G. S. Aslanyan, E. A. Tsirlina, and K. A. Yakimovich, Teplofiz. Vys. Temp. **17**, 495 (1979).
4. Yu. M. Baikov, T. Yu. Dunaeva, and V. B. Ptashnik, Fiz. Tverd. Tela **23**, 2504 (1981).
5. Yu. M. Baikov, G.Ya. Ryskin, Yu.P. Stepanov, *et al.*, Fiz. Tverd. Tela **11**, 3050 (1969).
6. Yu. M. Baikov, V. B. Ptashnik, and T.Yu. Dunaeva, Fiz. Tverd. Tela **20**, 1244 (1978).
7. V. A. Belov and N. I. Klychnikov, Teplofiz. Vys. Temp. **3**, 645 (1965).
8. M. Born and K. Huang, *Dynamical Theory of Crystal Lattices* (Oxford University Press, Oxford, 1954) [Russian translation edited by I. M. Lifshitz, published by Izdanja Inostr. Lit., 1958, p. 488].
9. M. V. Vol'kenshtein, *Structure and Physical Properties of Molecules* (Academy of Sciences of the USSR, Moscow, Leningrad, 1955), p. 638.
10. F. F. Voronov, V. A. Goncharov, O. V. Stal'gorova, *et al.*, Fiz. Tverd. Tela **8**, 1643 (1966).
11. Ya. I. Gerasimov and V. A. Geiderikh, *Termodynamika Rastvorov* [*Thermodynamics of Solutions*] (MGU, Moscow, 1980), p. 183.
12. J. O. Hirschfelder, C. F. Curtiss, and R. B. Bird, *Molecular Theory of Gases and Liquids* (Wiley, New York, 1954) [Soviet translation Izdanja Inostr. Lit., Moscow, 1961].
13. Zh. Goden, D. Kolomba, and G. Tonon, in *Problemy Lazernogo Termoyadernogo Sinteza* [*Laser Fusion*] (Atomizdat, Moscow, 1976), pp. 28–40.
14. L. V. Gurvich, G. A. Khachkuruzov, V. A. Medvedev, *et al.*, *Termodinamicheskie Svoistva Individual'nykh Veshchestv* [*Thermodynamic Properties of Single Substances*], 2 vols. (Academy of Sciences of the USSR, Moscow, 1961).
15. L. V. Gurvich, I. V. Veits, V. A. Medvedev, *et al.*, *Thermodynamic Properties of Single Substances* (Nauka, Moscow, 1978), Vol. 1, p. 495.
16. L. V. Gurvich, I. V. Veits, V. A. Medvedev, *et al.*, *Thermodynamic Properties of Single Substances* (Nauka, Moscow, 1982), Vol. 4, p. 470.
17. O. K. Davtyan, *Kvantovaya khimiya* [*Quantum Chemistry*] (Vyssh. Shkola, 1962), p. 783.
18. L. Girifalko, *Statistical Solid State Physics,* English translation (Mir, Moscow, 1975), p. 382.
19. A. V. Eletskii, L. A. Palkina, and B. M. Smirnov, *Yavleniya Perenosa v Slaboionizovannykh Gazakh i Plazme* (Atomizdat, Moscow, 1975), p. 333.
20. A. F. Kapustinskii, L. M. Shamovskii, and K. S. Bayushkina, Zh. Fiz. Khim. **10**, 4 (1937).
21. V. A. Kirillin, A. E. Sheindlin, and É. É. Shpil'rain, *Termofizika Rastvorov* [*Thermophysics of Solutions*] (Energiya, Moscow, 1980), p. 287.
22. C. Kittel, *Introduction to Solid State Physics,* 2nd ed. (Wiley, New York, 1956) [Russian translation, Fizmatgiz, Moscow, 1962].
23. C. Kittel, *Introduction to Solid State Physics,* 2nd ed. (Wiley, New York, 1956) [Russian translation edited by A. A. Gusev (Nauka, Moscow, 1978), p. 791].

24. V. S. Kogan and T. G. Omarov, Fiz. Tverd. Tela **7**, 933 (1965).

25. V. N. Kostryukov, Zh. Fiz. Khim. **35**, 1759 (1961).

26. O. A. Kraev, Teploenergetika **12**, 48 (1957).

27. O. A. Kraev, Zavod. Lab. **2**, 15 (1960).

28. N. M. Kulikov, Fiz. Tverd. Tela **20**, 2027 (1979).

29. L. D. Landau and E. M. Lifshitz, *Mekhanika* [*Mechanics*] (Nauka, Moscow), p. 208 [3rd ed., Pergamon, Oxford, 1976].

30. L. D. Landau and E. M. Lifshitz, *Statisticheskaya Fizika* [*Statistical Physics*] (Nauka, Moscow, 1964), p. 345 [2nd ed., Pergamon, Oxford, 1969].

31. G. Leibfried, "Microscopic Theory of Mechanical and Thermal Properties of Crystals," in *Handbuch der Physik*, edited by S. Flügge, Vol. 7, Part 1 (Springer-Verlag, Berlin, 1955), pp. 104-324 [Russian translation Fizmatgiz, Moscow, 1963].

32. G. Leibfried and W. Ludwig, "Theory of Anharmonic Effects in Crystals," Solid State Phys. **12**, 276 (1961).

33. V. E. Lyusternik, in *Teplofizicheskie Svoistva Veshchestv i Materialov* [*Thermophysical Properties of Elements and Materials*] (Standarts, Moscow), No. 10, p. 105.

34. E. Meizon, in *Kineticheskie Protsessy v Gazakh i Plazme* [*Kinetic Processes in Gas and Plasma*] (Atomizdat, Moscow, 1972), p. 52.

35. T. N. Mel'nikova and K. A. Yakimovich, Teplofiz. Vys. Temp. **17**, 975 (1979).

36. T. N. Mel'nikova, Teplofiz. Vys. Temp. **18**, 218 (1980).

37. T. N. Mel'nikova, Fiz. Tverd. Tela **22**, 588 (1980).

38. T. N. Mel'nikova and K. A. Yakimovich, Teplofiz. Vys. Temp. **18**, 305 (1980).

39. T. N. Mel'nikova, "Izotopicheskie Effekty i Termodynamicheskie Svoistva kristallov, Gidrid Litiya" ["Isotopic Effects and Thermodynamic Properties of Crystals, Lithium Hydride"], Candidate's Dissertation, Moscow, 1981.

40. C. E. Messer, E. L. Damon, P. K. Mayberry, *et al.*, in *Litii* [*Lithium*], a collection of papers, edited by V. E. Plyushchev [Russian translation Izdanja. Inostr. Lit., Moscow, 1959, p.175].

41. *Metal Hydrides*, edited by W. M. Mueller, J. P. Blackledge, and G. G. Libowitz (Academic Press, New York, 1968).

42. E. E. Nikitin, *Teoriya Elementarnykh Atomno-Molekulyarnykh Protsessov v Gazakh* [*Theory of Elementary Atomic-Molecular Processes in Gases*] (Khimiya, Moscow, 1970), p. 455.

43. M.Ya. Ovchinnikova, Zh. Eksp. Teor. Fiz. **49**, 275 (1965).

44. L. A. Palkina, B. M. Smirnov, and M. I. Chibisov, Zh. Eksp. Teor. Fiz. **56**, 340 (1969).

45. L. Pauling, *The Nature of the Chemical Bond and the Structure of Molecules and Crystals*, 2nd ed. (Cornell University Press, Ithaca, NY, 1940) [Russian translation, Goskhimizdat, M-L, 1947].

46. Z. T. Pinsker and R. N. Kurdyumova, Kristallogr. **3**, 501 (1958).

47. A. Ya. Polishchuk, É. É. Shpil'rain, and I. T. Yakubov, Physica C (a) **97**, 299 (1979).

48. A. Ya. Polishchuk, É. É. Shpil'rain, and I. T. Yakubov, Teplofiz. Vys. Temp. **17**, 1194 (1979).

49. A. Ya. Polishchuk, É. É. Shpil'rain, and I. T. Yakubov, Inzh.-Fiz. Zh. **38**, 429 (1980).

50. Yu. V. Privalov, "Investigation of the Sodium-Oxygen-Hydrogen System for Sodium-Coolant Technology," Candidate's Dissertation, IVTAN, Moscow, 1980.

51. I. Prigogine and R. Defei, *Chemical Thermodynamics*, Russian translation (Nauka, Novosibirsk, 1966).

52. I. B. Rabinovich, *Vliyanie Izotopii na Fiziko-Khimicheskie Svoistva Zhidkostei* [*Effect of Isotopy on the Physicochemical Properties of Liquids*] (Nauka, Moscow, 1968), p. 308.

53. C. Sarner, *Chemistry of Rocket Fuels*, Russian translation (Mir, Moscow, 1969), p. 488.

54. O. A. Skuratov, O. N. Pavlov, and I. V. Volkov, in *Proceedings of the Fifth All-Union Conference on Chemistry and Technology of Rare Alkali Elements* (Nauka, Moscow, 1977), p. 41.

55. B. M. Smirnov, Zh. Eksp. Teor. Fiz. **46**, 578 (1964).

56. B. M. Smirnov and M. I. Chibisov, Zh. Eksp. Teor. Fiz. **48**, 939 (1965).

57. B. M. Smirnov, *Atomnye Stolknoveniya i Elementarnye Protsessy v Plazme* [*Atomic Collisions and Elementary Processes in Plasma*] (Atomizdat, Moscow, 1968), p. 363.

58. B. M. Smirnov and M. I. Chibisov, Teplofiz. Vys. Temp. **9**, 513 (1971).

59. B. M. Smirnov, *Asimptoticheskie Metody v Teorii Atomnykh Stolknovenii* [*Asymptotic Methods in the Theory of Atomic Collisions*] (Atomizdat, Moscow, 1973), p. 294.

60. B. M. Smirnov and M. I. Chibisov, Teplofiz. Vys. Temp. **9**, 718 (1971).

61. *Handbook on Nuclear Physics*, Russian translation, edited by L. A. Artsimovich (Fizmatgiz, Moscow, 1963).

62. *Fundamental Physical Constants, Tables of Standard Reference Data* (Izdanja Standartov, Moscow, 1979), p. 7.

63. A. R. Ubbelohde, *Melting and Crystal Structure* (Clarendon, Oxford, 1965) [Russian translation, Mir, Moscow, 1969].

64. G. Huntington, Usp. Fiz. Nauk **74**, 303 (1961).

65. P. A. Cherkasov and L. N. Pankrat'eva, Zh. Fiz. Khim. **54**, 2778 (1980).

66. M. I. Shakhparonov, *Vvedenie v Molekulyarnuyu Teoriyu Rastvorov* [*Introduction to the Molecular Theory of Solutions*] (Gostekhizdat, Moscow, 1956), p. 507.

67. V. G. Shvidkovskii, *Nekotorye Voprosy Vyazkosti Rasplavlennykh Metallov* [*Viscosity of Molten Metals*] (Gostekhizdat, Moscow, 1955), p. 207.

68. É. É. Shpil'rain, K. A. Yakimovich, *et al.*, *Teplofizicheskie Svoistva Shchelochnykh Metallov* [*Thermophysical Properties of Alkali Metals*] (Izdanja Standartov, Moscow, 1970), p. 487.

69. É. É. Shpil'rain, D. N. Kagan, and G. A. Krechetova, Teplofiz. Vys. Temp. **9**, 1301 (1971).

70. É. É. Shpil'rain and K. A. Yakimovich, *Gidrid Litiya* [*Lithium Hydride*] (Izdanja Standartov, Moscow, 1971), p. 107.

71. É. É. Shpil'rain, K. A. Yakimovich, D. N. Kagan, *et al.*, in *Teplofizicheskie Svoistva Zhidkostei* [*Thermophysical Properties of Liquids*] (Nauka, Moscow, 1973), p. 7.

72. É. É. Shpil'rain, K. A. Yakimovich, and A. F. Tsitsarkin, Teplofiz. Vys. Temp. **11**, 1001 (1973).

73. É. É. Shpil'rain, K. A. Yakimovich, and A. F. Tsitsarkin, Teplofiz. Vys. Temp. **12**, 77 (1974).

74. É. É. Shpil'rain and K. A. Yakimovich, Teplofiz. Vys. Temp. **13**, 764 (1975).

75. É. É. Shpil'rain, K. A. Yakimovich, and V. A. Shereshevskii, Teplofiz. Vys. Temp. **13**, 872 (1975).

76. É. É. Shpil'rain, K. A. Yakimovich, V. A. Shereshevskii, *et al.*, Teplofiz. Vys. Temp. **14**, 405 (1976).

77. É. É. Shpil'rain, K. A. Yakimovich, and V. A. Shereshevskii, Teplofiz. Vys. Temp. **15**, 661 (1977).

78. É. É. Shpil'rain, K. A. Yakimovich, V. A. Shereshevskii, *et al.*, Teplofiz. Vys. Temp. **16**, 1204 (1978).

79. É. É. Shpil'rain, K. A. Yakimovich, and G. S. Aslanyan, Dokl. Akad. Nauk SSSR **253**, 857 (1979).

80. É. É. Shpil'rain and A. Ya. Polishchuk, Teplofiz. Vys. Temp. **17**, 285 (1979).

81. É. É. Shpil'rain, K. A. Yakimovich, A. F. Tsitsarkin, *et al.*, Teplofiz. Vys. Temp. **17**, 511 (1979).

82. É. É. Shpil'rain and K. A. Yakimovich, Teplofiz. Vys. Temp. **17**, 722 (1979).

83. É. É. Shpil'rain and A. Ya. Polishchuk, Teplofiz. Vys. Temp. **18**, 280 (1980).

84. É. É. Shpil'rain, K. A. Yakimovich, and E. A. Tsirlina, Inzh.-Fiz. Zh. **38**, 249 (1980).

85. K. A. Yakimovich and T. N. Mel'nikova, "Thermophysical Properties of Lithium Hydride and of Lithium-Containing Solutions," in partial fulfillment of PhD thesis.

86. K. A. Yakimovich, É. É. Shpil'rain, A. F. Tsitsarkin, *et al.*, Teplofiz. Vys. Temp. **17**, 314 (1979).

87. K. A. Yakimovich, *Thermodynamic Properties of Lithium Hydride, Deuteride, and Tritide and of the Lithium-Containing Solutions*, PhD thesis (IVTAN, Moscow, 1980), p. 39.

88. K. A. Yakimovich, T. N. Mel'nikova, and E. A. Tsirlina, Teplofiz. Vys. Temp. **18**, 92 (1980).

89. V. S. Yargin, N. I. Sidorov, and E. L. Studnikov, *Vyazkost i Teploprovodnost' Shchelochnykh Metallov v Gazovoi Faze* [*Viscosity and Thermal Conductivity of Alkali Metals in the Gas Phase*] (IVTAN, Moscow, 1978), p. 123 (review articles on the thermophysical properties of substances, 5th edition).

90. P. F. Adams, M. G. Down, P. Hubberstey, *et al.*, J. Less-Common Met. **42**, 325 (1975).

91. P. F. Adams, P. Hubberstey, and R. J. Pulham, J. Less-Common Met. **42**, 14 (1975).

92. P. F. Adams, P. Hubberstey, R. J. Pulham, *et al.*, J. Less-Common Met. **46**, 285 (1976).

93. S. A. Adelman and D. R. Herschbach, Molec. Phys. **33**, 763 (1977).

94. M. S. Ahmed, Nature **165**, 246 (1950).

95. M. S. Ahmed, Philos. Mag. **42**, 997 (1951).

96. J. L. Anderson, J. Nasise, K. Philipson, *et al.*, J. Phys. Chem. Solids **31**, 613 (1970).

97. E. N. Andrade, Philos. Mag. **17**, 497 (1934).

98. G. Benedek, Solid State Commun. **5**, 101 (1967).

99. J. M. Bijvoet and K. Karssen, Proc. Acad. Sci. Amsterdam, 1922, p. 25.

100. J. M. Bijvoet and K. Karssen, Acad. Amsterdam, Verslag., 1922, p. 31.

101. J. M. Bijvoet and K. Lonsdale, Philos. Mag. **44**, 204 (1953).

102. M. Blander, V. A. Maroni, E. Veleckis, *et al.*, Report ANL-7978, January 1973.

103. M. Blander, E. Veleckis, and E. Deventer, Report ANL-8023, October 1973.

104. H. Bode, Z. Phys. Chem. **13**, 99 (1931).

105. F. H. Bohn, H. Conrads, J. Darvas, *et al.*, in *Proceedings of the Fifth Symposium on Engineering Problems of Fusion Research* (Princeton University Press, Princeton, NJ, 1973), p. 117.

106. H U. Borgsted, Werkst. Korros. **28**, 529 (1977).

107. R. C. Bowman, J. Phys. Chem. **75**, 1251 (1971).

108. R. C. Bowman, J. Phys. Chem. Solids **34**, 1754 (1973).

109. M. A. Brediz, J. Chem. Phys. **46**, 4146 (1967).

110. D. E. Briggs, "Thermal Conductivity of Potassium Vapor," Dissertation, University of Michigan, Ann Arbor, Mich., 1968.

111. M. H. Brodsky and E. Burstein, J. Phys. Chem. Solids **28**, 1655 (1967).

112. R. S. Brokaw, Planet. Space Sci. **3**, 238 (1961).

113. W. Brückner, K. Kleinstück, and G. E. R. Schulze, Phys. Status Solidi **14**, 297 (1966).

114. R. H. Busey and R. B. Bevan, "Low-Temperature Heat Capacity of ^{7}Li Hydride and ^{7}Li Deuteride," ORNL-4706, 1971.

115. R. S. Calder, W. Cochran, D. Griffith, *et al.*, J. Phys. Chem. Solids **23**, 621 (1962).

116. J. Cazeneuve, Entropie **23**, 9 (1968).

117. A. Chretien, R. Kohlmuller, and P. Pascal, Nouveau Traité de Chimie Minéral, Paris, 1956.

118. L. T. Cowley, M. A. D. Fluendy, and K. P. Lawley, Trans. Faraday Soc. **65**, 2027 (1969).

119. L. Dass and S. C. Saxena, J. Chem. Phys. **43**, 1747 (1965).

120. K. Docken and J. Hinze, J. Chem. Phys. **57**, 4928 (1972).

121. D. T. Eash, J. Nucl. Mater. **37**, 358 (1970).

122. I. B. Fieldhouse, "Thermal Conductivity of Aircraft Structural and Reactor Materials. Thermodynamics and Transport Properties of Gases, Liquids, and Solids," ASME, 1959.

123. A. S. Filler and E. Burstein, Bull. Amer. Phys. Soc. **5**, 198 (1960).

124. C. R. Fischer, T. A. Dellin, S. W. Harrison, *et al.*, Phys. Rev. B **1**, 876 (1970).

125. D. Gerlich and C. S. Smith, J. Phys. Chem. Solids **35**, 1587 (1974).

126. T. R. P. Gibb, Report NEPA-1841, April 30, 1951 (see Refs. 127 and 190).

127. T. R. P. Gibb and C. E. Messer, "A Survey Report on Lithium Hydride," NYO-3957, May 2, 1954.

128. D. M. J. Goodal and G. M. McCracken, Proceedings of the Seventh Symposium on Fusion Technology, Grenoble, October 24–27, 1972, p. 97.

129. W. Gründler and W. Schulz, Z. Anorg. Allg. Chem. **410**, 293 (1974).

130. M. W. Guinan and C. F. Cline, J. Nonmetals **1**, 11 (1972).

131. S. R. Gunn and L. G. Green, J. Am. Chem. Soc. **187**, 4782 (1958).

132. S. R. Gunn, J. Phys. Chem. **71**, 1386 (1967).

133. P. Günther, Ann. Phys. **63**, 21 (1920).

134. A. Guntz, C. R. Acad. Sci. Paris, 122, 244 (1896).

135. A. Guntz, C. R. Acad. Sci. Paris, 123, 694 (1896).

136. R. Hultgren, P. D. Desai, D. T. Hawkins, *et al.*, *Selected Values of the Thermodynamic Properties of Elements* (American Society of Metals, Metals Park, Ohio, 1973).

137. J. E. Hanlon and A. W. Lawson, Phys. Rev. **113**, 472 (1959).

138. S. Haussühl and W. Skorczyk, Z. Kristallogr. **130**, 340 (1969).

139. R. F. S. Hearmon, Adv. Phys. **5**, 323 (1956).

140. K. Hensen, J. Inorg. Nucl. Chem. **31**, 919 (1969).

141. A. A. Hermann, E. Schumacher, and L. Wöste, J. Chem. Phys. **68**, 2327 (1978).

142. G. Herring, Rev. Mod. Phys. **34**, 631 (1962).

143. J. C. Hesson and H. Shimotake, in *Regenerative EMF Cells* (American Chemical Society, 149th Meeting, Washington, DC, 1967), p. 27.

144. F. K. Heuman and O. N. Salmon, "The Lithium Hydride, Deuteride, and Tritide Systems," KAPL-1667,1956.
145. R. O. Hickel, R. P. Cochran, C. M. Morse, *et al.*, "Experimental Investigation of Lithium Hydride and Water Absorbers as Possible Heat-Sink Materials for Hypersonic and Re-entry Vehicles." IAS paper No. 61-3, IAS 29th Annual Meeting, New York, January 23–25, 1961, p. 39.
146. J. H. Hildebrand and R. L. Scott, *The Solubility of Nonelectrolytes,* 3rd ed. (Dover, New York, 1964).
147. P. Hubberstey, P. F. Adams, and R. J. Pulham, in International Conference on Radiation Effects and Tritium Technology, Tennessee, 1975, CONF-750989, Vol. 3, pp. 117–124.
148. P. Hubberstey, P. F. Adams, R. J. Pulham, *et al.*, J. Less-Common. Met. **49**, 253 (1976).
149. P. Hubberstey, R. J. Pulham, and A. E. Thunder, Trans. Faraday Soc. **72**, 431 (1976).
150. C. B. Hurd and G. A. Moore, J. Am. Chem. Soc. **57**, 332 (1935).
151. R. P. Hurst, Phys. Rev. **114**, 746 (1959).
152. E. A. Hyleraas, Z. Phys. **63**, S.771 (1930).
153. H. R. Ihle and Ch.H. Wu, J. Inorg. Nucl. Chem. **36**, 2167 (1974).
154. H. R. Ihle and Ch.H. Wu, J. Chem. Phys. **63**, N 4, 1605 (1975).
155. H. R. Ihle and Ch.H. Wu, Proceedings of the Eighth Symposium on Fusion Technology, The Netherlands, June 17-21, 1974.
156. H. R. Ihle and Ch.H. Wu, Proceedings of the International Conference on Radiation Effects and Tritium Technology, Tennessee, 1975, CONF-750989, Vol. 3, p. 201.
157. H. R. Ihle and Ch.H. Wu, J. Phys. Chem. **79**, 22 (1975).
158. S. S. Jaswal and J. R. Hardy, Phys. Rev. **171**, 1090 (1968).
159. S. S. Jaswal, T. P. Sharma, and G. Wolfram, Sol. State Commun. **11**, 1151 (1972).
160. H. Jex, J. Phys. Chem. Solids **35**, 1221 (1974).
161. W. C. Johnson, Thesis of L. L. Hill, University of Chicago, Chicago, IL, 1938.
162. W. C. Johnson, Thesis of M. R. Perlow, Jr., University of Chicago, Chicago, IL, 1941.
163. C. E. Johnson, R. R. Heinrich, and C. E. Crouthamel, J. Phys. Chem. **70**, 244 (1966).
164. C. E. Johnson, S. E. Wood, and C. E. Crouthamel, J. Chem. Phys. **44**, 880 (1966).
165. C. E. Johnson, S. E. Wood, and C. E. Crouthamel, J. Chem. Phys. **44**, 884 (1966).
166. C. E. Johnson, S. E. Wood, and E. J. Cairins, J. Chem. Phys. **46**, 4168 (1967).
167. C. E. Johnson and R. R. Heinrich, in *Regenerative EMF Cells* (American Chemical Society, 149 Meeting, Washington, DC, 1967), p. 56.
168. H. Katsuta, T. Ishigai, and K. Furukawa, Nucl. Technol. **32**, 297 (1977).
169. K. K. Kelly, U.S. Bureau Mines Bulletin, Government Printing Office, Washington, DC, 1941, p. 434.
170. K. K. Kelly, U.S. Bureau Mines Bulletin 584, Government Printing Office, Washington, DC, 1960, p. 345.
171. A. J. Kirkham and B. Yates, Cryogenics **17**, 381 (1968).
172. F. Y. Krieger, "A Parametric Study of Certain Low-Molecular Weight. Compounds as Nuclear Rockets Propellants, IV Lithium Hydride." RM-2403 (RAND), August 1959, Rand Corp., Santa Monica, CA
173. P. H. Krupenie, E. A. Mason, and J. T. Vanderslise, J. Chem. Phys. **39**, 2399 (1963).

174. H. H. Landolt and R. Börnstein, *Zahlenwerte and Funktionen*, Bd. 11, 4 Teil, S.737–749 (Springer, Berlin, 1961).

175. J. L. Lang, "Thermodynamic Transport Properties of Gases, Liquids, Solids," Symposium paper, Lafayette, IN, ASME, 1959, p. 405.

176. D. Laplaze, M. Boissier, and R. Vacher, Solid State Commun. **19**, 445 (1976).

177. C. S. Lee, D. I. Lee, and C. F. Bonilla, Nucl. Eng. Des., **10**, 83 (1969).

178. R. E. Lo, "Some Properties of Lithium Hydride with Respect to Its Use as a Rocket Fuel," ORNL-tr-2471, November, 1965.

179. H. London, Z. Phys. Chem. (N.F.) **16**, S.302 (1958).

180. S. O. Lundqvist, Arkiv. för Fysik. Stockholm **8**, S.177 (1954).

181. V. A. Maroni, E. Veleckis, F. A. Cafasso, *et al.*, Proceedings of the Symposium on the Thermodynamics of Nuclear Materials, Vienna, IAEA-SM-190/51, 1975, p. 127.

182. V. A. Maroni, E. Veleckis, and E. H. Van Deventer, Proceedings of the Symposium on Tritium Technology Related to Fusion Reactor Systems, ERDA-50, CONF-741050, June 1975, p. 80.

183. V. A. Maroni, W. F. Calaway, E. Veleckis, *et al.*, Proceedings of the International Conference on Liquid Metal Technology in Energy Production, 1976, CONF-760503-Pl, p. 117.

184. S. P. Marsh, "Hugoniot Equations of State of Li^6H, Li^6D, $Li^n H$, and $Li^n D$," LA-4942, 1972.

185. G. M. McCracken, D. J. H. Goodal, R. T. P. Whipple, *et al.*, Fusion Reactor Design Problems. IAEA, Vienna, 1974, p. 439.

186. H. M. McCullough and B. Kopelman, Nucleonics **14**, 146 (1956).

187. C. E. Messer and T. R. P. Gibb, "A Survey Report on Lithium Hydride," NYO-8022, August 31, 1957.

188. C. E. Messer and L. G. Fasolino, J. Am. Chem. Soc. **77**, 4524 (1955).

189. C. E. Messer, "The Latent Heat of Fusion of Lithium Hydride from Oriogenic Measurements," NYO-8027, May 5, 1960.

190. C. E. Messer, "A Survey Report on Lithium Hydride," NYO-9470, October, 1960.

191. C. E. Messer, J. Mellor, J. A. Krol, *et al.*, J. Chem. Eng. Data **6**, 328 (1961).

192. C. E. Messer and I. S. Levy, Inorg. Chem. **4**, 543 (1964).

193. J. T. D. Mitchel and J. A. Booth, VII Symposium on Fusion Technology, Grenoble, France, 1972, p. 185.

194. K. Moers, Z. Anorg. Allg. Chemie **113**, 179 (1920).

195. L. Monchik, A. N. G. Pereira, and E. A. Mason, J. Chem. Phys. **42**, 3241 (1965).

196. C. D. Montgomery, Nucl. Eng. Des. **25**, 309 (1973).

197. A. Morita and K. Takahashi, Prog. Theor. Phys. **19**, 257 (1958).

198. C. R. Morse and R. O. Hickel, "Experimental Investigation of Lithium Hydride as a Heat-Sink Material," NASA, TND-1198, December, 1961.

199. B. Olinger and P. M. Halleck, Appl. Phys. Lett. **24**, 536 (1974).

200. L. Pauling, *The Nature of the Chemical Bond* (Pergamon, New York, 1960), p. 488.

201. K. Peters, Z. Anorg. Allg. Chem. **131**, 140 (1923).

202. B. T. Pickup, Proc. R. Soc. London A **333**, 69 (1973).

203. A. L. Piwinskii, E. M. Lilly, and G. S. Smith, Phys. Status Solidi A **32**, 2 (1975).

204. F. E. Pretzel, G. N. Rupert, C. L. Mader, *et al.*, J. Phys. Chem. Solids **16**, 10 (1960).

205. D. E. Pritchard and F. Y. Chu, Phys. Rev. A **2**, 1932 (1970).

206. T. R. Proctor and W. C. Stwalley, J. Chem. Phys. **65**, 2063 (1977).
207. R. J. Pulham, P. F. Adams, P. Hubberstey, *et al.*, International Conference on Radiation Effects and Tritium Technology, Tennessee, 1975, CONF-750989, Vol. 4, p. 144.
208. S. L. Quimby and P. M. Sutton, Phys. Rev. **91**, 1122 (1953).
209. F. Rossini *et al.*, "Selected Values of Chemical Thermodynamic Properties," NBS, Parts 1-2, 1952, p. 395.
210. H. Schinke and F. Sauerwalk, Z. Anorg. Allg. Chem. **287**, S.313 (1956).
211. R. B. Schulz, "Optimum Lithium Hydride Shield Vessel Design for Protection Against Meteoroid Punctures," NAA-SR-MEMO-TDR, 11790, 1966.
212. É. É. Shpil'rain, K. A. Yakimovich, A. F. Zizarkin, *et al.*, Proceedings of the Seventh Symposium on Thermophysical Properties, May 10–12, 1977, UBS, USA, p. 208.
213. É. É. Shpil'rain, K. A. Yakimovich, and A. F. Tsitsarkin, High Temp.-High Pressures **5**, 191 (1973).
214. D. K. Smith and H. R. Leider, J. Appl. Cryst. **1**, 246 (1968).
215. F. J. Smith, J. F. Land, G. M. Begun, *et al.*, "Equilibria in Hydrogen-Isotope-CTR-Blanket System," Report ORNL-5111, February, 1976.
216. F. J. Smith, J. B. Talbot, J. F. Laud, *et al.*, International Conference on Radiation Effects and Tritium Technology, Tennessee, 1975, CONF-750989, Vol. 3, p. 540.
217. F. J. Smith, La Gamma, A. M. Batistony, *et al.*, Proceedings of the Ninth Symposium on Fusion Technology, FRG, June 14–18, 1976, p. 28.
218. H. M. Smith and R. E. Webb, "Equilibrium Dissociation Pressures of Lithium Hydride and Lithium Deuteride," Report Y-2095, December 2, 1977.
219. F. J. Smith, J. D. Redman, R. A. Strehlow, *et al.*, Proceedings of the Symposium on Tritium Technology Related to Fusion Reactor Systems, N4, ERDA-50, October 1–2, 1974, CONF-741050, 1975, p. 12.
220. S. P. Srivastava, and R. S. Saraswat, J. Phys. Chem. Solids **36**, 351 (1975).
221. E. Starizky and D. L. Walker, Analyt. Chem. **28**, 1055 (1956).
222. D. R. Stephens and E. M. Lilly, J. Appl. Phys. **39**, 177 (1968).
223. D. R. Stull, JANAF Thermo-Chemical Tables, The Dow Chemical Co., Midlands, Mich., 1962, p. 576.
224. I. R. Tannenbaum and F. H. Ellinger, "The Thermal Expansion of Lithium Hydride in the Temperature Range 0–500 C," USAEC Report LAMS-1650, 1954.
225. Ch. Terras, "Elastic Constants of LiH Between 300 K and 77 K," CEA-R-4409, Commissariat a l'Energie Atomique, Bruyeresle-Chatel (France), Centre d'Etudes, March, 1973.
226. Ch. Terras and C. Moussin, C. R. Acad. Sci. Paris **272**, A815 (1971).
227. R. L. Tiede, J. Am. Ceramic Soc. **42**, 537 (1959).
228. L. Tronstad and H. Wergeland, Det Kongelige Norske Videnskabers Selskab Forhandlinger, 1937, Bd. 10, S.37.
229. A. R. Ubbelohde, Trans. Faraday Soc. **32**, 525 (1936).
230. P. Varotsos, Phys. Rev. B **9**, 1866 (1974).
231. P. Varotsos, J. Physique **37**, Coll. C7, 327 (1976).
232. L. Vegard and H. Dale, Z. Kristallogr. **67**, S.148 (1928).
233. E. Veleckis, "The Lithium-Lithium Deuteride System," ANL-8123, November 1974.
234. E. Veleckis, "The Lithium–Lithium Deuteride System," ANL-75-50, September 1975.

235. E. Veleckis, J. Phys. Chem. **81**, 6 (1977).
236. E. Veleckis, J. Nucl. Mater. **79**, 20 (1979).
237. E. Veleckis, E. H. Van Deventer, and M. Blander, J. Phys. Chem. **78**, 1933 (1974).
238. E. Veleckis, E. Van Deventer, and V. A. Maroni, "Physicochemical and Thermodynamic Studies of Lithium-Containing Systems," ANL-7923, August, 1972.
239. E. Veleckis and V. A. Maroni, International Conference on Radiation Effects and Tritium Technology Tennessee, 1975, CONF-750989, Vol. 3, p. 458.
240. E. Veleckis, R. M. Yonco, and V. A. Maroni, J. Less-Common. Met. **55**, 85 (1977).
241. E. Veleckis, R. M. Yonco, and V. A. Maroni, International Symposium of Thermodynamics of Nuclear Materials, July, FRG, IAEA-SM-236/56 January 29–February 2, 1979.
242. E. Veleckis, R. M. Yonco, and V. A. Maroni, "The Current Status of Fusion Reactor Blanket Thermodynamics," ANL-78-109, April, 1979.
243. J. L. Verbl, J. L. Warren, and J. L. Yarnell, Phys. Rev. **168**, 980 (1968).
244. W. Voight, Lehrbuch der Kristallphysik, Leipzig, 1928.
245. R. C. Vogel, M. Levenson, and V. H. Munnecke, Report ANL-6800, 1964.
246. J. M. Vogt, TAPCO Report NP-11888, October, 1961.
247. H. Weichselgartner, Proceedings of the Seventh Symposium on Fusion Technology, Grenoble, France, October 24-27, 1927, p. 187.
248. R. Weil and A. W. Lawson, J. Chem. Phys. **37**, 2730 (1962).
249. F. H. Welch, H. W. Northrup, W. H. Mink, *et al.*, "Viscosity of Liquid Lithium Hydride from 1252 °F to 1578 °F," TID-12738 (see NSA 15, 19881, 1961; CA 59, 8, 1843 g, 1963).
250. F. H. Welch, "Properties of Lithium Hydride III," XDC-61-5-67, 1961.
251. F. H. Welch, "Lithium Hydride Properties," DC-61-3-73, 1962.
252. F. H. Welch, Nucl. Eng. Des. **26**, **444** (1974).
253. J. C. Whitehead and R. Grise, Discuss. Faraday. Soc. **55**, 320 (1973).
254. H. W. Wilson, K. W. Beam, and W. J. Cooper, "Determination and Analysis of the Potentialities of Thermal Energy Storage Materials," ASD TR-187, June 30, 1961.
255. W. D. Wilson and R. A. Johnson, Phys. Rev. B **1**, 3510 (1970).
256. G. H. Wostenholm and B. Yates, Philos. Mag. **27**, 185 (1973).
257. C. H. Wu, J. Chem. Phys. **65**, 3181 (1976).
258. C. H. Wu and H. R. Ihle, J. Chem. Phys. **66**, 10 (1977).
259. A. Zalkin, "The Thermal Expansion of Lithium Hydride," USAEC Report, Lawrence Radiology Laboratory, UCRL-4239, 1953.
260. W. B. Zimmerman, and D. J. Montgomery, Phys. Rev. **120**, 405 (1960).
261. W. B. Zimmerman, Phys. Rev. B **5**, 4704 (1972).
262. E. Zintl and A. Harder, Z. Phys. Chem. **28B**, S.478 (1935).
263. B. Yates, G. H. Wostenholm, and J. L. Bingham, J. Phys. C **7**, 1769 (1974).
264. D. T. Hurd, An Introduction to the Chemistry of the Hydrides, New York-London, 1952.

Printed in the United States
By Bookmasters